Results and Problems in Cell Differentiation

Volume 66

More information about this series at http://www.springer.com/series/400

Leonora Buzanska
Editor

Human Neural Stem Cells

From Generation to Differentiation and Application

Springer

Editor
Leonora Buzanska
Stem Cell Bioengineering Unit
Mossakowski Medical Research Centre
Polish Academy of Sciences
Warsaw, Poland

ISSN 0080-1844 ISSN 1861-0412 (electronic)
Results and Problems in Cell Differentiation
ISBN 978-3-030-06671-0 ISBN 978-3-319-93485-3 (eBook)
https://doi.org/10.1007/978-3-319-93485-3

This Springer imprint is published by the registered company Springer Nature Switzerland AG
The registered company address is: Gewerbestrasse 11, 6330 Cham, Switzerland

Contents

Part I
Derivation and/or Generation of Human Neural Stem Cells

Chapter 1
Derivation of Neural Stem Cells from the Developing and Adult Human Brain

Claire M. Kelly and Maeve A. Caldwell

Abstract Neural stem cells isolated from the developing and adult brain are an ideal source of cells for use in clinical applications such as cell replacement therapy. The clear advantage of these cells over the more commonly utilised embryonic and pluripotent stem cells is that they are already neurally committed. Of particular importance is the fact that these cells don't require the same level of in vitro culture that can be cost and labour intensive. Foetal neural stem cells can be readily derived from the foetal brain and expand in culture over time. Similarly, adult stem cells have been explored for their potential in vitro and in vivo animal models. In this chapter we identify the progress made in developing these cells as well as the advantages of taking them forward for clinical use.

Neural stem cells may be derived from several sources, and the focus of attention in recent years has been on those from embryonic stem (ES) cells and induced pluripotent stem (iPS) cells. These cells can be readily differentiated down a neural lineage from where they can be further directed into the different cell types of the nervous system. One therapeutic approach for which these cells have been extensively explored is cell replacement therapy (CRT). CRT aims to replace the cells that have been lost due to disease process. Neurodegenerative diseases such as Parkinson's disease (PD) and Huntington's disease (HD), where there is focal cell loss, are ideal candidates for this type of approach. It has been shown that primary human foetal cells transplanted into the diseased brain can survive and integrate into the host brain, thereby recreating the lost circuitry and alleviating the motor symptoms of the disease. Proof of principle has been shown in clinical trials to date for both neurodegenerative diseases; however, there is an urgent need to identify an

C. M. Kelly
Biomedical Science, Cardiff School of Health Sciences, Cardiff Metropolitan University, Cardiff, UK
e-mail: ckelly@cardiffmet.ac.uk

M. A. Caldwell (✉)
Trinity College Institute for Neuroscience, Trinity College Dublin, Dublin, Ireland
e-mail: Maeve.Caldwell@tcd.ie

L. Buzanska (ed.), *Human Neural Stem Cells*, Results and Problems in Cell Differentiation 66, https://doi.org/10.1007/978-3-319-93485-3_1

alternative cell source that can be used to make this approach more viable. Currently, the source of cells is primary human foetal tissue taken within a restricted time window, around the time of birth of these neurons. Specifically, for PD this would be 4–6 weeks postconception and slightly later for HD, 8–12 weeks postconception. This restricted time window places significant constraints on the feasibility of large-scale clinical application. In addition several donors are required per patient, an issue compounded by the fact that bilateral transplants in PD require cells from approximately six foetuses and the time line for collecting this tissue is restricted to 7 days (thus causing logistical problems for coordinating cell collection, surgery and pathological screening of cells) and they are difficult to standardise. Hence, there is a need to identify a new source of cells that would make this possible. Indeed, for any type of cells to be considered as a cell replacement therapy, there are a number of critical issues that should be addressed: (1) the biology of the cells should be completely defined; (2) it should be possible to both expand and store these cells in clinically useful quantities; (3) they should have a reliable differentiation potential, i.e. their neurogenic potential must remain stable after passaging; (4) they must be able to restore function following transplantation; and (5) they must not undergo malignant transformation over time.

ES and iPS cells are being extensively explored for this purpose. ES and iPS cells are a pluripotent source of cells and thus require manipulation in vitro to direct them firstly to a neuronal fate and furthermore to a cell type-specific phenotype. An alternative approach is to seek to identify stem cells that are already committed to a neural lineage (i.e. tissue-specific) and, furthermore, from an even more restricted lineage, for example, striatal precursors from which it may be easier to drive an explicitly striatal phenotype, as required for HD. In addition, if these cells could survive cryopreservation, this would ease current practical constraints associated with scheduling the neurosurgery and would also permit at least some standardisation of the cells, which cannot currently be achieved for primary foetal tissue. Specifically, foetal tissue can only be reliably held in culture (using media to reduce metabolic processes, i.e. 'hibernation') for a short period of time (up to 8 days) which is an insufficient period of time to permit full quality control of the tissue (Hurelbrink et al. 2000). Furthermore, the multipotential nature of these cells means they are less likely to give rise to fast-growing tumours following grafting, a constant risk associated with pluripotent-derived neural cells. The focus of this chapter will be on those cells found in the developing and adult brain that have 'neural stem cell' characteristics.

Within the developing and adult brain, there are populations of neural cells that have stem cell-like characteristics. By definition 'neural stem cell' describes a multipotent stem cell that can self-renew and give rise to one or multiple neural or glial lineages. Furthermore, the terms 'neural progenitor cell' and 'neural precursor cell' refer to a lineage restricted to an unspecified neural cell or, if further down the developmental pathway, specified to a brain subregion. As well as being present throughout development, stem cell populations are present in adult tissues, where they may be continually active, such as stem cells that underlie the constant renewal of the skin, or may be largely quiescent but capable of being triggered to proliferate if the conditions are right, as for some populations in the adult CNS. It cannot be assumed, however, that tissue-specific adult stem cells are necessarily the same as their embryonic counterparts, and so they may require different conditions for self-renewal and may have different differentiation potentials.

1.1 Foetal Neural Stem Cells (NSCs)

All cells of the adult CNS arise from the neuroepithelium, a germinal layer that surrounds the ventricle of the embryonic brain (Larsen and Churchill 1998). In vivo fate mapping experiments in rodents have been undertaken to demonstrate that multipotential cells play a role in generating the phenotypic diversity of the mammalian CNS. The injection of replication-deficient retroviruses with a reporter gene into the forebrain ventricle of the embryo at a low level allowed individual cells to be infected and their multipotentiality to be assessed by examining their progeny in situ (Luskin et al. 1988). Based on this technique, it has been postulated that by E12–E14 (rodent embryonic day), most precursor cells in the mouse cortical germinal zone are fate-restricted and generate only neurons or glia (Grove et al. 1993; Luskin et al. 1988). However, it has also been shown by others that the label in some cases may be inherited by both neuronal and glial progeny which would suggest a common precursor cell (Price and Thurlow 1988). In utero fate mapping experiments have further classified the fate of cells in the striatum with cells from the MGE and LGE having their own distinct pattern of migration and differentiation (Wichterle et al. 2001). Furthermore, there is a need in all cases to verify the findings from such mapping experiments with gene expression markers of regional identity (Rowell and Ragsdale 2012).

Detailed study of the multipotentiality of precursor cells in the mammalian CNS is difficult given the technical constraints associated with the technique. Also, when carrying out such experiments, other factors that need to be taken into account include the possibility that two neighbouring unipotent cells may be infected with the label and thus lead to misleading interpretation of the clonality of the progeny, and as such migration of clonal cells may also be misinterpreted (Walsh and Cepko 1992, 1993). In vitro analysis of precursor cells using retrovirus labelling has helped to circumvent some of the problems just mentioned and has confirmed the presence of multipotential cells in the developing mouse cortex (Williams and Price 1995; Williams et al. 1991). Another method is to culture individual cells and follow their progeny, a technique that has been successfully used in the rodent CNS (Reynolds and Weiss 1992a, b, 1996; Gritti et al. 1996; Temple and Davis 1994). If single cells give rise to secondary clones, they must possess self-renewing capacity. The ability of clonal cultures to give rise to lineage-restricted daughter cells can be demonstrated by inducing them to differentiate, e.g. mitogen withdrawal and plating onto a permissive substrate. The offspring can then be analysed using antibodies specific to neural progeny, i.e. neurons, astrocytes and oligodendrocytes.

1.1.1 Mitogens Used for In Vitro Proliferation of hNSCs

Fibroblast growth factor-2 (FGF-2) is part of a large family of cytokines (22 identified to date) that are important in the control of precursor cell proliferation and differentiation in the embryo (Wagner 1991; Basilico and Moscatelli 1992). FGF family members signal through a family of four receptor kinases (FGFR1–FGFR4) (Johnson and Williams 1993), each of which is expressed at varying times in the developing rodent brain. The bioactivity of these FGFs is regulated by the heparan sulphate proteoglycan (HSPG). FGF-2 forms a complex with HSPG and the FGFR which requires sulphation and which is crucial for the receptor to be activated (Ornitz et al. 1992; Yayon et al. 1991). Heparin stabilises FGF-2 in culture media as well as acting directly with FGF-2 (Caldwell et al. 2004). Numerous studies have shown that FGF-2 promotes the proliferation of rodent NPCs in vitro (Kilpatrick and Bartlett 1993; Ray et al. 1993; Palmer et al. 1995; Gritti et al. 1999; Kelly et al. 2003). Furthermore, Kuhn and colleagues demonstrated that chronic intracerebroventricular infusion of FGF-2 increased the population of proliferating precursor cells in the SVZ of adult rats with a concurrent increase in the number of neurons migrating from the SVZ and reaching the olfactory bulb (Kuhn et al. 1997). In addition, Ambrosini et al. reported that hypoxia-induced apoptosis in human striatal precursor cell cultures could be prevented in the presence of FGF-2 (Ambrosini et al. 2015).

Epidermal growth factor (EGF) has also been shown to stimulate the division of rodent NPCs in vitro (Reynolds and Weiss 1992a; Kelly et al. 2005). EGF binds to the EGF receptor (EGFR) which is a tyrosine kinase encoded by the ErbB gene. Although it is not expressed as widely as the FGF family of receptors, EGFR is expressed in a pattern that is suggestive of a role in regulating proliferation of NPCs in the developing and indeed in the adult nervous system (Alroy and Yarden 1997). When EGF was chronically infused intracerebroventricularly, this also caused an increase in proliferation in the SVZ but, in contrast to FGF-2, increased the number of astrocytes reaching the olfactory bulb.

A number of rodent studies have highlighted the complex interactions between FGF-2 and EGF. Culturing rodent NPCs in FGF-2 promotes the appearance of EGF-responsive progenitor cells (Arsenijevic et al. 2001; Ciccolini and Svendsen 1998; Lillien and Raphael 2000; Santa-Olla and Covarrubias 1999). At earlier stages of development, stem cells are responsive to FGF-2 but not to EGF (Ciccolini and Svendsen 1998; Kilpatrick and Bartlett 1993, 1995; Lillien and Raphael 2000; Qian et al. 1997; Johe et al. 1996; Burrows et al. 1997). The acquisition of EGF responsiveness is associated with the appearance of a subpopulation of progenitor cells, which express relatively high levels of EGF receptors (Burrows et al. 1997; Kornblum et al. 1997). The presence of FGF-2 in the medium accelerates the appearance of EGF responsiveness (Lillien and Raphael 2000). Santa-Olla and Covarrubias (1999) suggested that FGF-2-responsive cells are themselves the precursors for EGF-responsive cells. Two conflicting theories have been proposed to account for the differences in EGF and FGF-2 responsiveness over time. One possibility is that two (or more) different stem cell populations, which have specific responses to different growth conditions, co-exist within the cultures (Kilpatrick and Bartlett 1995; Tropepe et al. 1999). Alternatively, cultures are comprised of FGF-2-responsive stem or precursor cells which acquire EGF responsiveness (Ciccolini and Svendsen 1998; Santa-Olla and Covarrubias 1999; Arsenijevic et al. 2001). In support of the latter theory, a series of studies have demonstrated that FGF-2-responsive NPCs later acquire EGF responsiveness by a mechanism that appears to involve upregulation of the EGFR by FGF-2 itself (Burrows et al. 1997; Lillien and Cepko 1992; Santa-Olla and Covarrubias 1999), which is antagonised by BMP-4 (Lillien and Raphael 2000).

Leukaemia inhibitory factor (LIF) is a member of the gp130 signalling family (which also includes interleukin-6 (IL-6), oncostatin M (OSM) and ciliary neurotrophic factor (CNTF)). In vitro studies of the developing central nervous system have demonstrated that the activation of gp130 by these cytokines promotes differentiation and/or survival of astrocytes (Koblar et al. 1998; Gadient et al. 1998), oligodendrocytes (Mayer et al. 1994) and specific neuronal subtypes (Marz et al. 1997).

Several other growth factors have the potential to enhance the neuronal differentiation of these cells down particular lineages, including nerve growth factor (NGF), insulin-like growth factor (IGF) and tumour necrosis factor (TNFα) (Arsenijevic et al. 2001; Cattaneo and Mckay 1990; Santa-Olla and Covarrubias 1995; Tropepe et al. 1997; Supeno et al. 2013). As has been shown for pluripotent cell-derived neural cultures, identifying an appropriate growth factor cocktail to the phenotype associated with each particular application may be a necessary prelude to using these cells for transplantation (Precious and Rosser 2012). Indeed, developments in our understanding of the developmental requirements of these cells and in establishing differentiation protocols for ES and iPS cells should now be considered for the culture conditions of these cells.

1.1.2 Longer-Term Growth of hNSCs

Rat-derived NPCs can only be expanded for relatively short periods of time in the presence of EGF and FGF-2 (approx. 35 days) (Ostenfeld et al. 2002); however, the addition of IGF-1 has been shown to lengthen this up to seven passages (>50 days) (Supeno et al. 2013). In contrast to this, mice and human NSCs can be expanded for extensive periods of time. Initiation of a human NPC culture is usually carried out in the presence of both EGF and FGF-2 (Carpenter et al. 1999; Svendsen et al. 1998; Vescovi et al. 1999) (Table 1.1). However, once the culture has been established, the requirement for FGF-2 is not so clear. Svendsen et al. omitted FGF-2 after the first 4 weeks and found no change in the rates of proliferation. The length of time it is possible to expand these cells varies between research groups. Ostenfeld et al. reported that human NPCs could be expanded for 40–50 population doublings (Ostenfeld et al. 2000). However, Carpenter et al. found that the addition of LIF to

Table 1.1 Long-term expansion of human neural progenitor cells

Brain region/donor age	Monolayer (ML)/ neurosphere (NS)	Media supplements	Growth factors	Expansion time	In vitro/in vivo differentiation	Reference
STR, VM, CTX/8–21 weeks	NS	N2	EGF ± FGF-2 ± LIF	>150 days	Neurons and astrocytes	Svendsen et al. (1998), Ostenfeld et al. (2000, 2002)
Telencephalon/14 weeks	ML	N2 and serum	EGF or FGF-2	1 year	Neurons, astrocytes and oligodendrocytes	Flax et al. (1998)
Telencephalon, CTX, whole brain, brain stem, spinal cord/ 6–14 weeks	ML	N2	EGF or FGF2	>2 years	Neurons, astrocytes and oligodendrocytes, no oligodendrocytes in vivo	Vescovi et al. (1999)
Forebrain/5–11 weeks	NS	N2	EGF, FGF-2, LIF	>1 year	Neurons, astrocytes and oligodendrocytes, no oligodendrocytes in vivo	Carpenter et al. (1999), Fricker et al. (1999), Englund et al. (2002a, b)
Forebrain	NS	N2	EGF, FGF-2, LIF	8 weeks	Neurons and astrocytes	Uchida et al. (2000)

STR striatum, *VM* ventral mesencephalon, *CTX* cortex

the culture media extended the time that cultures could be expanded to; up to 370 days with maximum proliferation rates have been maintained in culture media containing EGF, FGF-2 and LIF. Interestingly no differences have been reported in proliferation rates between tissues of different gestational ages (Vescovi et al. 1999; Carpenter et al. 1999). Others have reported the generation of neural stem cells from foetal tissue that have the capacity for indefinite culture (Conti et al. 2005), and they suggest the self-renewing capacity of these cells is associated with FGF-2. Following microarray analysis they found that there was transcriptional resetting of a subset of cells induced by FGF-2, thus promoting the NS-like state (Pollard et al. 2008).

1.1.3 Regional Differences

Ostenfeld et al. (2002) demonstrated that both human and rat NPCs grown under identical conditions are regionally specified. Human NPCs derived from the cortex and striatum produce greater numbers of neurons than NPCs from other regions including mesencephalon, cerebellum and thalamus (Kallur et al. 2006; Kim et al. 2006; Ostenfeld et al. 2002). Similar results were seen with rodent NPCs except that by 6-week proliferation, only striatal-derived cultures differentiated into significant numbers of neurons (Ostenfeld et al. 2002). A similar study carried out by Jain and colleagues compared regional NPCs from human and mouse. They demonstrated that there was regional variation in expansion rate which became significant at passage 4 with cortical NPCs showing significantly greater expansion than spinal cord (SC) or ventral mesencephalon (VM). In addition, the expansion rate for striatal and cerebellar NPCs was greater than VM and SC (Jain et al. 2003b). Taken together both these studies showed that there was greater proliferation with more rostrally derived NPCs, (in particular the cortex), than caudal areas of the neuraxis.

With regard to differentiation, human NPCs are capable of differentiation into neurons and astrocytes. In some studies the proportion of these human cells differentiating into neurons decreased with extensive periods in culture but still tend to remain higher than in murine cultures (between 20 and 37% (human) compared to 5 and 7% (murine)) (Carpenter et al. 1999; Kelly et al. 2007), whereas other studies have reported a stable capacity for neuronal differentiation (Svendsen et al. 1998; Vescovi et al. 1999). However, Carpenter et al. reported that the addition of LIF consistently differentiated cells into a higher proportion of neurons (Carpenter et al. 1999).

Jain et al. demonstrated that regional differences also exist upon differentiation; they reported greater numbers of neurons at early passage and that this decreased more significantly in the VM, CBM and SC (Jain et al. 2003b). Furthermore, the predominant neurotransmitter phenotype displayed was GABA, independent of region. This was in agreement with Ostenfeld et al. (2002) who demonstrated that human NPCs produce significantly more neurons when originating from the CTX or STR than other more caudal regions. Moreover, the predominant neurotransmitter phenotype was GABA.

More recently, Martin-Ibáῃez also reported regional variability in the proliferation and differentiation of human NPCs over the early stages in culture; however, this stabilised with passage, with forebrain, CTX and STR cultures being more readily expandable than cerebellar cultures. In addition, these cells retained their regional identity over time in culture (Martin-Ibanez et al. 2017).

Interestingly, in studies using rodent NPCs, molecular characterisation in vitro has shown that they retain a degree of their site-specific identity when environmental cues are absent but when cocultured with cells of different origin, they can adopt a new fate. For example, striatal NPCs in vitro will retain expression of site-specific genes such as IsletI and Er81 over time, but with neuronal differentiation, expressions of striatal specific neuronal markers such as DARPP-32 and Islet1 are lost, although they do express homeobox transcription factors DLX and MEIS2. Similarly, human NPCs characterised using quantitative high-throughput gene expression identified a temporal progression of gene expression over time in culture (Straccia et al. 2015). Over time in vitro NPCs lose their propensity to generate neurons which is a drawback for their use in regenerative medicine. One might propose that despite the continued proliferation of NPCs, the absence of developmental signals relating to positional identity over time will result in the loss of neurogenic potential. This has also been verified in transplantation studies in rodents carried out by us and others showing that NSCs after several passages in culture produce poorly surviving grafts with fewer neurons (Zietlow et al. 2005, 2012; Jain et al. 2003a) or grafts that are enriched in astrocytes (Burnstein et al. 2004; Anderson and Caldwell 2007). In contrast, short-term expansion of NSCs in culture can generate modest increases in cell number and produce integrating grafts with fibres projecting to the target sites of those cells (Armstrong et al. 2000; Kelly et al. 2007). Furthermore, the NPC grafts produced more profuse outgrowths than did their primary counterpart grafts, an encouraging finding for the potential of these cells. In contrast, when grafted to the neonatal brain, similar cells appear to respond to developmental signals and regional determinants by differentiating in a site-specific manner suggesting that they retain the capacity to respond to developmental signals if they are present (Titomanlio et al. 2011).

The case for astrocyte differentiation is more contradictory. Flax et al. (1998) were not able to generate astrocytes from human NPCs without a feeder layer of primary cortical cells, whereas Galli et al. (2000) reported that serum or cytokines needed to be added to the differentiation media for their successful generation. In contrast, other studies have reported their spontaneous generation (Caldwell et al. 2001; Jain et al. 2003b; Martin-Ibanez et al. 2017; Svendsen et al. 1998; Vescovi et al. 1999), which increases with passage number irrespective of region of origin (Jain et al. 2003a, b).

Caldwell et al. (2001) demonstrated that maintaining cell-to-cell contact by plating human NPCs as neurospheres rather than as a cell suspension substantially increased the number of neurons. Plating the neurospheres in the presence of either NT3 or NT4 significantly increased the proportion of neurons and reduced the proportion of astrocytes, while CNTF had the opposite effect. Vescovi et al.

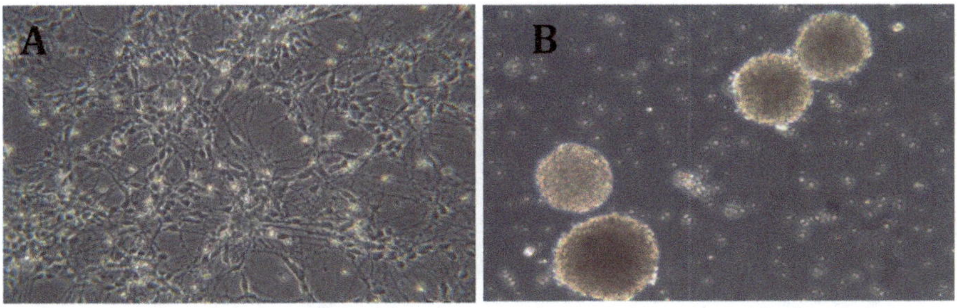

Fig. 1.1 Human foetal neural stem cells can expand in culture in two ways: (1) as a monolayer of cells as shown in (**a**), these cells adhere to the substrate on which they are growing and maintain close proximity to each other to support their proliferation over time in culture and (2) expand by forming free-floating spheres of cells ('neurospheres'), each of which contains several thousand cells, (**b**). Scale bar = 100 μm

(1999) used low concentrations of FGF-2 in the differentiation media to enhance neuronal differentiation (Fig. 1.1).

Despite the relatively low numbers generated following short-term expansion of these cells in comparison to the pluripotent cell sources, the ease with which these cells can be generated in comparison to the more laborious protocols used in ES- and iPS-derived neural differentiation, which are difficult to standardise, makes these cells an attractive alternative that should still be considered for CRT. Given that human foetal tissue will need to be collected for some time to come to supply cells for proof-of-concept and optimisation studies as outlined above, and secondly the advantages over hES and hiPS cells for regenerative medicine applications, makes further study on these cells a viable approach. The adaptability of the cells to respond to developmental signals could be considered as 'epigenetic memory', similar to that reported in adult somatic tissues, thus rendering these cells even more amenable to large-scale clinical application.

In summary, there is some evidence that NPCs retain the ability to default to a region specific-like phenotype following modest expansion in vitro and also that they are capable of projecting to target specific brain regions. It is important to acknowledge that NSCs are precursor cells, and in these experiments they have not undergone directed differentiation towards specific neuronal phenotypes. An important next step will be to subject NSCs to molecules known to be important for specific differentiation.

1.2 Adult Neural Stem Cells

Adult neural stem cells (ANSCs) are an example of an adult tissue-specific stem cell and, as their name suggests, are derived from the mature brain. Altman and colleagues provided the first clear evidence, using ^{3}H-thymidine autoradiography, that a

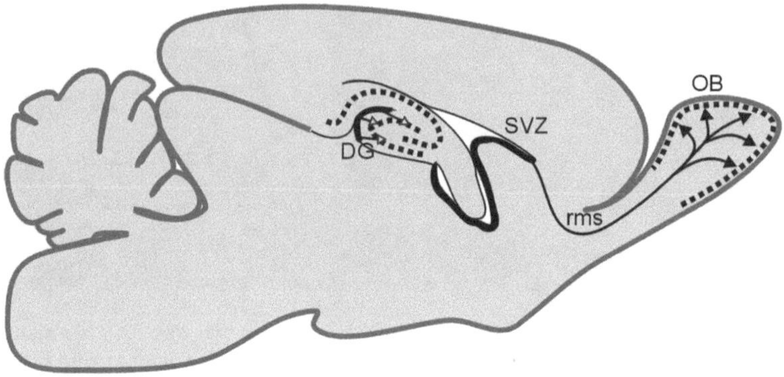

Fig. 1.2 Adult neural stem cells have been identified in the ventricular-subventricular zone (V-SVZ) and the dentate gyrus (DG) of the hippocampus. The V-SVZ cells migrate along the rostral migratory stream to the olfactory bulb, whereas newly formed neurons in the DG take up residence within in the granule cell layer (Dunnett et al. 2001)

low level of neurogenesis is ongoing in the dentate gyrus of adult rats (Altman and Das 1965). ANSCs have since been confirmed in the ventricular-subventricular zone (V-SVZ) of the lateral ventricles (Alvarez-Buylla et al. 2002), from where the newly formed neurons migrate via the rostral migratory stream (RMS) to the olfactory bulb (OB) (Lois and Alvarez-Buylla 1994). In addition, there is some evidence of the existence of an ANSC population in the sub-granular layers of the dentate gyrus of the hippocampus (Gage et al. 1995; Palmer et al. 1997; Reynolds and Weiss 1992b; Weiss et al. 1996; Gritti et al. 1996), but there is controversy as to whether this is a true ANSC or a more restricted progenitor cell (Bull and Bartlett 2005; Seaberg and Van der Kooy 2002). Nevertheless, whether this is an ANSC population or a more restricted progenitor pool, this dividing population gives rise to the newly formed neurons that repopulate the dentate gyrus (Gil-Perotín et al. 2013). More recently it has been reported that neural stem cells may also reside in other regions of the brain, albeit at an even lower concentration, including cortex (Gould et al. 1999; Rietze et al. 2000) and the medial-rostral part of the substantia nigra pars compacta in the lining of the cerebroventricular system of the midbrain (Zhao et al. 2003), although these reports remain controversial (Frielingsdorf et al. 2004; Taupin 2006b). In comparison to the spatially and temporally regulated niches of NSCs found in the developing brain in the adult brain, the neurogenic niches are restricted to specific regions. Furthermore, these cells have a heterogeneous nature coexisting in many different states and responding differently to physiological inputs (Taupin 2006a) (Fig. 1.2).

The NSCs of the V-SVZ are commonly referred to as type B1 cells and express glial-specific markers such as glial fibrillary acidic protein (GFAP), glutamate aspartate transporter (GLAST) and brain lipid-binding protein (BLBP) and thus are akin to brain astrocytes. Because these cells possess end feet on blood vessels and their location, they regularly are in contact with the brain ventricle, making them unique from other astrocytes found in brain. Furthermore, these cells can exist in two states, either quiescent or activated identified by the expression of nestin in the latter.

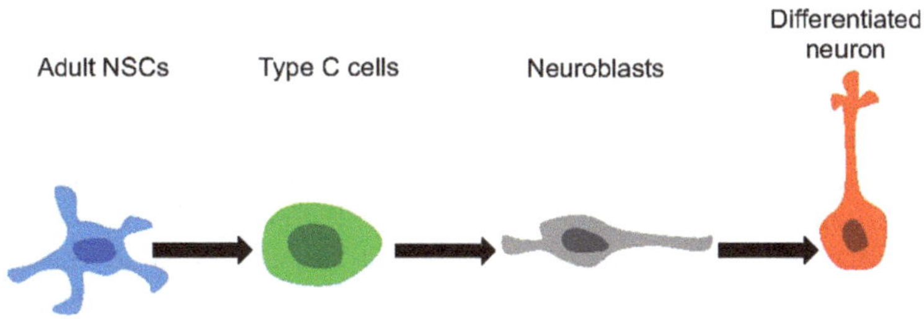

Fig. 1.3 Schematic representation of the differentiation of adult neural stem cells. These cells first divide and produce a type C transit-amplifying precursor which will then go on to produce migratory neuroblasts that will migrate along the rostral migratory stream before differentiating to a mature neuron in the olfactory bulb

The pathway taken by these cells is to first transition from the quiescent state to the more proliferative activated NSC that will then give rise to transit-amplifying precursors (type C cells) which then go on to generate neuroblasts (type A cells). It is these cells that then migrate along the RMS to the olfactory bulb OB (Lim and Alvarez-Buylla 2016) (Fig. 1.3).

The type B1 cells are understood to be derived from the neuroepithelial and radial glial cells that did not differentiate into astrocytes after birth. There is ongoing interest in determining the relationship between the quiescent and activated state with some suggesting they may be lineage related and others regarding the quiescent population as a reserve population that is recruited when needed (Chaker et al. 2016). Developments in techniques in recent times have been of significant benefit to the understanding of these cells. However, there is ongoing controversy surrounding the true multipotential nature of these cells. Using fluorescence-activated cell sorting (FACS), it has been shown that GFAP-positive single cells from the V-SVZ cultured on support astrocyte monolayers can generate colonies of neuronal and glial lineages. In contrast, using direct video time-lapse microscopy of NSCs from the V-SVZ failed to show both neuronal and glial lineages being generated. The extensive body of work reported on ANSCs is based primarily on rodent studies, but in 1998 neurogenesis and progenitor cells were described in discrete regions of the human brain (Eriksson et al. 1998; Curtis et al. 2012). Human ANSCs have been identified in small numbers deep within the postnatal human V-SVZ. However, upon histological examination it was found that the organisation of the V-SVZ of the human was very different to that reported in the rodent brain. Unlike the rodent brain where distinct populations of cells have been reported as described above, in the human brain, the V-SVZ has been found to have a gap layer (GAP) that is lacking in cells. Instead this layer consists of a dense network of interconnected processes from astrocytes and ependymal cells (Belenguer et al. 2016). Interestingly, it would appear unlikely that there is any real neurogenesis taking place in the adult brain but that neurogenesis as observed by the presence of DCX-positive migrating young neurons in the gap area is taking place in the infant brain <6 months of age.

There is evidence that progenitor cells exist in the human brain in both surgical specimens and post-mortem tissues. Palmer and colleagues reported that it was possible to isolate and expand cells from an 11-week-old post-natal male and a 27-year-old male. The cells from the younger donor could be expanded for 70 doublings and the ones from the older donor for 30 doublings. Neurons and astrocytes could be differentiated from both donors to a similar degree (Palmer et al. 2001). Over the course of their study, they expanded cells from 23 tissue samples from donors of different ages with the longest post-mortem interval being 20 h. Overall tissue samples from younger donors had more proliferating cells per gram, and these cells had a higher proliferative capacity.

The attraction of ANSCs as a donor supply for neural transplantation would be the possibility of autologous transplants, thus bypassing the immunological issues of graft rejection severe in the case of xenografts and not entirely benign even for allografts. Furthermore, it may eventually be possible to recruit such cells for endogenous repair without a requirement for their isolation and reimplantation. That is, it might be possible to stimulate the resident population of ANSCs to migrate to the site of degeneration, although adult stem cells remain difficult to isolate and grow in culture and the factors that would be required to enhance the proliferation of these cells and their differentiation into the particular phenotypes relevant to the site of degeneration remain unknown.

1.3 Good Manufacturing Practice Production

The EU directive 2004/23/EC applies to the use of human tissue and cells for use in clinical trials, and thus cells used for the purpose of cell replacement therapy in neurodegenerative disease as described here must adhere to the rules and regulations outlined in this directive. Namely, the donated tissues and cells must be procured, tested, processed, preserved and stored in accordance with validated and approved safety measures. The directive stipulates that the handling, storage and preparation of the cells meet with good manufacturing practice (GMP). In the UK the Human Tissue Authority (HTA) has implemented the UK Human Tissue Act which is the legal framework which sets out the regulatory framework for all matters concerning the removal, storage, use and disposal of human tissue for scheduled purposes. Relevant to the discussion here is the requirement for all work relating to the use of these cells to be carried out under GMP conditions. Taking this into consideration, the protocols used for the generation and propagation of these cells need to be adapted to avoid products that fall outside the remit of what is allowed under the directive. Therefore, alternatives that are nonanimal derived must be validated prior to progressing the use of these cells to clinical application.

1.4 Conclusion

Foetal- and adult-derived neural stem cells differ in their characteristics, and the potential of each for clinical application is varied. Human foetal neural stem cells are easily managed in culture; however, the conditions for maintaining these cells over time and specifying them require further optimisation. In contrast, adult neural stem cells mitigate the ethical and logistical issues of using other cell sources; however, as with foetal-derived cells, there are still many questions that need to be addressed in order to maximise the potential of this cell source.

References

Alroy I, Yarden Y (1997) The ErbB signaling network in embryogenesis and oncogenesis: signal diversification through combinatorial ligand-receptor interactions. FEBS Lett 410:83–86
Altman J, Das GD (1965) Autoradiographic and histological evidence of postnatal hippocampal neurogenesis in rats. J Comp Neurol 124:319–335
Alvarez-Buylla A, Seri B, Doetsch F (2002) Identification of neural stem cells in the adult vertebrate brain. Brain Res Bull 57:751–758
Ambrosini S, Sarchielli E, Comeglio P, Porfirio B, Gallina P, Morelli A, Vannelli GB (2015) Fibroblast growth factor and endothelin-1 receptors mediate the response of human striatal precursor cells to hypoxia. Neuroscience 289:123–133
Anderson L, Caldwell MA (2007) Human neural progenitor cell transplants into the subthalamic nucleus lead to functional recovery in a rat model of Parkinson's disease. Neurobiol Dis 27: 133–140
Armstrong RJ, Watts C, Svendsen CN, Dunnett SB, Rosser AE (2000) Survival, neuronal differentiation, and fiber outgrowth of propagated human neural precursor grafts in an animal model of Huntington's disease. Cell Transplant 9:55–64
Arsenijevic Y, Weiss S, Schneider B, Aebischer P (2001) Insulin-like growth factor-1 is necessary for neural stem cell proliferation and demonstrates distinct actions of epidermal growth factor and fibroblast growth factor-2. J Neurosci 27(18):7194–7202
Basilico C, Moscatelli D (1992) The FGF family of growth factors and oncogenes. Adv Cancer Res 59:115–165
Belenguer G, Domingo-Muelas A, Ferrón SR, Morante-Redolat JM, Fariñas I (2016) Isolation, culture and analysis of adult subependymal neural stem cells. Differentiation 91:28–41
Bull ND, Bartlett PF (2005) The adult mouse hippocampal progenitor is neurogenic but not a stem cell. J Neurosci 25:10815–10821
Burnstein RM, Foltynie T, He X, Menon DK, Svendsen CN, Caldwell MA (2004) Differentiation and migration of long term expanded human neural progenitors in a partial lesion model of Parkinson's disease. Int J Biochem Cell Biol 36:702–713
Burrows RC, Wancio D, Levitt P, Lillien L (1997) Response diversity and the timing of progenitor cell maturation are regulated by developmental changes in EGFR expression in the cortex. Neuron 19:251–267
Caldwell MA, He X, Wilkie N, Pollack S, Marshall G, Wafford KA, Svendsen CN (2001) Growth factors regulate the survival and fate of cells derived from human neurospheres. Nat Biotechnol 19:475–479
Caldwell MA, Garcion E, Ter Borg MG, He X, Svendsen CN (2004) Heparin stabilizes FGF-2 and modulates striatal precursor cell behaviour in response to EGF. Exp Neurol 188:408–420

Carpenter MK, Cui X, Hu ZY, Jackson J, Sherman S, Seiger A, Wahlberg LU (1999) In vitro expansion of a multipotent population of human neural progenitor cells. Exp Neurol 158: 265–278

Cattaneo E, Mckay R (1990) Proliferation and differentiation of neuronal stem cells regulated by nerve growth factor. Nature 347:762–765

Chaker Z, Codega P, Doetsch F (2016) A mosaic world: puzzles revealed by adult neural stem cell heterogeneity. Wiley Interdiscip Rev Dev Biol 5:640–658

Ciccolini F, Svendsen CN (1998) Fibroblast growth factor 2 (FGF-2) promotes acquisition of epidermal growth factor (EGF) responsiveness in mouse striatal precursor cells: identification of neural precursors responding to both EGF and FGF-2. J Neurosci 18:7869–7880

Conti L, Pollard SM, Gorba T, Reitano E, Toselli M, Biella G, Sun Y, Sanzone S, Ying QL, Cattaneo E, Smith A (2005) Niche-independent symmetrical self-renewal of a mammalian tissue stem cell. PLoS Biol 3:e283

Curtis MA, Low VF, Faull RL (2012) Neurogenesis and progenitor cells in the adult human brain: a comparison between hippocampal and subventricular progenitor proliferation. Dev Neurobiol 72:990–1005

Dunnett SB, Björklund A, Lindvall O (2001) Cell therapy in Parkinson's disease – stop or go? Nat Rev Neurosci 2(5):365–369

Englund U, Björklund A, Wictorin K (2002a) Migration patterns and phenotypic differentiation of long-term expanded human neural progenitor cells after transplantation into the adult rat brain. Brain Res Dev Brain Res 134(1-2):123–141

Englund U, Fricker-Gates RA, Lundberg C, Björklund A, Wictorin K (2002b) Transplantation of human neural progenitor cells into the neonatal rat brain: extensive migration and differentiation with longdistance axonal projections. Exp Neurol 173(1):1–21

Eriksson PS, Perfilieva E, Björk-Eriksson T, Alborn AM, Nordborg C, Peterson DA, Gage FH (1998) Neurogenesis in the adult human hippocampus. Nat Med 4:1313–1317

Flax JD, Aurora S, Yang C, Simonin C, Wills AM, Billinghurst LL, Jendoubi M, Sidman RL, Wolfe JH, Kim SU, Snyder EY (1998) Engraftable human neural stem cells respond to developmental cues, replace neurons, and express foreign genes. Nat Biotechnol 16:1033–1039

Fricker RA, Carpenter MK, Winkler C, Greco C, Gates MA, Björklund A (1999) Site-specific migration and neuronal differentiation of human neural progenitor cells after transplantation in the adult rat brain. J Neurosci 19(14):5990–6005

Frielingsdorf H, Schwarz K, Brundin P, Mohapel P (2004) No evidence for new dopaminergic neurons in the adult mammalian substantia nigra. Proc Natl Acad Sci USA 101:10177–10182

Gadient RA, Lein P, Higgins D, Patterson PH (1998) Effect of leukemia inhibitory factor (LIF) on the morphology and survival of cultured hippocampal neurons and glial cells. Brain Res 798: 140–146

Gage FH, Ray J, Fisher LJ (1995) Isolation, characterization, and use of stem cells from the CNS. Annu Rev Neurosci 18:159–192

Galli R, Borello U, Gritti A, Giulia Minasi M, Bjornson C, Coletta M, Mora M, De Angelis MGC, Fiocco R, Cossu G, Vescovi AL (2000) Skeletal myogenic potential of human and mouse neural stem cells. Nature 3:986–991

Gil-Perotín S, Duran-Moreno M, Cebrián-Silla A, Ramírez M, García-Belda P, García-Verdugo JM (2013) Adult neural stem cells from the subventricular zone: a review of the neurosphere assay. Anat Rec (Hoboken) 296:1435–1452

Gould E, Reeves AJ, Graziano MSA, Gross CG (1999) Neurogenesis in the neocortex of adult primates. Science 286:548–552

Gritti A, Parati EA, Cova L, Frolichsthal P, Galli R, Wanke E, Faravelli L, Morassutti DJ, Roisen F, Nickel DD, Vescovi AL (1996) Multipotential stem cells from the adult mouse brain proliferate and self-renew in response to basic fibroblast growth factor. J Neurosci 16:1091–1100

Gritti A, Frolichsthal-Schoeller P, Galli R, Parati EA, Cova L, Pagano SF, Bjornson CR, Vescovi AL (1999) Epidermal and fibroblast growth factors behave as mitogenic regulators for a

single multipotent stem cell-like population from the subventricular region of the adult mouse forebrain. J Neurosci 19:3287–3297
Grove EA, Williams BP, Li DQ, Hajihosseini M, Friedrich A, Price J (1993) Multiple restricted lineages in the embryonic rat cerebral cortex. Development 117:553–561
Hurelbrink CB, Armstrong RJ, Barker RA, Dunnett SB, Rosser AE (2000) Hibernated human fetal striatal tissue: successful transplantation in a rat model of Huntington's disease. Cell Transplant 9:743–749
Jain M, Armstrong RJ, Elneil S, Rosser AE, Barker RA (2003a) Migration and differentiation of transplanted human neural precursor cells. Neuroreport 14:1257–1262
Jain M, Armstrong RJ, Tyers P, Barker RA, Rosser AE (2003b) GABAergic immunoreactivity is predominant in neurons derived from expanded human neural precursor cells in vitro. Exp Neurol 182:113–123
Johe KK, Hazel TG, Muller T, Dugich-Djordjevic MM, Mckay RD (1996) Single factors direct the differentiation of stem cells from the fetal and adult central nervous system. Genes Dev 10: 3129–3140
Johnson DE, Williams LT (1993) Structural and functional diversity in the FGF receptor multigene family. Adv Cancer Res 60:1–41
Kallur T, Darsalia V, Lindvall O, Kokaia Z (2006) Human fetal cortical and striatal neural stem cells generate region-specific neurons in vitro and differentiate extensively to neurons after intra-striatal transplantation in neonatal rats. J Neurosci Res 84:1630–1644
Kelly CM, Zietlow R, Dunnett SB, Rosser AE (2003) The effects of various concentrations of FGF-2 on the proliferation and neuronal yield of murine embryonic neural precursor cells in vitro. Cell Transplant 12:215–223
Kelly CM, Tyers P, Borg MT, Svendsen CN, Dunnett SB, Rosser AE (2005) EGF and FGF-2 responsiveness of rat and mouse neural precursors derived from the embryonic CNS. Brain Res Bull 68:83–94
Kelly CM, Precious SV, Penketh R, Amso N, Dunnett SB, Rosser AE (2007) Striatal graft projections are influenced by donor cell type and not the immunogenic background. Brain J Neurol 130:1317–1329
Kilpatrick TJ, Bartlett PF (1993) Cloning and growth of multipotential neural precursors: requirements for proliferation and differentiation. Neuron 10:255–265
Kilpatrick TJ, Bartlett PF (1995) Cloned multipotential precursors from the mouse cerebrum require FGF-2, whereas glial restricted precursors are stimulated with either FGF-2 or EGF. J Neurosci 15:3661
Kim HT, Kim IS, Lee IS, Lee JP, Snyder EY, Park KI (2006) Human neurospheres derived from the fetal central nervous system are regionally and temporally specified but are not committed. Exp Neurol 199:222–235
Koblar SA, Turnley AM, Classon BJ, Reid KL, Ware CB, Cheema SS, Murphy M, Bartlett PF (1998) Neural precursor differentiation into astrocytes requires signaling through the leukemia inhibitory factor receptor. Proc Natl Acad Sci USA 95:3178–3181
Kornblum HI, Hussain RJ, Bronstein JM, Gall CM, Lee DC, Seroogy KB (1997) Prenatal ontogeny of the epidermal growth factor receptor and its ligand, transforming growth factor alpha, in the rat brain. J Comp Neurol 380:243–261
Kuhn HG, Winkler J, Kempermann G, Thaal L, Gage FH (1997) Epidermal growth factor and fibroblast growth factor-2 have different effects on neural progenitors in the adult rat brain. J Neurosci 17:5820–5829
Larsen WJ, Churchill L (1998) Essentials of human embryology. Elselvier, New York
Lillien L, Cepko C (1992) Control of proliferation in the retina: temporal changes in responsiveness to FGF and TGF-alpha. Development 115:253–266
Lillien L, Raphael H (2000) BMP and FGF regulate the development of EGF-responsive neural progenitor cells. Development 127:4993–5005
Lim DA, Alvarez-Buylla A (2016) The adult ventricular-subventricular zone (V-SVZ) and olfactory bulb (OB) neurogenesis. Cold Spring Harb Perspect Biol 8(5)

Lois C, Alvarez-Buylla A (1994) Long-distance neuronal migration in the adult mammalian brain. Science 264:1145–1148

Luskin MB, Pearlman AL, Sanes JR (1988) Cell lineage in the cerebral cortex of the mouse studied in vivo and in vitro with a recombinant retrovirus. Neuron 1:635–647

Martin-Ibanez R, Guardia I, Pardo M, Herranz C, Zietlow R, Vinh NN, Rosser A, Canals JM (2017) Insights in spatio-temporal characterization of human fetal neural stem cells. Exp Neurol 291:20–35

Marz P, Herget T, Lang E, Otten U, Rose-John S (1997) Activation of gp130 by IL-6/soluble IL-6 receptor induces neuronal differentiation. Eur J Neurosci 9:2765–2773

Mayer M, Bhakoo K, Noble M (1994) Ciliary neurotrophic factor and leukemia inhibitory factor promote the generation, maturation and survival of oligodendrocytes in vitro. Development 120:143–153

Ornitz DM, Yayon A, Flanagan JG, Svahn CM, Levi E, Leder P (1992) Heparin is required for cell-free binding of basic fibroblast growth factor to a soluble receptor and for mitogenesis in whole cells. Mol Cell Biol 12:240–247

Ostenfeld T, Caldwell MA, Prowse KR, Linskens MH, Jauniaux E, Svendsen CN (2000) human neural precursor cells express low levels of telomerase in vitro and show diminishing cell proliferation with extensive axonal outgrowth following transplantation. Exp Neurol 164: 215–226

Ostenfeld T, Joly E, Tai YT, Peters A, Caldwell M, Jauniaux E, Svendsen CN (2002) Regional specification of rodent and human neurospheres. Brain Res Dev Brain Res 134:43–55

Palmer TD, Ray J, Gage FH (1995) FGF-2-responsive neuronal progenitors reside in proliferative and quiescent regions of the adult rodent brain. Mol Cell Neurosci 6:474–486

Palmer TD, Takahashi J, Gage FH (1997) The adult rat hippocampus contains primordial neural stem cells. Mol Cell Neurosci 8:389–404

Palmer TD, Schwartz PH, Taupin P, Kaspar B, Stein SA, Gage FH (2001) Cell culture: progenitor cells from human brain after death. Nature 411(6833):42–43

Pollard SM, Wallbank R, Tomlinson S, Grotewold L, Smith A (2008) Fibroblast growth factor induces a neural stem cell phenotype in foetal forebrain progenitors and during embryonic stem cell differentiation. Mol Cell Neurosci 38:393–403

Precious SV, Rosser AE (2012) Producing striatal phenotypes for transplantation in Huntington's disease. Exp Biol Med 237:343–351

Price J, Thurlow L (1988) Cell lineage in the rat cerebral cortex: a study using retroviral-mediated gene transfer. Development 104:473–482 PM:3151483

Qian X, Davis AA, Goderie SK, Temple S (1997) FGF-2 concentration regulates the generation of neurons and glia from multipotent cortical stem cells. Neuron 18:81–93

Ray J, Peterson M, Schinstine M, Gage FH (1993) Proliferation, differentiation, and long-term culture of primary hippocampal neurons. Neurobiology 90:3602–3606

Reynolds BA, Weiss S (1992a) A multipotent EGF-responsive striatal embryonic progenitor cell produces neurons and astrocytes. J Neurosci 12:4565–4574

Reynolds BA, Weiss S (1992b) Generation of neurons and astrocytes form isolated cells of the adult mammalian central nervous system. Science 255:1707–1710

Reynolds BA, Weiss S (1996) Clonal and population analyses demonstrate that an EGF-responsive mammalian embryonic CNS precursor is a stem cell. Dev Biol 175:1–13

Rietze R, Poulin P, Weiss S (2000) Mitotically active cells that generate neurons and astrocytes are present in multiple regions of the adult mouse hippocampus. J Comp Neurol 424:397–408

Rowell JJ, Ragsdale CW (2012) BrdU birth dating can produce errors in cell fate specification in chick brain development. J Histochem Cytochem 60:801–810

Santa-Olla J, Covarrubias L (1995) Epidermal growth factor (EGF), transforming growth factor-Ã (T GF-Ã), and basic fibroblast growth factor (bFGF) differentially influence neural precursor cells of mouse embryonic mesencephalon. J Neurosci Res 42:172–183

Santa-Olla J, Covarrubias L (1999) Basic fibroblast growth factor promotes epidermal growth factor responsiveness and survival of mesencephalic neural precursos cells. J Neurobiol 40:14–27

Seaberg RM, Van der Kooy D (2002) Adult rodent neurogenic regions: the ventricular subependyma contains neural stem cells, but the dentate gyrus contains restricted progenitors. J Neurosci 22:1784–1793

Straccia M, Garcia-Diaz Barriga G, Sanders P, Bombau G, Carrere J, Mairal PB, Vinh NN, Yung S, Kelly CM, Svendsen CN, Kemp PJ, Arjomand J, Schoenfeld RC, Alberch J, Allen ND, Rosser AE, Canals JM (2015) Quantitative high-throughput gene expression profiling of human striatal development to screen stem cell-derived medium spiny neurons. Mol Ther Methods Clin Dev 2:15030

Supeno NE, Pati S, Hadi RA, Ghani AR, Mustafa Z, Abdullah JM, Idris FM, Han X, Jaafar H (2013) IGF-1 acts as controlling switch for long-term proliferation and maintenance of EGF/FGF-responsive striatal neural stem cells. Int J Med Sci 10:522–531

Svendsen CN, Ter Borg MG, Armstrong RJ, Rosser AE, Chandran S, Ostenfeld T, Caldwell MA (1998) A new method for the rapid and long term growth of human neural precursor cells. J Neurosci Methods 85:141–152 ISI:000077608900004

Taupin P (2006a) Adult neural stem cells, neurogenic niches, and cellular therapy. Stem Cell Rev 2:213–219

Taupin P (2006b) Neural progenitor and stem cells in the adult central nervous system. Ann Acad Med Singap 35:814–820

Temple S, Davis AA (1994) Isolated rat cortical progenitor cells are maintained in division in vitro by membrane-associated factors. Development 120:999–1008. PM:7600974

Titomanlio L, Bouslama M, Le Verche V, Dalous J, Kaindl AM, Tsenkina Y, Lacaud A, Peineau S, El Ghouzzi V, Lelievre V, Gressens P (2011) Implanted neurosphere-derived precursors promote recovery after neonatal excitotoxic brain injury. Stem Cells Dev 20:865–879

Tropepe V, Craig CG, Morshead CM, Van der Kooy D (1997) Transforming growth factor-a null and senescent mice show decreased neural progenitor cell proliferation in the forebrain subependyma. J Neurosci 17:7850–7859

Tropepe V, Sibilia M, Ciruna BG, Rossant J, Wagner EF, Van der Kooy D (1999) Distinct neural stem cells proliferate in response to EGF and FGF in the developing mouse telencephalon. Dev Biol 208:166–188

Uchida N, Buck DW, He D, Reitsma MJ, Masek M, Phan TV, Tsukamoto AS, Gage FH, Weissman IL (2000) Direct isolation of human central nervous system stem cells. Proc Natl Acad Sci USA 97(26):14720–14725

Vescovi AL, Parati EA, Gritti A, Poulin P, Ferrario M, Wanke E, Frolichsthal-Schoeller P, Cova L, Arcellana-Panlilio M, Colombo A, Galli R (1999) Isolation and cloning of multipotential stem cells from the embryonic human CNS and establishment of transplantable human neural stem cell lines by epigenetic stimulation. Exp Neurol 156:71–83

Wagner JA (1991) The fibroblast growth factors: an emerging family of neural growth factors. Curr Top Microbiol Immunol 165:95–118

Walsh C, Cepko CL (1992) Widespread dispersion of neuronal clones across functional regions of the cerebral cortex. Science 255:434–440. PM:1734520

Walsh C, Cepko CL (1993) Clonal dispersion in proliferative layers of developing cerebral cortex. Nature 362:632–635. PM:8464513

Weiss S, Dunne C, Hewson J, Wohl C, Wheatley M, Peterson AC, Reynolds BA (1996) Multipotent CNS stem cells are present in the adult mammalian spinal cord and ventricular neuroaxis. J Neurosci 16:7599–7609

Wichterle H, Turnbull DH, Nery S, Fishell G, Alvarez-Buylla A (2001) In utero fate mapping reveals distinct migratory pathways and fates of neurons born in the mammalian basal forebrain. Development 128:3759–3771

Williams BP, Price J (1995) Evidence for multiple precursor cell types in the embryonic rat cerebral cortex. Neuron 14:1181–1188. PM:7605631

Williams BP, Read J, Price J (1991) The generation of neurons and oligodendrocytes from a common precursor cell. Neuron 7:685–693. PM:1931054

Yayon A, Klagsbrun M, Esko JD, Leder P, Ornitz DM (1991) Cell surface, heparin-like molecules are required for binding of basic fibroblast growth factor to its high affinity receptor. Cell 64: 841–848

Zhao M, Momma S, Delfani K, Carlen M, Cassidy RM, Johansson CB, Brismar H, Shupliakov O, Frisen J, Janson AM (2003) Evidence for neurogenesis in the adult mammalian substantia nigra. Proc Natl Acad Sci 100:7925

Zietlow R, Pekarik V, Armstrong RJ, Tyers P, Dunnett SB, Rosser AE (2005) The survival of neural precursor cell grafts is influenced by in vitro expansion. J Anat 207:227–240

Zietlow R, Precious SV, Kelly CM, Dunnett SB, Rosser AE (2012) Long-term expansion of human foetal neural progenitors leads to reduced graft viability in the neonatal rat brain. Exp Neurol 235:563–573

Chapter 2
Human Somatic Stem Cell Neural Differentiation Potential

David J. Eve, Paul R. Sanberg, Leonora Buzanska, Anna Sarnowska, and Krystyna Domanska-Janik

Abstract Human somatic stem cells can be identified and isolated from different types of tissues and are grouped here based on their developmental maturation and ability to undergo neural differentiation. The first group will represent afterbirth somatic tissues, which are perinatal stem cells including placental blood and tissue, amniotic fluid and tissue, and umbilical cord blood- and umbilical cord tissue-derived cells. The second group of cells discussed in this chapter is the adult stem cells, generally those in a transient period of development, thus placing them in the special position of transitioning from the perinatal to young somatic tissue, and they include the menstrual blood-, the peripheral blood-, and the bone marrow-derived stem cells.

Keywords Placenta · Amnion · Umbilical cord · Blood · Bone marrow · Neural differentiation

2.1 Afterbirth Somatic Tissues

2.1.1 Placenta

The gestational tissue known as the placenta originates from both the fetus and the mother. The maternal portion of the placenta is the decidua or endometrial lining into which the fetal blastocyst implants. The trophoblast develops from the outer layer of the blastocyst to form the outer layers of the placenta that can be divided into the amnion and chorion. Within the placenta, the fetus is bathed in amniotic fluid, and stem cells have been isolated from all of these fetally derived sources, as well as from

D. J. Eve (✉) · P. R. Sanberg
Department of Neurosurgery and Brain Repair, Center for Excellence in Aging and Brain Repair, University of South Florida, Tampa, FL, USA

L. Buzanska · A. Sarnowska · K. Domanska-Janik (✉)
Stem Cell Bioengineering Unit, Mossakowski Medical Research Centre Polish Academy of Sciences, Warsaw, Poland

L. Buzanska (ed.), *Human Neural Stem Cells*, Results and Problems in Cell Differentiation 66, https://doi.org/10.1007/978-3-319-93485-3_2

the maternally derived decidua (In 't Anker et al. 2004). The amnion and amniotic fluid-derived cells will be discussed in the next section, while the decidua- and chorion-derived cells will be discussed below.

2.1.1.1 Maternal Decidua-Derived Cells

The majority of placental studies focus on the fetal portion, though a few do explore the potential of the maternal decidua as a source of stem cells and the latter normally involves term placenta (afterbirth). For instance, Sabapathy et al. (2012) isolated maternal placental tissue and, following digestion, cultured plastic adherent fibroblast-like cells in a mesenchymal expansion medium. The cells were passaged multiple times before reaching confluence, and their multipotency, including adipogenic, chondrogenic, and osteogenic differentiation, was examined, as well as their cell surface markers, confirming that they were of the mesenchymal stem cell (MSC) lineage. Chromosomal integrity was also verified at multiple passages. Additionally, Sabapathy et al. performed neural or retinal differentiation using short-term incubation with α-minimum essential medium (αMEM) and either 5 mM β-mercaptoethanol (βME) or 50 μM taurine and 1 mM βME. Morphological changes toward a neural or retinal phenotype were observed. Staining with antibodies for microtubule-associated protein 2 (MAP 2), neuronal nuclei (NeuN), glial fibrillary acidic protein (GFAP), and neural filament were used to confirm that the cells were expressing neural-specific proteins, and polymerase chain reaction (PCR) analysis was performed to validate the presence of the photoreceptor genes calbindin2 and recoverin in the retinal-like cells. Despite these findings, the use of βME to induce neural differentiation is believed to be flawed causing changes to the actin cytoskeleton rather than triggering transdifferentiation (see Bone Marrow section for more information). While this study demonstrated that cells expressing MAP 2 and NeuN were generated in vitro, no electrophysiological evidence was provided confirming that these cells possessed a depolarized membrane or that they were able to transmit action potentials or secrete neurotransmitters. In addition, Sabapathy et al. (2012) do not describe whether normally cultured placental stem cells also express these factors. Hayati et al. (2011) observed protein expression by immunostaining for neurogenic markers such as GFAP, neurofilament, neuron-specific enolase (NSE), vimentin, and nestin in placental-derived stem cells under normal culture conditions (Dulbecco's modified Eagle's medium (DMEM), Ham's F12, 10% fetal bovine serum (FBS), 1 × GlutaMAX, 50 μg/ml vitamin C, and 1 × antibiotics) suggesting that these cells either possess neural stem cell-like properties without any need for neural induction or have the inherent plasticity to undergo neural differentiation in the right conditions. Nestin expression was also confirmed in at least half the placental-derived cells cultured by Semenov et al. (2010) using non-hematopoietic stem cell expansion medium (Miltenyi Biotec). However, there is a caveat with regard to nestin expression, since it is also expressed by endothelial cells (as well as other cell types) and so is not an exclusive neural stem cell marker (Koning et al. 2016). Another problem was mentioned earlier, and that is the use of βME in the media (described further in Bone Marrow section).

In a two-step process, Lu et al. (2012) generated dopamine (DA) neuron-like cells by inducing neurospheres from decidua-derived MSCs and then culturing the potential neural stem cells isolated from the neurospheres. Placental tissue was obtained at the birth of male babies so that the maternal source of the MSCs could be confirmed using karyotyping by the absence of the Y chromosome. Adherent MSCs were cultured in DMEM/F12 containing 20 ng/ml epidermal growth factor (EGF), 20 ng/ml human basic fibroblast growth factor (hbFGF), and 1:50 B27 supplement. Within 8–10 days, free-floating neurospheres were observed containing cells that expressed nestin and cluster of differentiation 133 (CD133), proposed markers of neural stem cells. However, placental decidua-derived MSCs under normal culture conditions also expressed slightly lower levels of nestin and negligible CD133. Fewer neurospheres were observed with late passaged cells (P19) compared with cells from an earlier passage (P3). Subsequent culturing of disassociated cells from the neurospheres in DMEM/F12 supplemented with 200 ng/ml sonic hedgehog (SHH), 100 ng/ml FGF8, 50 ng/ml brain-derived neurotrophic factor (BDNF), and 10 mM forskolin (known to increase cAMP) for 7–10 days led to cells exhibiting long branching processes and expressing NSE, GFAP, and tyrosine hydroxylase (TH). TH was only expressed after this second culturing step, while expression of GFAP and NSE in normal placental MSCs or neurospheres was not investigated, and so it is unclear whether expression was induced or already present. The cells were also able to secrete DA, but no electrophysiological evidence was provided.

While the majority of decidua studies involve term placenta, Park et al. (2013) cultured cells obtained from first trimester placenta as well as term placenta. They found that cells from both sources resembled MSCs and expressed "pluripotency-coupled" genes such as octamer-binding transcription factor 4 (OCT-4), Nanog, activin, and reduced expression protein 1 (Rex1), though the genes were more highly expressed in the first trimester cells. Culturing high concentrations of placenta MSCs in DMEM, 10% FBS, 0.1 mM βME, and 1% nonessential amino acids (NEAA) on polyethyleneimine resulted in spheroid formation, which on replating led to a single layer of differentiating nestin-expressing cells. The first trimester and term cells expressed markers of neural progenitors such as doublecortin (DCX) and growth-associated protein 43 (GAP43), but only the first trimester cells expressed sex-related Y-chromosome box 1 (SOX1) and neurogenic differentiation (NeuroD) suggesting they had a greater degree of plasticity than the term cells.

A myofibroblast-like cell has also been identified within the decidua that expresses pluripotency markers, such as stage-specific embryonic antigen 1 (SSEA-1), OCT-4, TRA-1-81, and the hematopoietic cell marker CD34. They are also able to undergo osteogenic, adipogenic, and chondrogenic differentiation (Strakova et al. 2008). Culturing these cells on gelatin, collagen, and collagen-carbon nanotube composite matrices in DMEM, 10% FBS, and 1% NEAAs was revealed to lead to neural cell commitment marker expression (SOX1 and nestin) within 8 h particularly when using the collagen-carbon nanotube composite matrix

(Sridharan et al. 2013). By day 6 on the composite matrix, the cells were positive for NeuN or neuronal βIII tubulin (using Tuj-1 antibody), and this was dependent on the β-1 integrin-mediated β-catenin signaling pathway. Use of an electrospun aligned collagen fibril matrix had similar effects and provided further evidence that the neural differentiation was related to the β-1 integrin-mediated β-catenin signaling pathway (Li et al. 2014). Remodeling of the 3-D collagen fibril matrix by the cells was also observed, suggesting that neural differentiation could be promoted by enhancing collagen-β-1 integrin interactions. However, there is some debate as to the specificity of neuronal βIII tubulin as a neuronal marker since it has also been identified in melanocytes, along with nestin, NGF, BDNF, neurotrophin 3 (NT-3), MAP 2, TH, and neurotrophic receptors (Locher et al. 2014). Therefore the absence of the detection of the melanocyte marker Melan-A may be useful to demonstrate that these cells are likely to be neuronal.

In two studies, Huang et al. (2009, 2010) explored whether decidua-derived stem cells could survive under hypoxic conditions in vitro, since hypoxic or ischemic conditions are likely to exist within damaged tissue such as the brain. They determined that the cells were viable under these conditions and were still able to differentiate into osteoblasts, chondrocytes, and adipocytes, but unfortunately they did not explore their neurogenic abilities. Since they appear to survive under hypoxic or ischemic conditions, these cells may be beneficial in the treatment of damaged brain tissue. Wu et al. (2015a) labeled the cells with [methyl-^{14}C]thymidine (^{14}C-TdR) and injected the cells via the caudal vein into female BALB/C nude mice. The distribution of the cells was tracked for up to 180 days. No cells were found to have migrated into the brain, though this could be due to the lack of any specific migratory signal, since no brain injury was present. It is therefore unclear whether these cells with their potential neurogenic properties could be beneficial in treating brain disorders when administered intravenously. Predifferentiation to NSCs or neural progenitors may have increased the likelihood of migration to the brain (as well as cues from injured brain tissue with a potentially impaired blood-brain barrier), and so further study is required. Fisher-Shoval et al. (2012) transplanted decidua-derived MSCs intracerebroventricularly into the experimental autoimmune encephalomyelitis mouse model of multiple sclerosis. The cells exerted a therapeutic effect by prolonging cell survival, and the cells survived for at least 5 days at the injection site and within the spinal cord. It is unclear whether the surviving cells were differentiating into neural stem cells and neural progenitors or whether they were remaining as MSCs due to the very short time frame. Additional long-term studies would be necessary to help elucidate this.

2.1.1.2 Fetal Chorion-Derived Cells

In addition to the maternally derived placental stem cell studies, there are several utilizing the fetal chorion-derived placental stem cells. These cells can be harvested at term, during villocentesis, or abortion for prenatal diagnosis (first to second trimester), or from nonviable tissue (first or second trimester) from either the chorion membrane layer or the chorionic villi (CV). One of the first studies utilized the explant culture method to culture chorionic villi from the term placenta (Igura et al. 2004). The cells harvested from the explants were either spindle-shaped or large flat cells, expressed MSC, but not hematopoietic markers and could undergo osteogenic, chondrogenic, and, to a limited extent, adipogenic differentiation, suggesting that they could be classified as MSCs. In line with some of the studies described earlier, neuroglial markers such as nestin, NSE, βIII tubulin, GFAP, and myelin basic protein (MBP) were also detected. Twenty-four hours after culturing in neural induction media (DMEM/F12, B27 supplement, 100 μM butylated hydroxyanisole [BHA], 0.5 mM 3-isobutyl-1-methylxanthine [IBMX], 1 mM dibutyryl cyclic adenosine monophosphate [db-cAMP], 1.5% dimethyl sulfoxide [DMSO], and 20 ng/ml bFGF), the cells changed morphologically to a neural lineage and sent out projections that interacted with multiple neighboring cells in a network-like fashion. The neural markers described above were also still expressed along with TH, suggesting that the cells may be able to secrete DA. However no electrophysiological data was provided to confirm that functioning neurons had been generated and the cells were not followed long term.

Portmann-Lanz et al. (2006) quantified the proportion of first trimester CV and term placental chorion layer MSC-like cells that appeared to undergo neuronal differentiation when cultured in DMEM, 10% FBS, and 30 μM all-trans-retinoic acid. They observed that approximately 20–30% of the cultured cells expressed CD133, nestin, or heavy neurofilament 200 (NF200), which they suggest are indicative of NSCs, neural progenitors, and mature neurons, respectively. Morphologically these cells possessed long thin processes and growth cones, but no electrophysiological or neurotransmitter expression data was performed to verify that these cells were neurons. In a subsequent study, the same authors used three different methodologies to select and differentiate these cells into the neural lineage (Portmann-Lanz et al. 2010). Firstly, they cultured them for four passages in Neurobasal medium that included 2% B27 supplement, 2 mM GlutaMAX™, 20 ng/ml human epidermal growth factor (EGF), 20 ng/ml hFGF-2, and 2 μg/ml heparin in the absence of serum, and the cells formed neurospheres, or in α-MEM with 10 ng/ml FGF-2, 10 ng/ml EGF, and 20% serum, where the cells were adherent. A third method involved magnetically selecting cells that expressed SSEA-4 to obtain a pure stem cell population. The neurospheres and adherent cells were terminally differentiated for a week on collagen-coated plates in serum-containing (10%) Neurobasal media along with 1 μM all-trans-retinoic acid and 15 ng/ml hBDNF, while the SSEA-4$^+$ cells were initially cultured in the above media supplemented with 10 ng/ml hEGF and 10 ng/ml hFGF-2, followed by a further week in media that excluded hEGF and hFGF-2. Approximately 33% of the cells cultured utilizing the first two methods died within 3 days, and those that survived either rapidly formed neurospheres or became thinner and longer cells. Within 24 h in differentiation media, the cells underwent neurite-like branching, with the projections extending and the cell body shrinking over time, and the SSEA-4$^+$ cells adopted a similar phenotype when cultured in the same media. Within a week, over half the cells were morphologically neuronal and expressed early postmitotic markers such as βIII tubulin and turned on after division (TOAD)/unc33-like protein (Ulip)/collapsin-response-mediator protein (CRMP) 4 (TUC-4). Up to 5% of control placental cells also expressed βIII tubulin, but did not show any neuronal morphology. More mature neuronal markers such as NeuN and medium neurofilament were also expressed by 10% of the neuron-like cells from neurospheres or adherent culture. Nestin expression increased from 5 to 25% following differentiation suggesting that NSCs were present. A proportion of the cells also expressed different neurotransmitters. For instance, 15% of the neurosphere- or adherent-cultured cells expressed DA, while 10% and 3% of the magnetically selected cells expressed serotonin and glutamate. No cells expressed adult neuronal markers such as MAP 2 or Tau.

Up to 20% of the cells were immature oligodendrocytes following differentiation, with less than 1% being present in nondifferentiated cell cultures. The oligodendrocyte transformation was lower from the first trimester chorion tissue compared with the term tissue. MBP^+ cells were also observed, but no cells expressed GFAP, suggesting no astrocytes were present. Poloni et al. (2008) confirmed that oligodendrocytes could be obtained from first trimester chorion tissue (obtained by villocentesis) following culture in Neurocult^{TM} neuronal proliferation and differentiation media (Stem Cell Technologies, Cambridge, MA, USA) after neurosphere formation. They also detected nestin before and after differentiation. While these results suggest that neural stem cells and immature neurons expressing some neurotransmitters were generated, long-term culture and electrophysiological evidence are necessary to determine whether the cells can proliferate and mature and whether they are really neuronal.

In a separate study using term placenta, the heterogenous chorionic cells were enriched by magentic cell sorting for c-KIT, a stem cell factor (also known as CD117), and neurogenic differentiation was induced via culturing with α-MEM, 5% FBS, and 3 μg/ml all-trans-retinoic acid (Resca et al. 2013). Passaging in α-MEM of both an enriched and an unenriched c-KIT^+ cell population revealed that c-KIT expression decreased in the unenriched, but remained in the enriched population. The c-KIT population after neural differentiation demonstrated more βIII tubulin-expressing cells as well as a higher number of oligodendrocytes (revealed by 2′,3′-cyclic-nucleotide 3′-phosphodiesterase [CNPase] staining). An absence of $GFAP^+$ cells was again seen, suggesting that these cells do not differentiate into astrocytes.

Several of the above studies suggest that undifferentiated chorionic stem cells may have neurogenic properties due to their expression of nestin and other neural markers within a subpopulation of the cells. This was explored further by Calzarossa et al. (2013) who demonstrated that the heterogeneous population of cells from first trimester villocentesis samples following culture in AmnioMAX II medium (Invitrogen, Carlsbad, CA, USA) expressed NSC and early progenitor markers such as nestin (86%) and SOX-2 (38%) as well as markers of neuronal precursors and immature neurons or neuroblasts such as DCX (13%), GFAP (18%), O4 (oligodendrocyte marker; <10%), and galactosylceramidase (GalC; <10%), even though no neural phenotype was observed. Electrophysiological evidence showed a negative resting potential (−40 mV), but no additional specific neuronal activity was observed such as depolarization or action potentials. The cells did secrete the neurotrophin BDNF, but not glial-derived neurotrophic factor (GDNF) or neurotrophin-3 (NT-3). Interestingly their conditioned media were neuroprotective when SH-SY5Y cells were pretreated for 2 h prior to 6-hydroxydopamine (6-OHDA)-induced apoptosis. However, when Calzarossa et al. (2013) grew these cells in an NSC-specific media of unreported composition, they formed neurospheres that were unable to proliferate, and the cells soon died, suggesting that they were not "classical" NSCs.

The neurogenic potential of first trimester and term chorion-derived MSCs was explored by Jones et al. (2012). They used DMEM supplemented with 0.5% FBS, 2 mM L-glutamine, and the flavone glycoside baicalin (0.1%), along with membrane insert coculture with C17.2 mouse NPCs. Both first trimester and term chorion-derived MSCs expressed βIII-tubulin, MAP 2, and N-methyl D-aspartate (NMDA) receptor NR1. A higher level of "pluripotency-related" genes were expressed by the first trimester cells, and these cells were more effective in restoring bone quality and plasticity in the osteogenesis imperfecta mouse model or enhanced wound healing compared with the term cells. The authors did not directly compare their ability to undergo neural differentiation in vivo, but their other studies may suggest that the first trimester cells could be most effective.

Similar to the decidua-derived cells, culture of term chorion-derived cells on gelatin-coated tissue culture plates and electrospun nanofibrous gelatin scaffolds was shown to lead to neural differentiation (Faghihi et al. 2016). However, in this study, the culture media contained 20% serum as well as differentiating factors such as 10 ng/ml bFGF, 250 mM IBMX, and 100 mM βME for initial expansion and 10 μM all-trans-retinoic acid and 100 ng/ml sonic hedgehog (SHH) followed by 100 ng/ml BDNF for differentiation. By the second week, long axon-like projections were evident, along with immunostaining for choline acetyltransferase (ChAT) and SMI-32 (or heavy neurofilament) suggesting the cells may be differentiating into motor neurons. However no functional or electrophysiological analysis was performed to confirm these findings.

Several of the compounds used to differentiate stem cells to neural stem cells or neurons lead to activation of cAMP-dependent protein kinases and/or increase intracellular cAMP, which then activate protein kinase A (Tio et al. 2010). Growth factors that further support differentiation and survival are frequently also included. Talwadekar et al. (2016) isolated cells from the central part of the term placenta (likely to be chorionic villi) and cultured them in Neurobasal medium, supplemented with 2% B27, 1% N2, 40 ng/ml bFGF, 20 ng/ml EGF, 1 mM db-cAMP, 0.5 mM IBMX, and 200 mM L-glutamine, and they observed a gradual increase in neurite formation by 6 days and βIII tubulin expression over 10 days. They observed nestin expression in the majority of their cells prior to differentiation as well as low levels of MAP 2, NF200, and S100β, all of which increased significantly following differentiation. However an increased senescence and reduction in βIII tubulin was observed by 15 days, suggesting that long-term culturing may not be viable. However, the authors do not mention that the cells were passaged, and therefore the cells may have become confluent with rampant contact inhibition, and so further clarification of why the cells became senescent is required, since the other studies mentioned above do not appear to show this, but did explicitly involve passaging.

The authors also added the histone deacetylase (HDAC) inhibitor, valproic acid (VPA; 1 mM), to the media, since HDAC transcriptionally represses numerous genes, and so removal of this repression by HDAC inhibition could potentially promote differentiation. Enhanced differentiation was observed with higher neurite formation and expression of neuronal lineage markers by 10 days—unfortunately the authors did not investigate what effect was seen at 15 days. Mitochondrial membrane potential was also enhanced suggesting greater differentiation. VPA was also observed to decrease notch-1 signaling, which has been implicated in the maintenance of proliferation rather than differentiation.

Chorion-derived MSCs have been transplanted into a number of animal models including those for stroke. In 2009, Yarygin et al. (2009) transplanted chorion-derived MSCs labeled with Dragon-fluorescent-tagged magnetic particles intravenously into male Wistar rats 2 days after they were given a temporary middle cerebral artery occlusion. The cells were followed immunocytochemically for 19 days and by magnetic imaging for 28 days. The cells migrated toward the infarct via the vasculature, and the infarct size was decreased by 12 days after transplant. Adjacent to the infarct, $NeuN^+$-transplanted cells were observed, while by 19 days, a few $GFAP^+$ transplanted cells could be detected within the infarct. This suggests that differentiation of the cells may be occurring in vivo toward a neuronal and astroglial phenotype, respectively. Migration toward the neurogenic niches, subventricular zone, and dentate gyrus of the hippocampus was also evident which coincided with potentiation of endogenous neurogenesis, suggesting that the reduced infarct size was more likely due to a paracrine effect on the endogenous cells rather than the transplanted cells differentiating toward a neuronal or astroglial phenotype. No functional or electrophysiological analysis was performed to confirm neuronal differentiation of the transplanted cells.

Park et al. (2011) isolated cells from the chorionic villi of first trimester placenta, and 30–50% of the cells expressed nestin, and the majority expressed pluripotency markers. However one possible cause for concern is the same authors refer to these cells as decidua-derived in a future paper, and so there may be some confusion over their exact source (Park et al. 2013). After culturing them on polyethylenimine (PEI)-coated plates for 3 days, the cells formed neurospheres, and after expansion on a fresh culture plate, they formed a rosette-like spread of neural progenitor-like cells. The cells were then expanded for several passages. A neurogenic differentiation media containing 10 μM retinoic acid and N2 supplement, followed by exposure to serum-free conditioned media from a dopaminergic neuroblastoma (MN9D) culture, was also used to differentiate the neurospheres. Nestin expression increased to nearly 100% of the sphere-forming cells, while the pluripotency marker expression decreased. Neural markers such as DCX, SOX1, GAP43, and NeuroD increased during neural progenitor differentiation. Terminal differentiation with the conditioned media resulted in paired box protein 6 (Pax6), MAP 2, GFAP, and TH expression by Western blotting and PCR.

Bilateral striatal transplantation of the neural precursors into neonatal rats, 2 weeks after the rats were subjected to a hypoxic-ischemic insult, improved locomotor activity. The cells survived over 8 weeks and coexpressed NeuN and human nuclear antigen and TH and human nuclear antigen, suggesting that the progenitor cells were differentiating into dopaminergic cells. In a second study, the authors also transplanted the neural progenitors into the striatum of rats that had received 6-OHDA 30 days prior to denervate the dopaminergic nigrostriatal projections (Park et al. 2012). Amphetamine-induced rotational behavior was significantly reduced 2 weeks following transplantation, and this persisted for 22 weeks. DA reuptake was significantly increased compared with pretransplant values suggesting that DA was being secreted. The colocalization of human nuclear protein and NeuN and human mitochondria and TH within the grafted cells 12 weeks after transplant implies that the transplanted cells survived and differentiated into dopaminergic-secreting cells.

A few studies appear to use cells isolated from the placenta as a whole without any discrimination between the different layers, suggesting that they may be a heterogeneous population of fetal amnion and chorion and maternal decidua-derived cells. In their first study Yen et al. (2005) differentiated their MSC-like cells, which also expressed "pluripotency" genes, into neuron-like cells, that expressed MAP 2 and NSE, by culturing in media supplemented with 1 μM retinoic acid for 6–14 days. Since retinoic acid is cytotoxic at high concentrations, the authors explored in a subsequent study whether 0.5 mM IBMX or neonatal rat brain cell cocultures were as effective as 1 μM retinoic acid for differentiating their placental-derived cells (Yen et al. 2008). They observed NSE expression within the cells by all three methods, suggesting that some degree of neural differentiation does occur. The authors also determined that an inhibitor of Rho kinase (ROCK), Y-27632 (10 μM), was also able to induce changes toward a neural phenotype including cytoskeletal alterations such as the formation of neurite-like projections, as well as increased nestin and MAP 2 expression, in a greater proportion of their placental-derived cells than 0.5 μM retinoic acid (Wang et al. 2013a). The undifferentiated placental cells were also transplanted into the rat cortex prior to occlusion of the middle cerebral artery and their location and effect on the ischemia-induced behavioral deficits recorded (Wu et al. 2015b). Improvement in the behavioral tests was observed by day 5, but no migration of the transplanted cells from the injection site was apparent. The size of the infarct and degree of microglial activation were also significantly decreased. However no examination of whether the cells were differentiating was performed, though the lack of migration may suggest that their effect was primarily paracrine rather than differentiation and cell replacement. Additional studies involving transplantation of mixed placental-derived cells also do not study whether the cells differentiated in vivo and so are beyond the subject of this book chapter and will not be discussed here.

2.1.2 Amnion and Amniotic Fluid

The amnion is fetus-derived and can be divided into epithelial and mesenchymal layers, both of which contain stem cells (AECs and AMSCs, respectively). These cells may be secreted into the amniotic fluid which has also been shown to contain stem cells. There are therefore three different types of potential cells that can be classified as amnion: epithelial, mesenchymal, and fluid. Amniotic fluid can be obtained during amniocentesis in the second trimester or at term, while the epithelial and mesenchymal layers of the amnion can theoretically be obtained at any stage of development (first or second from miscarriage or aborted tissue, or at term).

2.1.2.1 Amniotic Epithelial-Derived Cells

The amniotic epithelial layer lines the amniotic cavity and so is in direct contact with the amniotic fluid. This layer can be scraped away from the other amniotic layers, and the cells cultured in Roswell Park Memorial Institute (RPMI) 1640 medium supplemented with 10% FBS. Immunocytochemical analysis shows that a high proportion of these cells express neural markers such as neurofilament, MAP 2, vimentin, and a smaller proportion express GFAP, CNPase, and MBP (Sakuragawa et al. 1996). This suggests that many of the cells are potentially neurogenic, while a smaller proportion of them are potentially astrocytic or oligodendrocytic. The same authors also demonstrated that there is a heterogeneous expression of choline acetyltransferase (ChAT) by these cells, implying that they could have neurogenic potential to replace the damaged or missing ChAT-expressing neurons that are typical of Alzheimer's disease and other disorders (Sakuragawa et al. 1997). They also showed that approximately 10% of these cells express TH, secrete DA, and reduce apomorphine-induced rotation following striatal transplantation into 6-OHDA-treated rats (Kakishita et al. 2000). However, while some TH-positive cells remained at the graft site, no morphological evidence of differentiation was apparent, as the cells remained oval or round in shape. Endogenous cell sprouting was not evident over the 2-week period after transplantation, suggesting any improvement was a direct effect of the transplanted amniotic epithelial cells (AECs).

The above studies suggest that AECs have an intrinsic neural progenitor-like capacity, and so can this be enhanced by use of neural differentiating medium? Serum-free Neurobasal medium supplemented with B27 was utilized by Okawa et al. (2001) for the culture and differentiation of rat AECs, and after 4 days, they observed a change in morphology from isolated round or oval cells to pyramidal cells connected by processes. These cells were highly positive for nestin (cells cultured only in DMEM and FBS were also slightly positive). Transplantation of these cells into the hippocampus of gerbils that had been subjected to ischemia and reperfusion showed some migration of cells, which was not seen in nonischemic animals. Some of the cells were positive for neurofilament, and the cells that migrated were strongly stained for MAP 2 but poorly stained for nestin, suggesting that the migrating cells may have differentiated into neurons (or at least MAP 2^+ cells).

Differentiation factors such as retinoic acid, bFGF, and other neurotrophic factors have also been used. For instance, Miki et al. (2005) initially cultured AECs at high density in DMEM, 10% FBS, 2 mM *L*-glutamine, 1% NEAAs, 55 μM βME, 1 mM sodium pyruvate, and 10 ng/ml EGF, and they observed cells that remained free-floating, cells that were loosely attached as spheroids, and cells that strongly adhered to the plastic of the culture flasks or dishes after 3 days. AECs expressed "pluripotency" genes such as SSEA-3, SSEA-4, Oct-4, and nanog, with the highest expression present in the spheroid-forming cells. Culturing the fresh AECs in media supplemented with 50 μM retinoic acid and 10 ng/ml bFGF led to an increase in nestin and glutamate dehydrogenase (GAD) expression over 7 days by the cultured cells. A neuronal morphology was observed along with GFAP and CNPase expression suggesting cells were being induced toward glial and neuronal cells.

Using the same initial media described above, a separate study explored the contribution of retinoic acid and bFGF to the differentiation of these cells in culture (Niknejad et al. 2010). They isolated the floating cells and spheroid-forming cells and transferred them to nonadherent dishes for 5 days. The aggregated cells were then dissociated and plated on gelatin-coated plates with media supplemented with (or without) 10 ng/ml bFGF, 1 μM retinoic acid, and 100 ng/ml Noggin for 21 days. Noggin inhibits bone morphogenetic proteins (BMPs), which inhibit neurogenesis, meaning that Noggin should promote neurogenesis. In the absence of bFGF, the expression of neural markers, such as βIII tubulin, was decreased, while NSE and NeuN appeared to be unaffected. Removal of EGF led to a significant decrease in viability showing that it is crucial for AEC survival. Retinoic acid had a similar effect to bFGF, with the greatest expression of βIII tubulin observed when both factors were present. Inhibition of BMPs by Noggin also promoted neural differentiation with the highest βIII tubulin expression observed when Noggin, retinoic acid, and bFGF were used together. The absence of Noggin actually promoted expression of nonneural markers, e.g., the endodermal marker GATA-6. Since FBS is xenogeneic for human cells and there is some suggestion that serum may also inhibit neurogenesis, alternatives were also explored. Unfortunately removal of serum completely, significantly reduced cell viability which could be partially restored by use of insulin/transferrin/selenium (ITS; 1%). However use of tenfold less serum (1% instead of 10%) and ITS restored the cell viability to 10% serum values and was also shown to significantly increase βIII tubulin expression. As described earlier, Okawa et al. (2001) and others have used serum-free Neurobasal media to differentiate AECs suggesting that FBS is not necessary under the right conditions.

The use of different substrates for culturing may also have a bearing on differentiation potential. Fresh AECs were cultured on poly-D-lysine-/laminin-coated coverslips in neural basal A medium containing 50 μM all-trans-retinoic acid, 10 ng/ml FGF4, and supplements N2 and B27 (Ilancheran et al. 2007). At the beginning of culturing, the majority of the cells expressed nestin, MAP 2, and GFAP. Cell numbers were seen to decline rapidly when cultured in neural differentiation media and by 4 weeks. The majority of the cells still expressed GFAP, but only 5 and 10% expressed nestin and MAP 2, respectively. The MAP 2^+ cells resembled neurons with large central bodies and long thin projections, while the $GFAP^+$ cells resembled astrocytes with a similar but distinct morphology.

Additional supplementary compounds have also been explored for neural differentiation. AECs express melatonin 1 but not melatonin 2 receptors, and melatonin may be neuroprotective by promoting endogenous neurogenesis (Kaneko et al. 2011). Use of a standard culture media consisting of DMEM, 10% FBS, 2% B27, and 10 ng/ml bFGF supplemented with 100 μM melatonin for 5 days led to melatonin 1 receptor-expressing AECs to exhibit a neuronal phenotype, and βIII tubulin was significantly higher than in cells treated in the absence of melatonin. Using a slightly different basal media (DMEM/F12, 10% FBS, 10 ng/ml EGF, and 55 μM βME), the addition of 50 μM retinoic acid and 0.01–100 μM melatonin and 18 days of culturing, bovine AECs were seen to express high levels of βIII tubulin, MAP 2, and GFAP (Gao et al. 2016). Wingless-related integration site 4 (WNT4) knockout AECs exhibited decreased expression suggesting that WNT4 may be involved in the neural differentiation of AECs. Combined retinoic acid, melatonin, and WNT4 supplementation induced the highest level of neural progenitor gene and protein expression. Axonal-like projections were also observed with regard to these cells suggesting that neural progenitors or immature neuronal cells were formed. Neurally differentiated and undifferentiated AECs were transplanted into the spinal cord of immunosuppressed mice 6 days after a spinal cord contusive injury. After 2 weeks, many surviving neural cells were evident, but only a few undifferentiated AECs remained. In some ways this is surprising since AECs only express low levels of human leukocyte antigen ABC (HLA ABC) and are believed to be immunosuppressive, so they are not expected to trigger an immune rejection. Behavioral improvement was also observed, suggesting that transplantation of neurally differentiated AECs may be beneficial in treating spinal cord injury.

2.1.2.2 Amnion Mesenchyme-Derived Cells

Between the epithelial layers of the amnion and the chorion lies a mesenchymal layer from which amniotic mesenchymal stem cells (AMSCs) can be obtained. Sakuragawa et al. (2004) were again among the first to attempt to differentiate AMSCs into neural stem cells. They initially cultured the cells in DMEM/F12 supplemented with 10 ng/ml human leukemia-inhibitory factor (LIF), 0.2 mM βME, and 10% FBS. A high proportion of the cells expressed Musashi-1 and nestin, with a modest population also expressing βIII tubulin and GFAP, suggesting that they possess a propensity to become neural stem cells. Neural differentiation occurred by culturing the cells in DMEM, 20% FBS, and 10 ng/ml bFGF for 24 h followed by DMEM supplemented with 100 μM BHA, 10 μM forskolin, 2% DMSO, 5 U/ml heparin, 5 nM K252a (staurosporine analogue), 25 mM KCl, 2 mM valproic acid, 1 × N2, 10 ng/ml bFGF, and 10 ng/ml platelet-derived growth factor-BB (PDGF-BB) for another 24 h. This significantly increased the expression of βIII tubulin, GFAP, and neurofilament-M (medium; NF-M) and did not significantly affect nestin or Musashi-1. More long-term culture under these conditions would help to confirm whether the cells did adopt a neural stem cell/neuroglial phenotype and whether the cells were functional.

A similar neural differentiation protocol consisting of DMEM, 2 mM valproic acid, 15 mM betaine, 2.5 mM taurine, 175 μM BHA, 27 nM selenium, 20 nM progesterone, 10 μM forskolin, 10 nM K252a, 5 U/ml heparin, 1 mM sodium pyruvate, 50 μM α-thioglycerol, 20 nM bathocuproinedisulfonic acid, and 10 ng/ml bFGF after an initial incubation in DMEM, 1% FBS, and 10 ng/ml bFGF was utilized by Marcus et al. (2008) to differentiate rat AMSCs. Prior to differentiation they also observed expression of neural-related genes such as nestin, βIII tubulin, GFAP, Olig1, MAP 2, and NF-M. Within 24 h of the addition of the differentiation media, the majority of the AMSCs adopted a neuron-like phenotype, including compact cell bodies and elaborate processes that formed networks. These cells also increased their expression of neuronal genes such as NF-M, and they began to express tau as well as downregulating expression of NSC markers such as SOX2 and NeuroD. This may mean that neural differentiation of the AMSCs utilizing the media in this study may result in terminally differentiated cells rather than NSCs or progenitors, though the cells are likely to pass through a progenitor stage at an earlier time frame. Of course no electrophysiological data or measurement of neurotransmitter release was made, so there is no confirmation that a mature neuronal phenotype was obtained. These cells were followed for 7 days compared with the 1 day of the previous study.

Other media compositions have also been tested. For instance, Neurobasal medium, supplemented with 50 μM all-trans-retinoic acid, 10 ng/ml FGF-4, and N2 and B27 supplements, was utilized by Fatimah et al. (2013). They observed that the cells again began to adopt a neural-like phenotype, and some of the cells accumulated to form neurospheres. Stemness gene expression was shown to decline with serial passaging and differentiation, though nestin expression did not decline. In vivo, cells are frequently located within an extracellular matrix (ECM), and cell culture flasks/dishes are frequently coated with an extracellular matrix-like layer to foster attachment. Hu et al. (2013) explored whether an acellular matrix derived from the amnion aids the growth and differentiation of AMSCs. Initially the AMSCs were cultured on the matrix in DMEM/F12, 20 mg/ml B27, and 20 ng/ml bFGF for 2 days, followed by 1 day in DMEM/F12, 20 mg/ml B27, 1 μM retinoic acid, and 20 ng/ml β-NGF before a further 3 days in DMEM/F12, 10% FBS, and 20 ng/ml bFGF. They observed no significant change in nestin expression (though there was a trend toward a decrease) but a significant increase in NSE expression during neural differentiation. Synaptophysin expression also increased, suggesting that the AMSCs may be differentiating to terminally differentiated neurons. Additional studies are required to determine whether NSCs are formed first, since the AMSCs express nestin prior to differentiation and no other NSC markers were investigated. A decrease in GFAP expression was also observed suggesting a possible reduction in differentiation toward astrocytes. In the absence of the acellular matrix, nestin expression was significantly higher and synaptophysin significantly lower than that observed in its presence, which implies that the matrix could be potentiating synapse formation and terminal differentiation. The expression of homeobox 9 (Hb9; also known as motor neuron and pancreas homeobox 1) suggests that the cells may be differentiating into a motor neuron precursor.

After transplanting AMSCs (or any other stem cell), it would be helpful to track what happens to the cells. One possible method of labeling cells is the use of superparamagnetic iron oxide particles (SPIOs). Any labeling method must be shown to not affect the cells' viability or ability to differentiate. Zeng et al. (2011) investigated this in vitro using a single or multiple labeling steps. AMSCs were neuronally differentiated using DMEM, 10 ng/ml bFGF, 50 ng/ml NGF, and 10 μM retinoic acid for 7 days. Within 3 days the AMSCs began to adopt a neural-like phenotype, and nestin levels decreased, while NSE expression increased suggesting that the cells were maturing into neurons since no GFAP expression was present. SPIOs at concentrations greater than 14 μg/ml also appeared to reduce the cell viability, though the presence of SPIO did not affect the differentiation of the cells.

Another means of labeling stem cells is the use of DiI. AMSCs labeled with DiI were shown to begin to express GFAP and βIII tubulin, 45 days after transplantation into a permanent middle cerebral artery occlusion (MCAo) rat stroke model, while GFAP and NeuN were expressed by the cells 21 days after transplant in a temporary MCAo stroke model (Kholodenko et al. 2012). The cells were also observed to migrate into the ischemic region and hippocampus. This shows that in vivo, some of the cells will differentiate into neurons and astrocytes.

Differentiated and nondifferentiated AMSCs could also be transplanted to determine their ability to repair brain injury. Yan et al. (2013) used media composed of Knockout™ DMEM/F12 (Invitrogen), 20 ng/ml EGF, 20 ng/ml bFGF, 1:50 StemPro® NSC SFM supplement (Invitrogen), and 1:100 GlutaMAX™-1 supplement to differentiate AMSCs in ultralow attachment flasks. The cells formed neurospheres within 3 days and were cultured for 10 days. The neurospheres were then replated onto poly-L-lysine and laminin-coated plates and cultured in Neurobasal medium containing 0.5 μM retinoic acid, 1% FBS, 5% horse serum, and 1% N2 supplement to confirm that they could be terminally differentiated. The cells within the neurospheres and the nondifferentiated cells were transplanted into rats 4 days after they had been subjected to a traumatic brain injury, and 28 days later, their survival and differentiation state was analyzed. Prior to differentiation, few nestin-, SOX2-, and Musashi-expressing cells were detected, but this significantly increased after neurosphere formation. Following terminal differentiation, a proportion of the cells expressed βIII tubulin, GFAP, and GalC; however not all the cells adopted the expected morphology of cells expressing these factors (i.e., neurons, astrocytes, and oligodendrocytes, respectively). The greatest cognitive and functional recovery was observed in rats treated with the neurosphere-derived cells, and around 7% of the transplanted cells survived by 28 days. The transplanted cells expressed βIII tubulin, GFAP, and GalC, but were not of the expected morphology, suggesting that full differentiation into neurons, astrocytes, and oligodendrocytes may not have occurred.

Another study involving transplantation of hAMSCs for treating ischemia in rats used DMEM, 2% DMSO, and 100 μM BHA to differentiate the cells for 21 days (Li et al. 2012). AMSCs expressing NSE and GFAP were used, and the cells underwent morphological changes to resemble neural cells. Nondifferentiated AMSCs were transplanted 2 weeks after ischemia, and the rats followed for 8 weeks. A significant improvement in the neurological score/behavioral tests was observed, and the cells were observed around the injection site and within the lesioned area. Unfortunately no study to identify whether the transplanted cells had differentiated was performed even though they have the ability to do so according to the in vitro cultures. Multiple injections of hAMSCs into mice modeling amyotrophic lateral sclerosis (ALS) showed benefit, but no evidence of neural differentiation in vivo, based on the lack of βIII tubulin and GFAP staining (Sun et al. 2014). The cells were shown to differentiate in vitro (based on NF-M expression) when cultured in α-MEM, 10% FBS, 100 nM dexamethasone, 500 nM linoleic acid, 10 ng/ml PDGF, and 10 ng/ml bFGF for 29 h. In an intracerebral hemorrhage rat model, transplantation of hAMSCs 24 h after injury led to improvements in behavioral testing, as well as evidence of neurogenesis (by bromodeoxyuridine [BrdU] and DCX expression) in the subventricular zone (SVZ) and angiogenesis 27 days later (Zhou et al. 2016). The transplanted cells did not express βIII tubulin or GFAP suggesting that they were not undergoing neural differentiation, and the majority were found in the vicinity of the perihematomal region, implying that they were not the cells undergoing neurogenesis in the SVZ, though confirmation of the lack of BrdU and DCX expression would have helped to confirm the endogenous nature of the increased neurogenesis.

2.1.2.3 Amniotic Fluid-Derived Cells

The amniotic fluid bathes the growing embryo and is likely to contain cells from the embryo itself, the amnion, and potentially other placental regions throughout the gestational period, as well as water and nutrients. These cells are therefore primarily fetal in origin. Amniotic fluid samples can be obtained relatively safely during amniocentesis (normally performed for genetic testing in the second trimester), amnioreduction (from the beginning of the third trimester if fluid levels and hence pressure become too high within the amniotic sac), and cesarean section at term. The amniotic fluid is likely to contain more cells as the fetus and extraembryonic tissue develop toward term. Due to the number of potential different sources for cells within the amniotic fluid, a heterogeneous mixture can be found including terminally differentiated cells which comprise the majority of cells, as well as mesenchymal stem cells (AFMSCs), epithelial-derived, and fibroblastic cell types (Hoehn et al. 1974, 1975; In 't Anker et al. 2003). Additionally a multipotent stem cell type that expresses CD117 (or c-KIT), but is lineage negative (CD117(c-KIT)$^{+}$/lin^{-} cells), has been classified as amniotic fluid stem cells (AFSCs), though they only comprise approximately 1% of a population of amniotic fluid-derived cells that adhere after 7–10 days of culture (De Coppi et al. 2007). These cells, isolated from both human and rodent amniotic fluids, could be differentiated along the neural lineage (according to nestin expression) following culture in DMEM, 2% DMSO, 200 μM BHA, and 25 ng/ml NGF for 2 days and further culture in DMEM and 25 ng/ml NGF for 6 days. Following differentiation the cells were transplanted intracerebroventricularly into neonatal mouse brains and survived for at least 2 months. During this time, they migrate to periventricular regions of the brain including the hippocampus and olfactory bulb. However, no investigation of whether the cells were functionally integrated into the neural network or differentiated was performed. A second two-step differentiation protocol, adapted from that previously used to generate dopaminergic neurons from embryonic stem cells

(Perrier et al. 2004), involved culturing the cells in fibronectin-coated culture plates in DMEM/F12 supplemented with 10 ng/ml bFGF and N2 for 8 days, followed by N2 medium supplemented with 200 ng/ml SHH, 100 ng/ml FGF8, 20 ng/ml BDNF, and 0.2 mM ascorbic acid. Additionally the above media in the absence of SHH and FGF8 and the addition of 20 ng/ml GDNF, 1 mM db-cAMP, and 1 ng/ml transforming growth factor type β3 (TGFβ3) was also applied. These cells expressed G-protein-gated inwardly rectifying potassium channel 2 (GIRK2) and exhibited evidence of barium-sensitive voltage clamping as well as the expression of nestin. While this evidence may suggest that the cells were differentiating toward a neuronal phenotype, other interpretations are also possible, since neither nestin nor GIRK2 are exclusively expressed by neurons (Toselli et al. 2008), so additional electrophysiological evidence of sodium channels and development of synapses is necessary.

Since there are several potential sources for amniotic fluid stem cells, a number of studies have attempted to determine which sources contribute to the pool of cells. Jezierski et al. (Jezierski et al. 2010) determined that cells obtained early (i.e., first to second trimester) expressed epithelial cell markers and so are likely to arise from epithelial sources such as the AECs and other fetal epithelial locations. This was found to decrease significantly when looking at cells obtained during the third trimester. A subpopulation was also found to express low levels of early neural genes such as nestin, pax6, and CD133, as well as pluripotency markers such as SOX2, OCT4, and NANOG. Approximately 1% of the amniotic fluid cells also expressed c-KIT and higher levels of the pluripotency genes. Amniotic fluid cells that expressed SOX2 (or were subcloned to express SOX2 using a plasmid) appeared to differentiate into cells that adopted a neuronal morphology and expressed neurofilament and NSE as well as βIII tubulin and MAP 2 after culturing in DMEM with 0.5% FBS and N2 supplements for a week. Cells that did not express SOX2 did not differentiate under these conditions. Since NSCs express SOX2 (Episkopou 2005), the presence of SOX2 prior to neural differentiation may prejudice cells toward the neural lineage, and so this could be a method to promote the yield of NSCs and ultimately neural lineage cells.

Donaldson et al. (2009) attempted to differentiate two AFSC lines utilizing two different differentiation protocols that had previously demonstrated neuronal differentiation to DA neurons in embryonic stem cells (ESCs), MSCs, or NSCs. The first two-step protocol involved serum-free media supplemented with 20 ng/ml bFGF for 1 week, followed by 1 week with 10 ng/ml aFGF, 20 μM DA, 200 nM tissue plasminogen activator (TPA), 250 μM IBMX, and 50 μM forskolin. The initial step increased nestin from its initial moderate expression, while the latter step increased βIII tubulin from its low-level expression, but no evidence of additional DA cell markers or cell differentiation was observed. The second protocol involved neurosphere formation induced by the addition of 200 ng/ml Noggin to the media for 5–7 days, followed by plating on a polyornithine/fibronectin coating in media containing 1–500 μM cAMP and 200 μM ascorbate. Again the first step increased nestin expression, though βIII tubulin expression was not increased by the second step and no evidence of neuronal differentiation or expression of DA neuronal markers was detected. The authors therefore tried a number of other differentiation media previously reported to differentiate other cell types and again were unsuccessful in inducing DA neuronal markers or other evidences of neuronal differentiation in AFSCs in their hands.

Two different neural differentiation media protocols were also used by Maraldi et al. (2014). Their first protocol utilized αMEM supplemented with 10% FBS and 20 μM retinoic acid for 4 weeks, while their second protocol was 10% FBS and 1 mM βME for 24 h followed by 10 mM βME, 2% DMSO, and 200 μM BHA for up to 3 weeks. Initially the c-KIT^{+} cells also expressed OCT4, but the expression of both proteins declined with differentiation. GFAP and CNPase expression increased along with peripheral myelin protein 22 (PMP22) and S-100 (myelin and glial or Schwann cell markers, respectively). The second protocol lacking serum was only effective for a minority (2/5) of amniotic fluid samples; however approx. 20% of cells in both protocols were seen to also resemble neurons and express βIII tubulin, NeuN, and MAP 2, as well as synapsins, suggesting synaptic vesicle formation was occurring. Cells survived for at least 6 weeks throughout the brain following transplantation of nondifferentiated AFSCs into the lateral ventricles of neonatal rats, based on human mitochondrial antibody staining. CNPase was colocalized with human mitochondrial staining in over 30% of cells suggesting that they were differentiating into oligodendrocytes, while approximately 25% coexpressed βIII tubulin or synapsin with human mitochondria. The presence of synapsin suggests the neuron-like cells maybe forming synapses, though further study is required to confirm this. However, it has recently been shown that transplanted (and endogenous) cells can transfer mitochondrial proteins or even whole mitochondria to host cells (Hayakawa et al. 2016; Hsu et al. 2016; Mahrouf-Yorgov et al. 2017), and since human mitochondria staining was used here, the double-labeled cells could possibly be host cells that have taken up human mitochondria. Use of a different human-specific stain would therefore be necessary to exclude this possibility.

Several studies of AFSCs did not utilize cell sorting and so cultured a mixture of cells. In 2004, AFSCs were plated on culture dishes and expanded in nutrient mixture Ham's F10 medium supplemented with 10% FBS and L-glutamine (Prusa et al. 2004). Neurogenic differentiation media reduced the FBS concentration (from 10 to 2%) and included 1.25% DMSO. Prior to culture, the cells did not express neural markers; however expansion in normal media led to the expression of nestin, neurofilament, CNPase, and CD133 within a subpopulation of the cells, suggesting that normal culturing practices may induce expression of these factors in AFSCs. After 2 weeks in neurogenic differentiation media, the cells expressed higher levels of these markers, and approximately 1% of the AFSCs appeared to be sending out neurite-like projections as they adopted a neurogenic phenotype. No electrophysiological or synaptic protein expression analysis was performed to confirm this finding.

In a separate study, cells were cultured for a week in Amniomed (Euroclone, UK), an amniocyte-specific media (Bossolasco et al. 2006). The media were then removed and spun down and the cells replated in Amniomed, or DMEM + 20% FBS, or M199 media supplemented with 10% FBS and 20 μg/ml endothelial cell growth factor (ECGF) for at least two passages. Neuroglial differentiation was performed by culturing in either NS-A basal media alone or supplemented with 50% culture media from mouse NSCs, or Knockout™ DMEM + Knockout serum replacement, for 2 weeks. A heterogeneous mixture of cells with diffuse βIII tubulin and GFAP protein expression was observed, and all the cells expressed nestin prior to differentiation, while at the mRNA level, GAP43, NSE, and MAP 2 were also observed. However, after neuroglial differentiation, no morphological changes were observed, and only a modest increase in GFAP and βIII tubulin expression was detected suggesting that the media cultures were unable to initiate differentiation.

Culture of amniocytes and their immortalization by transfection of Ad5 E1 resulted in epithelial-like cells that could be differentiated to take on a neural morphology and express neural markers by plating on polyornithine-laminin-coated plates and culturing in DMEM/F12 supplemented with 100 ng/ml FGF8 and 1 μM retinoic acid, or using DMEM/Hams F12, 10% FBS, 20 ng/ml EGF, 20 ng/ml bFGF, and N2 for 5 days (Arnhold et al. 2008). Neural markers included the early neuronal markers nestin, human natural killer 1 (HNK-1), βIII tubulin, and α-internexin, the glial precursor and astrocyte marker vimentin, the oligodendrocyte precursor marker A2B5 (oligodendrocyte precursor cells), and astrocyte marker GFAP. The authors found that nonimmortalized cells did not express neural markers, even after use of the aforementioned neural differentiation media in contrast to previously described studies. This may be because the cells in this case were described as predominantly epithelial-like or the differentiation step was too short. However, in a later study by the same authors, in which they used CD117 magnetic cell sorting to generate both $CD117^+$ and $CD117^-$ cell populations, they did observe expression of neural markers such as βIII tubulin, α-internexin, GFAP, and HNK-1 after use of differentiation media, and this was greater in the $CD117^-$ cell population compared to either the $CD117^+$ or unsorted cell populations (Arnhold et al. 2011). They also found that transplantation of amniotic fluid cells transfected with GFP into the rat striatum led to neural-like cells being observed that partially expressed GFAP after 3 weeks. The $CD117^+$ population was primarily mesenchymal in nature and comprised approximately 2–5% of the total population, while the $CD117^-$ population was described as being primarily epithelial. This is a higher number than previously reported by de Coppi et al. (2007). Growth characteristics were also similar between all three cell populations.

Since neurons have a resting potential and undergo depolarization, the presence of active sodium channels can act as an indicator of a neural phenotype. In 2009, Mareschi et al. (2009) cultured adherent AFSCs, obtained via amniocentesis, for 2 or more passages, before culturing in neural progenitor maintenance media (NPMM; Lonza) for 3 weeks. Neurosphere-like aggregates were observed within the first 24 h. After 3 weeks, variable neural marker expression was observed with approximately 75% of cells being positive for nestin, NSE, MAP 2, and/or GFAP. Voltage clamp studies demonstrated the presence of tetrodotoxin-sensitive Na^+ channels in some of the positively labeled cells. These cells were also shown to express the genes for several of the Na channel subunits found in neurons (Nav1.1, Nav1.2, and Nav1.3). However the density of these channels was 100-fold less than you would expect in mature neurons, suggesting the cells may be on the way to maturing into functioning neurons, i.e., a neural stem cell, neural precursor cell, or immature neuron-like cell, potentially supported by their expression of nestin.

Frequently only a subset of the AFSCs will actually start expressing factors or adopt characteristics that suggest that they are differentiating into a neural phenotype. Wei et al. (2014) looked at a number of markers to see if they were predictive of the ability to undergo neural differentiation. Adherent AFSCs obtained from 23 patients during amniocentesis were cultured in α-MEM, 20% FBS, and 4 ng/ml bFGF. At some undefined point, the culture media were changed to neural differentiation medium (Cellular Engineering Technologies Inc., Coraville IA) for 2–4 days. NeuN, nestin, βIII tubulin, and TH staining was observed in a proportion of the cells which were undergoing morphological changes that included sending out axon-like projections. Based on the expression levels of these markers, the AFSCs could be divided into good and poor neurogenic groups. Expressions of the NSC predictive markers zinc finger protein 521 (Zfp521), OCT6, SOX1, SOX2, SOX3, and SOX9 were analyzed, and SOX9 levels were found to be predictive of whether the AFSCs were going to be good or poor neurogenic-differentiating cells. It is unclear whether following the cells for a longer period of time would have shown continued neural differentiation. In addition, as previously discussed, many of the neural markers investigated (nestin, βIII tubulin, and TH) as well as the supposed predictor molecule SOX9 are also expressed by melanocytes (Locher et al. 2014), and so without additional evidence of neurogenic differentiation, it is unclear whether the AFSCs really were undergoing neural differentiation.

In addition to directly differentiating AFSCs to neural progenitors, there are alternative means of obtaining neural progenitors. For instance, Jiang et al. (2014) reprogrammed AFSCs isolated from second and third trimester amniotic fluid with the Yamanaka factors (OCT4, SOX2, Krüppel-like factor 4 [KLF4], and c-MYC) using Sendai viral vectors to induced pluripotent stem cells (iPSCs). These cells could then be differentiated into neural progenitor cells by culturing in neural induction medium (DMEM/F12, GlutaMAX, NEAA, and N2) for 2 days until neurosphere-like structures were formed. The neurospheres were transferred to polyornithine-laminin-coated plates to enhance rosette formation and then replated on polyornithine-laminin-coated plates in neural stem cell media (Neurobasal media, B27, NEAA, GlutaMAX, and FGF-2). These cells were cultured for eight passages and shown to express a number of early neural markers such as SOX2, SOX3, PAX6, nestin, Musashi RNA-binding protein 1 (MSI1), and polysialylated-neural cell adhesion molecule (PSA-NCAM). Removal of both FGF-2 and NEAA and replating the cells onto poly-L-ornithine- and laminin-coated plates for up to 70 days were shown to induce neuronal differentiation to mature neurons, while the use of DMEM, N2, GlutaMAX, and 1% FBS for 4 weeks led to astrocytic differentiation. Of course this procedure is very time consuming with at least a month required to generate only small numbers of iPSCs, due to the low reprogramming efficiency. Time would therefore be required to culture and passage these cells further to generate sufficient quantities to differentiate (Jiang et al. used passage 8 iPSCs (Jiang et al. 2014)) as well as allow the time for neural differentiation (passage 8 NPC-iPSCs). The authors suggested that you can obtain a sizeable population of neural progenitors within 8 weeks from a small sample of amniotic fluid, and so this method may be considered too time consuming than other more direct methods. In addition, no confirmation of functioning neurons as an end point was performed.

Transplantation of AFSCs into both normal and ischemic rat striata revealed long-term survival and some degree of neural differentiation (Cipriani et al. 2007). Two hundred thousand MSCs extracted from the AFSC cultures by use of plastic adherence and culturing were transplanted into the striatum of immunosuppressed rats 7 days after ischemic injury (or sham). Cell survival and differentiation were studied 10, 30, and 90 days after transplantation by immunohistochemistry, and human nuclei-positive cells were observed at all time points. Cells coexpressing human nuclei and DCX were observed at 10 days, but not at later time points, while no cells coexpressing human nuclei and βIII tubulin were observed at any time point. This suggests that an early neuronal precursor cell may have been generated, but it did not survive or mature to 30 days in vivo. By contrast, human nuclei staining did colocalize with GFAP at all time points, suggesting that astrocytic differentiation and survival did occur. This is in contrast to the previously mentioned Maraldi et al. (2014) study in which 700,000 c-KIT$^+$ AFSCs were injected into the right lateral ventricle of newborn rats, and the cells studied 6 weeks later. The cells were stained for human mitochondria and neural markers and revealed no GFAP colocalization, but some cells coexpressed βIII tubulin or synapsin or CNPase and human mitochondria suggesting that neuronal and oligodendrocytic, but not astrocytic differentiation, had occurred. This discrepancy may have arisen due to the host rats being neonatal rather than adult, the increased number of cells administered, the selection of a slightly different cell population from the AFSC cultures, or, as mentioned earlier, mitochondrial transfer.

2.1.3 The Umbilical Cord

Human umbilical cord forms a conduit between the developing fetus and the placenta, allowing for the exchange of gases, metabolites, and nutritive factors between umbilical and maternal blood. The cord is formed about the 26th day of pregnancy and reaches a length of 30 to 50 centimeters at birth time (Sarugaser et al. 2005). During prenatal development, the umbilical cord is physiologically and genetically part of the fetus developed from the remnants of the yolk sac and from allantois. It is covered with an amniotic epithelium and contains two arteries and one vein surrounded by a unique kind of connective tissue, formed largely from mucopolysaccharides, that is known as Wharton jelly. The gelatinous connective tissue, containing the microfibrils and collagen fiber glycoproteins network, was first described by Thomas Wharton (1656). Collagen fibers build the type of skeleton surrounded by the main vessels on which glycosaminoglycans and hyaluronic acid are arranged (Wang et al. 2004). Umbilical cord blood is the blood of fetal origin found in the vessels of the umbilical cord and fetal part of the placenta.

2.1.3.1 Human Umbilical Cord Blood (hUCB)

The umbilical cord blood is characterized as a heterologous population containing three types of stem cells with unique molecular and cellular properties: hematopoietic stem cells (HSCs), mesenchymal stem cells (MSCs), and non-hematopoietic multipotent stem cells similar to the preMSC being described in the next chapter and expressing some of the markers typical for pluripotent stem cells, such as OCT4, SOX2, and NANOG (Habich et al. 2006; Ali and Al-Mulla 2012). The presence in the cord blood of different types of stem cells, together with the enrichment of more primitive stem/progenitor cells that are able to produce relatively long-term repopulating stem cells in vivo (as compared to adult sources of stem cells such as bone marrow or peripheral blood), makes hUCB attractive for applications in regenerative medicine. In addition cord blood is still of primitive ontogeny, little exposed to immunologic challenges, and induces less frequent and less severe acute and chronic GVHD than adult sources of stem cells, which contain a higher number of activated T cells (Gluckman and Rocha 2009). In addition hUCB can be non-controversially and noninvasively obtained. While the use of cord blood transplantations for hematological disorders is a routine procedure for the reconstitution of the ablated BM since the first transplant in 1972 (to treat a case with lymphoblastic leukemia, (Ende and Ende 1972)), the potential use of umbilical cord blood in the treatment of other pathologies, especially those with a neurological background, is still under debate for their plausibility. The possible applications of the cord blood in the treatment of the neural disorders were highlighted by the first evidence in the early 2000s that cord blood stem cells can cross the tissue and even germ-line barriers to attain in vitro neural features. These pioneering studies came out in parallel from the Labs of Krystyna Domanska-Janik in Warsaw, Poland, and Paul Sanberg in Tampa, USA (Buzanska et al. 2001, 2002; Sanchez-Ramos et al. 2001; Zigova et al. 2001).

In order to differentiate hUCB stem cells into neural phenotypes, Buzanska et al. (2002) used in their study non-hematopoietic $CD34^-$ adherent, mononuclear cell population grown in Iscoves' modified Dulbecco's medium (IMDM) supplemented with 10% FBS for 6 weeks. The hematopoietic $CD34^+$ stem cells were eliminated by immunomagnetic sorting at the beginning of the selection process, and the cells were grown in the presence of 10 ng/ml EGF in DMEM as the mixed cultures with adherent and floating cells. Cells from both subfractions were clonogenic and nestin-positive and, in the presence of retinoic acid, as well as in coculture with rat cortical primary cells, were differentiated into neuronal, astrocytic, and oligodendroglial-like cells (Buzanska et al. 2002). This group was also successful in establishing a nonimmortalized, spontaneous NSC line (human umbilical cord blood neural stem cells: hUCB-NSCs) by sequential passaging of only the floating cells from the mitogen-expanded culture (Buzanska et al. 2005, 2006). The cells retain their normal chromosomal pattern and an unchanged capacity to proliferate and self-renew. By establishing special culture conditions (Buzanska et al. 2006), hUCB-NSCs could be maintained at different developmental stages (Fig. 2.1) and then used for pharmacological and toxicological (Buzanska et al. 2009) screening.

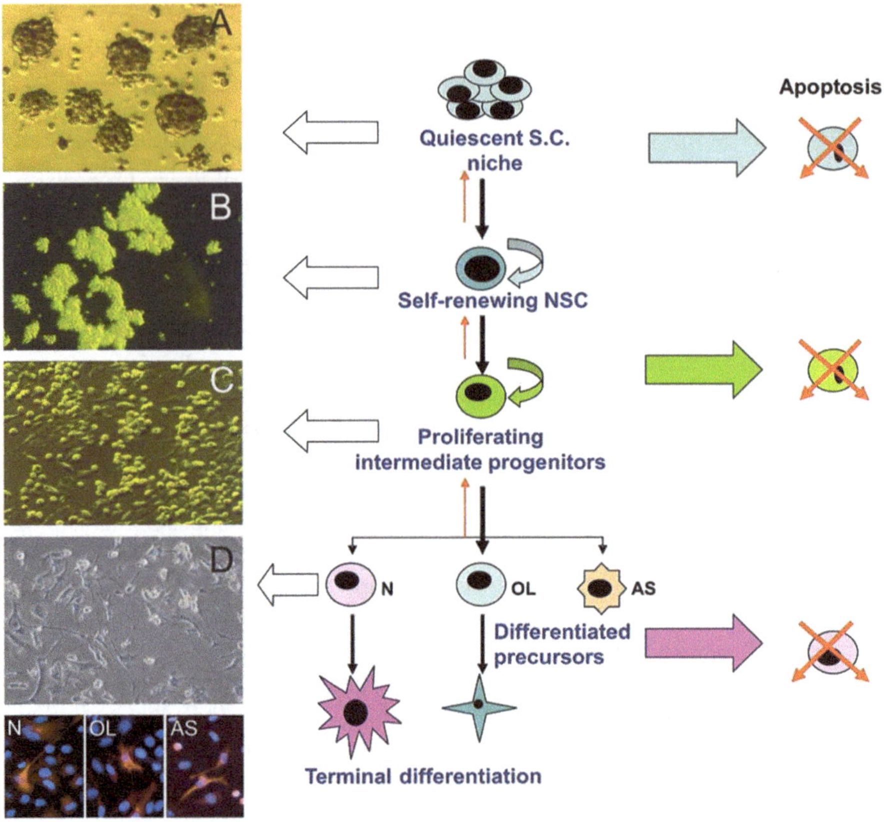

Fig. 2.1 Developmental steps of hUCB-NSCs toward neural lineages. Phase contrast (**a–d**). (**a**) hUCB-NSC-derived neurospheres-quiescent stem cell niche; (**b**) self-renewing, floating hUCB-NSCs in serum-free medium; (**c**) intermediate, neural-committed hUCB-NSCs grown in low serum medium at floating and adherent fraction. (**d**) hUCB-NSC-derived neuronal, astrocytic, and oligodendroglial precursors differentiated in media supplemented with neuromorphogens. Immunocytochemistry: (N) ßIII tubulin, (OL) GalC, and (AS) GFAP for neuronal, oligodendrocytic, and astrocytic cells, respectively (Buzanska et al. 2005)

In the culture with no serum added, cells were able to form free-floating, undifferentiated spheres, resembling the spheres obtained from the human central nervous system, which in the presence of serum easily attach, spread out, and differentiate spontaneously. An analysis of such spheres has revealed the presence of cells expressing proteins characteristic of differentiating neurons (MAP 2), astrocytes (GFAP), and oligodendrocytes (GalC) on the edges of the sphere, while the core remained proliferating, Ki67- and nestin-positive (Fig. 2.2; Buzanska et al. 2005; Domanska-Janik et al. 2008).

Further directed differentiation in the presence of neuromorphogens (RA and dBcAMP) led to the expression of more advanced neuronal markers: eventually even proteins typical of functional neurons as revealed by immunocytochemistry

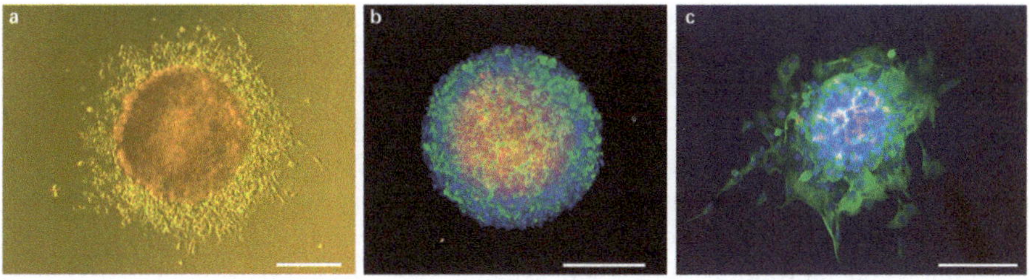

Fig. 2.2 (**a**) Neurospheres derived from hUCB attach to the bottom of the culture dish in the presence of serum. (**a**, **c**) Cells from the surface region of the attached neurosphere spread out and differentiate mainly toward neurons positive for βIII tubulin (**c**; green); however, some glial cells positive for GFAP are present as well (**c**; red). Neurospheres derived from hUCB before adhesion are cultured and propagated as floating cell clusters in SF medium supplemented with EGF. (**b**) Undifferentiated floating neurospheres comprise at least two regions of neural stem/progenitors: inner cells positive for nestin (red) and outer cells positive for GFAP (green). (**b**, **c**) Cell nuclei are stained with Hoechst (blue). Scale bar: 100 μm (Buzanska et al. 2006a)

and transcriptomic data; however patch clamping experiments did not prove the presence of action potentials in morphologically mature neurons (Sun et al. 2005; Buzanska et al. 2006). The functionality of the neuronal network obtained from hUCB-NSCs was demonstrated in the three-dimensional, scaffold-based culture of hUCB-NSCs by registering active field potentials on multielectrode array (MEA) chips (Jurga et al. 2009).

To answer the question on the dynamic of the fate change of hUCB into the neural phenotype, the molecular and immunocytochemical characterization of the freshly isolated mononuclear $CD34^-$ during the first stages of neural commitment was performed. The obtained data revealed the unexpected transient presence of cells expressing pluripotent markers: Oct3/Oct4, Sox2, and Rex1 (Habich et al. 2006). During 24 h of culture of the freshly isolated hUCB $CD34^-$ mononuclear fraction, the frequency of Oct3/Oct4 immunopositive cells increased together with the appearance of $CD133^+$ cells and parallel enlargement of the "side population." Concomitantly, cultured cells started to form aggregates, which after adhesion differentiated spontaneously expressing pro-neural genes, such as *SOX2*, *OTX1*, *nestin*, *GFAP*, and *NF-200*. Further culture of these neurosphere-like adherent aggregates revealed spontaneous expression of neural markers, such as βIII tubulin, *MAP 2*, *GFAP*, *S100b*, *DCX*, and *GalC*. The supplementation with neuromorphogens (BDNF and RA) increased the percentage of differentiated cells in the culture. Thus, as will be discussed further in this chapter, we postulated the native presence of more primitive pluripotent-like cells preexisting in the hUCB mononuclear fraction and contributing in vitro to the phenotypic shift of the hUCB rather than direct conversion of hematopoietic precursors due to expression of overlapping genetic program. Such a view was further confirmed by epigenetic studies on the hUCB-NSC cell line at different developmental stages, where substantial demethylation of the genes linked to pluripotency was observed only in the floating, nondifferentiated population (Habich et al. 2013). The data of the contribution of pluripotent-like cells to the stem cell population existing in human cord blood was supported by the findings of other groups (Kogler et al. 2004; McGuckin et al. 2005; Kucia et al. 2007; Haris and Rogers 2007; Ali and Al-Mula 2012). McGuckin et al. (2004) by isolating a discrete lineage negative (LinNeg) stem cell population (0.1% of CB mononucleated cell [MNC] population) was able to expand primitive nonadherent hematopoietic progenitors (up to 47-fold) and simultaneously produce slow-dividing adherent cells expressing GFAP with neuroglial progenitor cell morphology. In a subsequent study, this group has isolated cord blood-derived embryonic-like stem cells, expressing TRA-1-60, TRA-1-81, SSEA-4, SSEA-3, and Oct-4 (McGuckin et al. 2005). HUCB as a source of embryonic-like stem cells was also reported by Kucia et al. (2007). In that study a two-step isolation procedure of cord blood-derived stem cells was applied: removal of erythrocytes by hypotonic lysis combined with multiparameter sorting, which included such parameters as $CXCR4^+$, lin^-, $CD45^-$, $CD34^+$, or $CD133^+$ in combinations—$CXCR4^+$, lin^-, $CD45^-$ or $CD34^+$, lin^- $CD45^-$, or $CD133^+$ lin^- $CD45^-$. This resulted in isolation of very small embryonic-like cells (CB-VSELs), which are very small (3–5 μm) and express nuclear embryonic transcription factors Oct-4 and Nanog and surface embryonic antigen SSEA4.

The results from the preclinical study of Domanska-Janik group have indicated that intracarotid artery infusion into rats of $CD34^-$ hUCB directed or not in vitro toward a neural phenotype ameliorates neurological deficits associated with stroke-like, ouabain-induced brain injury showing reduction in lesion volume and significant behavioral improvement (Janowski et al. 2004; Gornicka et al. 2006; Kozlowska et al. 2007). All these data and that coming out from other groups (Nan et al. 2005; Chen et al. 2006) suggest that cells with pluripotency or neural progenitor properties are contributing substantially to the observed hUCB therapeutic effects. However, it is still not clear how transplanted stem cells can participate in the CNS repair process. In the study of Gornicka-Pawlak et al. (2011) where intra-arterial infusion of hUCB-derived cells at different stages of their neural conversion was evaluated, the most effective in functional restoration was freshly isolated mononuclear cells (D-0) as compared to neurally directed progenitors (D-3) and neural stem cells derived from UCB. What's more, none of these different fractions of transplanted cord blood-derived cells were detected in the rat's brains over 30 days after treatment. This suggests other mechanisms underlying the positive effects of intra-arterially infused hUCB than direct transvascular neural cell supplementation.

HUCB has been proven beneficial as the source of therapeutic cells in several preclinical models of stroke; thus the first approved autologous, intraventricular transplantation of hUCB into the human brain was undertaken by a group of clinicians and scientists under the coordination of Prof. Krystyna Domanska-Janik (Jozwiak et al. 2010). The 18-month-old patient diagnosed with a vegetative state, due to global cerebral ischemia, underwent cell transplantation to the lateral ventricle of autologous hUCB-derived stem cells that were neurally directed in vitro and labeled with superparamagnetic iron oxide (SPIO). MRI evaluation revealed that the cells were detected in the patient's brain with a decreasing signal until 4 months and was undetectable after 33 months. The prospective clinical evaluation revealed that the treatment was safe with signs of functional improvement in the patient (Janowski et al. 2014).

The pioneering research on the derivation of neural stem cells from hUCB carried out in the group of Paul Sanberg was performed on cryopreserved, nonselected mononuclear cells expanded in the presence of mitogens (EGF and bFGF) and neural induction "N2" medium (Sanchez-Ramos et al. 2001). Differentiation of the expanded cells in the presence of RA and NGF resulted in the increased immunoreactivity of Musashi-1, βIII tubulin, and GFAP in the differentiating population. Microarray data revealed upregulation of genes involved in neural development, including pleiotrophin (*PTN;* a neurite outgrowth-promoting protein), *GFAP*, neuronal pentraxin II (*NPTX2*), neuronal growth-associated protein 43 (*GAP-43*), *MAP 2,* and downregulation of genes associated with development of blood lines. Further studies by this group have proved that 20% of hUCB cells transplanted into the neonatal rat brain survive, and some of them can differentiate into GFAP and βIII tubulin positive cells (Zigova et al. 2002). Willing et al. (2003) administered hUCB cells intravenously into the femoral vein or directly into the striatum of rats with permanent MCAO to compare the effect of the route of cell administration for the behavioral recovery of rats. Intravenous delivery was proven to be more effective for long-term functional benefits to the stroked animal. For the treatment of traumatic brain injury (TBI) which is associated with neuroinflammation, Acosta et al. (2014) proposed combination therapy with intravenously delivered hUCB cells and granulocyte colony-stimulating factor (GCSF). Positive synergistic effects by diminishing neuroinflammation while enhancing endogenous neurogenesis and reducing hippocampal cell loss were observed.

Among the first publications showing the commitment of human cord blood to neural features was also the short report by Ha et al. (2001). The authors indicated protein expression of neural markers (nestin, NeuN, NF-M, MAP 2, and GFAP) in the cord blood nonselected mononuclear population after culturing of cells for 20 days in 20% FBS without additional stimulation with neuromorphogenes, but no quantitative data from this study was reported. The same group has confirmed functional improvement of spinal cord injury (SCI) in rats after transplantation of hUCB MNCs in the presence of BDNF into the injured spinal cord (Kuh et al. 2005), and they have demonstrated in situ expression of neuronal markers by the transplanted cells.

The MSC fraction isolated from human cord blood is another kind of stem cell which has been shown to differentiate in vitro into neural phenotypes and plays a beneficial role in transplantation into neurally deficient rodents. In the study of Jeong et al. (2004), adherent cells expressing MSC-related antigens such as SH2, CD13, CD29, and α-smooth muscle actin (αSMA) have been isolated from the mononuclear cell fraction of human UCB and differentiated into neural cells expressing βIII tubulin, TrkA, GFAP, and CNPases as demonstrated by immunofluorescence and RT-qPCR analyses.

A new, two-step induction protocol for improving the differentiation of hUCB-derived mesenchymal stem cells into neural progenitor cells was proposed by

Rafieemehr et al. (2015). In this protocol culturing of the mononuclear fraction of hUCB in preinduction medium with RA, bFGF, and EGF for 2 days was followed with NGF, IBMX, and ascorbic acid in basal medium for 6 days. Molecular verification by quantitative real-time PCR proved that this protocol significantly increased the expression of *GFAP*, *MBP*, and *MAP-2* genes.

Since the ability of UCB-MSCs to generate neurons appears to be different among batches, some of them require very simple neuronal induction protocols, whereas others need extensive exposure to combinations of growth factors in a stepwise protocol, Divya et al. (2012) decided to verify whether human UCB-MSCs contain multiple types of progenitors with varying neurogenic potentials. The authors identified two different populations of progenitors within the UCB-MSCs: one expressing pluripotent stem cell markers (Oct4, Nanog, Sox2, ABCG2) and nestin (neuroectodermal marker) and the second expressing typical MSC markers; however both populations expressed CD29 and CD105, indicating their MSC lineage.

The first population was capable of expanding and differentiating into neurons, while the second required extensive exposure to a combination of growth factors. This data support the hypothesis of a pluripotent-like origin of the tissue-derived stem cells that differentiate into the neuronal lineage.

Last, but not least, is the role of the microenvironment on the differentiation of tissue-derived stem cells into neuronal lineages. In that regard the recent study of Kheirandish et al. (2017) revealed that hypoxic preconditioning was effective in enhancing the proliferation capacity of hUCB-MSCs and can trigger expression of some of the neural genes, such as *nestin*, *NT3*, and *GFAP*. These data are in line with the demonstrated beneficial influence of low-oxygen (physiologically normoxic) conditions on the survival, proliferation, and differentiation capacity of hMSCs isolated from other tissues (Lech et al. 2016).

While the population of circulating hematopoietic stem cells ($CD34^+$) isolated from hUCB was shown in several studies to be beneficial in the partial restoration of cortical tissue or functional recovery in stroke animals (Taguchi et al. 2004; Peterson 2004), there was no evidence in vitro and in vivo for neuronal differentiation of these cells. Recently, along with the onset of reprograming technology, cord blood $CD34^+$ cells were directly converted to induced neural stem cells (iNSCs) by ectopic expression of Oct4 transcription factor (Liao et al. 2015). In the presence of specific differentiation factors (e.g., forskolin, BDNF, GDNF, SHH, or T3), the iNSCs expressing nestin and Musashi-1 are able to differentiate into multilineage mature neural cells, displaying patch clamp recorded action potentials. Upon engraftment to NOD/SCID mice, CB-iNSCs were able to survive at least 3 months and consistently maintained their differentiation capacity in vivo, thus confirming their feasibility for possible application to treat neurological disorders.

2.1.3.2 Human Umbilical Cord Tissue (Wharton Jelly)

One potential source of cells with the ability to undergo neural differentiation in vitro is the umbilical cord mesenchymal stem cells isolated from Wharton jelly (WJ-MSCs) (Davies et al. 2017).

The main bulk of cord MSCs are derived from perivascular and subcutaneous tissue as well as from the subendothelial umbilical vein and artery regions. There are data suggesting that WJ-MSCs are more primitive and have greater potential for proliferation and differentiation than those obtained from other regions of perinatal tissues (Karahuseyinoglu et al. 2007; Sarnowska and Domańska-Janik 2017). Regardless of certain local differences between particular regions, the umbilical cord seems to be one of the best and most efficient sources of MSCs. Secco's group (Secco et al. 2008) has shown that MSC derivation from umbilical cord sources has a very high, almost 100% efficiency, whereas isolation from umbilical cord blood sources is only about 10% efficient.

Like MSCs from other sources, those obtained from the umbilical cord display the typical, fibroblast-like oblongate shape, with the cell body adhering to the plastic surface. In vitro, they can undergo more than 30 passages with a population doubling time of approximately 24 h with stable morphology and karyotype when cultured under the most suitable 5% oxygen concentration (Lech et al. 2016). Cell cycle analysis of the third passage proved that the overwhelming majority of the cells (88.86%) are in the G0/G1 growth phase and only 5.69% in the G2/M phase. The antigens typical of all MSCs are found on WJ-derived cells including CD10, CD13, CD29, CD44, CD73, CD90, CD105, CD106, SH2, SH3, and SH4. Accordingly, they do not express the markers typical for hematopoietic cells: CD14, CD31, CD34, CD38, CD45, and HLA-DR.

Moreover, WJ-MSCs as a heterogeneous but primitive population also express pluripotent stemness-related cell markers: SSEA3 and SSEA4, Tra-1-60, Oct4, Sox-2, Rex1, and Nanog (Dezawa 2016; Drela et al. 2014). Also other markers typical for primitive stem cells, such as leukemia-inhibitory receptor, embryonic stem cell 1 gene, and reverse transcriptase telomerase (Troyer and Weiss 2008), were described. It has been shown that WJ-MSCs have longer telomeres and express greater telomerase activity compared to MSCs obtained from other sources. This primitive cell subpopulation is favored in the hypoxic/physioxic culture condition (lower than 5% oxygen in Drela et al. 2014), where the cells acquired the typical stemness-related phenotype characterized by comparatively high proliferation rate, longevity, and augmented scope of cell differentiation especially toward therapeutically important neural and endothelial cells, responsible directly for neurogenesis and angiogenesis, respectively (Lech et al. 2016; Obtulowicz et al. 2016).

Another notable feature, especially important in the clinical application of WJ-MSCs, is the lack of HLA class II expression in the presence of HLA-G, a pattern which usually is not expressed in stem cells derived from adult tissues (Troyer and Weiss 2008). This specific HLA system would allow WJ-MSCs to be used for allogenic transplantations.

The above basic characteristic may differ significantly depending on the applied method of isolation procedure. MSCs could be obtained from Wharton's jelly tissue with the enzymatic (Wang et al. 2004) or mechanical (Mitchell et al. 2003) method. In our experiments we have compared these two fractions (Lech et al. 2016). The first one was more efficient and grew faster. Despite comparable expression levels of typical mesenchymal markers (CD73, CD90, CD105, CD166, vimentin, collagen, and fibronectin), the mechanically isolated cells were more stable in culture and had a shorter population doubling time (PDT), higher ability to CFU-F colony formation, and a lower number of senescent cells. Moreover, along with the longer cultivation period, the cells showed significantly higher expression of neural/neuronal markers: nestin, βIII tubulin, GFAP, and NF-200. Then, the chosen method of cell isolation may substantially affect the cell's ability to neural differentiation and their pattern of neurotrophic factor secretion, which all together may influence their neuroprotective properties. Therefore the efficiency of the procedure cannot be the only determinant for the method of MSC isolation.

MSC transplantation to the rodent brain, without any additional inductive or supportive treatments, would be a rather short living procedure. Therefore, their time-related neural differentiation in vivo would be even more questionable. However, it has been reported that WJ-MSCs transduced with transgene nuclear receptor-related 1 (Nurr-1) and then treated with a defined cocktail of combined factors (including all-trans-retinoic acid, neuroregulin 1 [glial growth factor 2; GGF-2], bFGF, PDGF, and forskolin) may effectively differentiate toward neurons. After transplantation the cells successfully colonize injured spinal cord tissue and actively participate in the repair process (Yang et al. 2017). The other way to enhance specific neural differentiation is pretreatment with forskolin. In WJ-MSC culture it promotes a strong upregulation of the neurotrophic Trk receptors and induces transdifferentiation of WJ-MSCs specifically in the dopaminergic direction. The dopaminergic phenotype was confirmed by immunocytochemistry and Western blot analysis that revealed the significant induction of Nurr1, NeuroD1, and TH protein expression (Paldino et al. 2014). The percentage of dopaminergic neurons obtained from WJ-MSCs depended on the specific induction protocols, i.e., the use of transcription and/or growth factors as transgenes including EGF, SHH, and NGF, and the selection of specific conditioned media derived from amniotic epithelial cells, choroidal conditioned medium (Boroujeni et al. 2017), or cerebrospinal fluid (Aliaghaei et al. 2016). In the latter case, when differentiated cells were transplanted into the striatum of rats, they survived for the longest time and effectively suppressed neuronal apoptosis in the injured host tissue.

Interestingly the three-dimensional scaffold cell culture induces even more advanced neural differentiation of WJ-MSCs than the conventional two-dimensional system (Hosseini et al. 2015). Moreover the influence of the nanoscaffold culture in the presence of retinoic acid and sonic hedgehog further enhanced WJ-MSC differentiation especially in the motor neuron direction (Bagher et al. 2016). Such cells additionally supplemented with heparin and bFGF expressed even higher levels of motor neuron RNA and protein biomarkers. After such inductive processing, transplanted neurons started to synthesize cholinergic neurotransmitters which is decisive for the therapeutic effectiveness of a transplant (Liu et al. 2013).

As mentioned above, the WJ-MSCs have a unique potential for spontaneous neural differentiation; however this ability is seriously hampered by the number of passages that the cells pass through in culture. At the beginning the isolated WJ-MSC population is very heterogeneous, and the cells do not express any neural markers. Also the clonogenic subfraction of genuine MSCs would comprise only a relatively small subpopulation of the overall WJ cells dominated mainly by endothelial cells (ECs). Later, these cells are disposed, whereas mesenchymal cells take over the whole monolayer (Qiao et al. 2008), in which the neural cell phenotypic markers will start to be detected (Lech et al. 2016).

Surprisingly, our functional experiments showed that the ability for neuroprotection is much higher in the primary cohort of the WJ-migrating cells (passage O). Along with further passaging, the cell phenotype is changing from mesenchymal to neural-like features, but their neuroprotective effect is successively diminishing. Thus, it seems that the native ability of the primary culture to secrete the mixture of still poorly defined, different cytokines is preferentially responsible for the observed neuroprotection rather than the much later appearing neural-like cell differentiation (Dabrowska et al. 2018). This conclusion is supported also by the previously gathered preclinical and approved clinical data, showing that the naïve, stemness transcription factor (SRTF) expressing MSC cultures, being capable of time-locked proliferation, migration, and ultimately later neural differentiation, is more efficient in various therapeutic transplantation models (Castillo-Melendez et al. 2013; Dabrowska et al. 2018; Gornicka-Pawlak et al. 2011).

One strategy used to expand this type of undifferentiated, naive transplantation material in culture is the transient formation of the spheroids reported in different types of MSC cultures (Fotia et al. 2015), but they are most easily obtained from the prenatal tissues like umbilical cord blood (Habich et al. 2006; Habich and Domanska-Janik 2011; Monti et al. 2017), or Wharton jelly (Zhu et al. 2017). The WJ-derived spheroids appeared within 4–5 days after seeding, while the formation of bone marrow stromal cell (BMSC)-derived spheres needs 7–9 days by using the same induction methods (Hermann et al. 2004). The cells could be forced to grow as free-floating spheres by plating on nonadhesive surfaces with occasional agitation. Enhanced proliferation and the well-preserved stemness of the cultured spheres were confirmed by expression of the stemness transcriptional factors Oct4A, Sox2, Nanog, and Rex1 with additional stimulation following cultivation in the stem niche-specific 5% O_2 atmosphere. Inside the clusters, cells express genes characteristic for the ectodermal, mesodermal, and endodermal fate (Monti et al. 2017; Adamzyk et al. 2016). These cells appear to be rejuvenated by transient growth in these nonadhesive 3D, floating clusters to undifferentiated, embryonic-like phenotypes. However, after secondary adhesion to adherent surfaces promoted by 21% O_2 atmosphere, their ability for outspreading monolayer formation and subsequent neural-like differentiation is restored. Further neuronal differentiation could be accelerated by forskolin (FK) treatment. Appearance of neural/neuronal markers, e.g., MAP 2, βIII tubulin, TUC-4, and NF-L in MSCs, was observed at 4 days after FK exposure (Bonilla-Porras et al. 2017). Importantly, the neurosphere-like aggregates, even after growth in the plain Neurobasal medium, when transplanted into injured spinal cord, may spontaneously differentiate into oligodendrocyte-like cells. Moreover their pretreatment with 10 ng/ml recombinant human BDNF or FK prompts them farther to undergo bordered neural differentiation as confirmed by MAP-2, βIII tubulin, TUC-4, and NF-L expression and functional ability to improve axonal regeneration (Zhang et al. 2009; Bonilla-Porras et al. 2017). From the practical standpoint, the new technology of neurosphere culture would be essential for expanding undifferentiated, primitive stem cells to undergo multilineage differentiation and allogenic transplantation in accordance with the medical demands.

As already stated, while the freshly isolated cells from Wharton jelly (at O passage) express numerous markers typical for either pluripotent, mesenchymal, or endothelial cells such as the SRTF, fibronectin, and VEGFR-1, they are temporarily devoid of any neural markers. Shortly after the first passages, they start to spontaneously express early neural markers, e.g., βIII tubulin or NF200 (Datta et al. 2011; Lech et al. 2016; Dabrowska et al. 2018). After a longer culture period, the number of MSC- and pluripotency marker-expressing cells decreases (Baer et al. 2010), whereas the bulk of the cells acquire a more ramified morphology and express characteristics of mature neural cells, including neuron-specific MAP 2 and enolase (Mitchell et al. 2003), GFAP for astrocytes, and CNPase for oligodendrocyte identification (Drela et al. 2016; Fink et al. 2013; Tracy et al. 2011). Additionally these cells can be further induced toward more advanced and specific neural lineages by supplementation with the cocktails of various neurotrophic factors and other neuroprotective small molecules in vitro (Paldino et al. 2014; Zhang et al. 2009).

Considerable augmentation of WJ-MSC multilineage differentiation can also be promoted by changing the culture atmosphere from the physiological 5% O_2 to 21% O_2 concentration. This causes inhibition of cell proliferation, increased cell aging, and a rapid switch from the stemness-specific, almost anaerobic, to functionally hyperoxic metabolism (Lech et al. 2016; Sandvig et al. 2017; Dabrowska et al. 2018). The mechanism of this response is still not clear but triggers selective death and elimination of the part of MSCs. Simultaneously, stimulation of epigenetic cell remodeling leads to still poorly defined, adaptive changes in the rest of the surviving cells. The relationship observed between the stages of culture oxygenation, proliferation, and differentiation seems to be controlled by the hypoxia-inducible factor alpha (HIFα) signaling and is negatively regulated by the histone deacetylase inhibitor TSA in a dose-dependent manner (Drela et al. 2014).

Another way to enhance the WJ cell propensity to undergo neural differentiation would be to coculture them with nerve tissue ex vivo (Dabrowska et al. 2018). The cells differentiate preferentially into oligodendroglial NG2-positive precursors but also into NF200- and MAP-2-expressing neurons (Mitchell et al. 2003; Hosseini et al. 2015). Such results give a hope for possible spontaneous WJ-MSC differentiation after transplantation into devastated, injured, or demyelinated CNS tissues in vivo. However, direct data confirming MSC differentiation toward the neural phenotype and then integration within the host injured human nervous tissue in situ are still lacking. The other important but also unresolved question is at what stage of predifferentiation do the WJ-MSCs need to be for optimal transplantation? In the treatment of neurological disorders, some of the preclinical data gathered using rats and mice have confirmed that the neural and/or endothelial committed MSCs can survive and integrate well after transplantation in the host nervous tissue (Jablonska et al. 2010; Jurga et al. 2011; Chen et al. 2013; Sarnowska et al. 2013).

2.2 Adult Somatic Tissues

2.2.1 Blood

Transitioning from afterbirth to somatic tissues, we can briefly mention menstrual blood before tackling peripheral blood, since it is likely to contain endometrial and potentially other stem cells. Their ability to undergo neural differentiation has only been studied in a few cases, which are discussed below.

2.2.1.1 Menstrual Blood

The endometrium can be divided into two layers: the functionalis which regenerates and is expelled every menstrual cycle and the basalis layer which contains the cells to regenerate the functionalis layer (reviewed in Rodrigues et al. (2016)). These cells can be expelled during menstruation and therefore can be found in menstrual blood. Three different groups identified stem cells in menstrual blood around 10 years ago and showed that they were multipotent (Meng et al. 2007; Hida et al. 2008; Patel et al. 2008). Since then, the isolation of stem cells from menstrual blood has been increasingly performed, with the company Cryo-Cell (Oldsmar, FL) setting up a menstrual stem cell banking service. The three groups may have isolated different stem cell populations, since the cells isolated by Patel et al. (2008), named menstrual blood-derived stromal stem cells (MenSCs), are positive for c-KIT and SSEA-4, while the cells isolated by the other two groups, coined as endometrial regenerative cells (ERCs) (Meng et al. 2007) and menstrual blood-derived mesenchymal cells (MMCs) (Hida et al. 2008), were not identified as being positive for these markers. Patel et al. (2008) cultured cells, obtained from menstrual blood by a proprietary isolation procedure, for five passages and then isolated the MenSCs by using c-KIT magnetic bead cell sorting. OCT4 and SSEA-4 expression was also observed by these cells. Different induction media were shown to be able to differentiate these cells into a number of different cell lineages, including neural. For the latter, the cells were plated on fibronectin in DMEM/F-12, 2 mM GlutaMAX, 1 × N2 supplement, and 10 ng/ml bFGF. After 4 days the cells were passaged, and 10 ng/ml PDGF and 20 ng/ml EGF were added for 5 days, and a further 7 days without EGF. Around 50% of the cells were shown to express O4 and GalC suggesting oligodendroglial differentiation, GFAP suggesting astrocytic differentiation, and MAP 2 and βIII tubulin suggesting neuronal differentiation. Neural marker RNA expression, such as nestin, NCAM, and Nurr1, was also detected.

The ERCs derived by Meng et al. (2007) were obtained by culturing the Ficoll-separated mononuclear cells found in menstrual blood in DMEM, 1% amphotericin B, 1% glutamine, and 20% FBS. The cells were cultured for 2 weeks, and then single cells were plated to generate clonal colonies. These cells were shown by flow cytometry to be of a nonhematopoietic stem cell phenotype as well as expressing OCT4. Culturing in specific induction media again resulted in multiple different cell lineages. Neural differentiation was tested by culturing adherent cells in NPMM neural induction medium (Cambrex) supplemented with 0.2 mM GlutaMAX and 20 ng/ml hFGF-4 for 21 days. The cells expressed nestin and GFAP suggesting some degree of neuronal differentiation may have occurred.

The MMCs of Hida et al. (2008) were mesenchymal-like, and following coculture with murine fetal cardiomyocytes, the MMCs were seen to differentiate into cardiac precursor cells. Their ability to undergo neural differentiation was not explored.

In a separate study, the MenSCs were shown to be able to differentiate into neural stem cells in vitro following culture in DMEM/F12, N2, and FGF-2 for a week, and the same media supplemented with retinoic acid for a further 3 weeks (Borlongan et al. 2010). The cells expressed nestin, MAP 2, and GFAP, but did not express NeuN, suggesting that full maturation to NeuN-expressing neurons did not occur. In addition, following either intravenous administration, or direct transplantation, of MenSCs into the striatum of rats who underwent an MCAo, no evidence of neural differentiation was observed by the surviving cells after 14 days, even though behavioral benefit and reduced lesion size was detected, suggesting the cells may have exerted a paracrine effect rather than differentiation/cell replacement.

Azedi et al. have performed two studies on MenSCs, whereby they differentiated them to glial-like cells (Azedi et al. 2014) or neuron-like cells (Azedi et al. 2017). In both instances they first generated neurosphere-like cells by culturing MenSCs in P4-8F medium containing 20 ng/ml EGF and 20 ng/ml FGF-2 for 3 weeks. Within the first 3 h, morphological changes were seen resembling a neural phenotype, but the cells reverted to their fibroblast morphology within days. After 10–14 days, the cells coalesced into small spheres that became more neurosphere-like with continued culture. The spheres likely contain NSCs, and after trypsinization the cells comprising the spheres were cultured as individual cells on glass cover slips in Neurobasal media containing 1% FBS, 5% horse serum, 1% N2 supplement, and 0.5 μM all-trans-retinoic acid plus either 10 ng/ml BDNF for neuronal (Azedi et al. 2017) or 10 ng/ml PDGF for glial (Azedi et al. 2014) differentiation. After 12–16 days, the neuronal differentiation resulted in cells expressing MAP 2 and GABA B receptor proteins and K^+ and Ca^{2+} channel mRNA, as well as being electrophysiologically similar to mature neurons (Azedi et al. 2017). In contrast, the glial-differentiated cells expressed Olig-2 and MBP, suggesting the cells had become oligodendrocytes (Azedi et al. 2014).

Mesenchymal cells isolated directly from the endometrium (human endometrial-derived stem cells; HEDSCs), which may also be found in menstrual blood, have also been investigated for neuronal differentiation (Wolff et al. 2011). The cells were first treated for 48 h with DMEM, 10% FBS, 1% antibiotics, 2 mM L-glutamine, 10 ng/ml rhFGF, 10 ng/ml rhEGF, and N2 supplement B. This was followed a further 96 h by DMEM, antibiotics, 2 mM L-glutamine, N2 supplement B, 200 μM BHA, 1 mM db-cAMP, 0.5 mM IBMX, and 1 μM all-trans-retinoic acid. Long axon-like projections and pyramidal cell bodies were observed following differentiation. The cells were nestin- and TH-positive, and they possessed a barium-sensitive K^+ current similar to the G-protein coupled inwardly rectifying K^+ current that are present on neurons. This would suggest that the cells had differentiated into dopaminergic-like neurons in vitro. Undifferentiated and differentiated cells have been transplanted into the striatum of mice modeling PD, and some cells were found to have survived after 5 weeks. The undifferentiated cells migrated to the substantia nigra where they were shown to express nestin and TH, suggesting that they had differentiated into a neurogenic phenotype. The differentiated cells were not observed to migrate and remained at the site of transplantation in the striatum. Increased DA and dihydroxyphenylacetic acid (DOPAC) levels were observed suggesting that the cells were able to help restore the loss of DA.

There is also some debate as to whether the endometrial stem cells truly arise from the endometrium or whether stem cells originating from the bone marrow are involved. Some cells isolated from the endometrium (and so also potentially in menstrual blood) have been shown to originate from the bone marrow in bone marrow transplant patients by taking advantage of HLA mismatching between the donor and host bone marrow (Taylor 2004). Also Wang et al. (2012) compared ERCs and BMSCs with respect to microRNA, gene expression, cytokine and growth factor secretion levels, and their immunosuppressive abilities. Differences were observed suggesting that they were different populations. One important distinction is that ERCs do not express STRO-1, a BMSC marker. Consequently there may be several different stem cell populations present in endometrial tissue and by extension menstrual blood.

These few studies suggest that differentiation to an NSC-like cell may also be possible with menstrual blood-derived stem cells.

2.2.1.2 Peripheral Blood

The peripheral blood can be easily isolated from subjects of any age or gender and contains a mixture of different cells, including erythrocytes; platelets; granulocytes such as eosinophils, basophils, and neutrophils; MNCs such as monocytes and lymphocytes; and, in much smaller quantities, HSCs and nonhematopoietic stem cells, such as MSCs. These cells originate from the bone marrow and are released into the blood stream. The majority of studies focus on the MNC fraction, including the stem cells. This is normally obtained from the blood by a centrifugal isolation technique such as Ficoll centrifugation. The specific cell populations can be further isolated by additional isolation steps and manipulation of the culturing conditions. Culturing the MNC fraction of human peripheral blood in DMEM and 10% FBS on fibronectin-coated plates for 7–10 days leads to a relatively pure population of adherent monocyte-derived cells that exhibit multipotent capabilities (MOMCs) (Kuwana et al. 2003). These cells can be differentiated into neural-like cells by coculturing with primary rat neurons in DMEM, 10% FBS, 2 mM L-glutamine, and ITS supplement for 4 weeks (Kodama et al. 2006). Direct contact between the cells was not required (demonstrated by use of a Transwell chamber), suggesting that factors secreted by the rat neurons triggered the change rather than cell fusion. This is further supported by the absence of any neural-like cells possessing more than one nucleus. Within 3 days of coculture, the cells were expressing early neural transcription factors such as mammalian achaete-scute homolog 1 (Mash1), neurogenin 2 (Ngn2), NeuroD, and nestin, factors that free-floating monocytes did not express even after coculture. Within 14 days of coculture, the cells possessed an interconnecting network of axon-like structures, and after 21 days, MAP 2, βIII tubulin, NeuN, and neuron-specific RNA-binding protein Hu expression was also evident, suggesting neuronal differentiation may have occurred. Rat GFP-MOMCs have been transplanted into the striatum of rats 7 days after they were given a stroke by MCAo. The number of surviving GFP^+ cells peaked around 2 weeks and then declined. While no neural differentiation was observed, neovascularization and $CD31^+/GFP^+$ cells were observed suggesting that endothelial differentiation had occurred (Hattori et al. 2012).

In a study exploring the potential neural differentiation of the peripheral blood lymphocyte population, an induced pluripotent stem cell step was utilized. Tsai et al. (2015) expanded the MNC layer from several patients in AIM-V medium (Invitrogen) supplemented with 300 IU/ml recombinant human IL-2 on plates bound with 10 ng/ml anti-CD3 antibody for 3 days. The latter induces T-cell proliferation and activation. The obtained T cells were then nucleotransfected with integration-free expression plasmids for the four Yamanaka factors and Epstein-Barr nuclear antigen 1 (EBNA1). Following transfection, the cells were grown on mouse embryonic feeder (MEF) cells in DMEM/F12 with 20% knockout serum replacer, 0.1 mM NEAA, 1 mM L-glutamine, 0.1 mM βME, 10 ng/ml rhbFGF, and antibiotics for 4 weeks. The T-cell-derived iPSCs were then differentiated into neural precursor cells using a typical ESC/iPSC neural differentiation medium such as DMEM/F12, containing glucose, ITS, 20 nM progesterone, 60 μM putrescine, 2 mM glutamine, 3 mM sodium bicarbonate, 5 mM HEPES, 2 μg/ml heparin, 20 ng/ml EGF, and 20 ng/ml bFGF for 14 days. Culture for a further week in the absence of EGF, but with the addition of 20 ng/ml SHH, 10 ng/ml BDNF, and 100 nM all-trans-retinoic acid, led to further neural differentiation. The cells formed $SOX2^+/nestin^+$ neurospheres with extensive neurite outgrowth, suggesting generation of NSCs. Over time, the cells began to express MAP 2, GFAP, and NCAM implying that some degree of neuronal differentiation may have occurred.

In another study, episomal, non-integrating vectors encoding the Yamanaka factors, as well as SV40, Lin-28, Nanog, and EBNA1, were nucleotransfected into MNCs from a single male patient and then plated on MEF cells in 50% DMEM/F12, 50% Neurobasal media, 1 × N2, 2 × B27, 1% GlutaMAX, 10 ng/ml rhLIF, 3 μM glycogen synthase kinase 3β (GSK3β) inhibitor CHIR99021, and 2 μM TGFβ receptor inhibitor SB431542 for 20 days (Tang et al. 2016). Expansion on poly-*D*-lysine/laminin-coated plates was then performed, followed by removal of CHIR99021, SB31542, and rhLIF, and the addition of 1 μM retinoic acid and

1 μM *Toxoplasma gondii* surface antigen (SAG1) to the media for a further 2 weeks. The induced NSCs obtained were able to proliferate and were positive for Ki67. In addition they expressed SOX1, SOX2, and nestin protein, and fatty acid-binding protein 7 (*FABP7*), *SOX2*, hairy and enhancer of split 5 (*HES5*), and *SOX1* mRNA. Further neural differentiation was also demonstrated by culturing on poly-*D*-lysine/laminin-coated-glass coverslips in DMEM/F12, 1 × N2, 1 × B27, 1 × NEAA, 100 ng/ml cAMP, 10 ng/ml GDNF, 10 ng/ml BDNF, 10 ng/ml IGF-1, and 10 ng/ml NT-3. The resulting cells were MAP 2^+/NeuN$^+$/postsynaptic density 95 (PSD95)$^+$/vGlut1$^+$ action potential-capable neurons and GFAP$^+$ astrocytes. The above media supplemented with 1 μM retinoic acid, 20 ng/ml PDGF-AB, 10 ng/ml bFGF, and 1 μM SAG1 for 2 weeks resulted in PDGFRα$^+$ oligodendrocyte precursor cells. Replacement of RA and bFGF with 60 ng/ml triiodothyronine (T3) for a further 6 weeks led to the detection of O1$^+$ oligodendrocytes. However compared to the time taken to perform in vitro transdifferentiation, neural or other means of induction appear to be highly time consuming.

In addition, the CD34$^+$ HSC population can also be isolated using CD34$^+$ antibody-linked fluorescent- or magnetic-activated cell sorting (CD34$^+$ FACS or MACS). The cells can be isolated directly from the MNCs and cultured in media containing 100 ng/ml thrombopoietin, 100 ng/ml fms-like tyrosine kinase 3 ligand, 100 ng/ml stem cell factor (SCF), 20 ng/ml IL-6, and 20 ng/ml IL-7 for 24 h (Wang et al. 2015), before the free-floating cells are replated in fresh media for a further 5 days and then induced using a CytoTune-iPSC Sendai virus reprogramming mix which will lead to sphere formation and aggregation, and the cells will become adherent. These cells can then be cultured in DMEM/F12, 1 × N2, 0.1% BSA, 20 ng/ml bFGF, and 120 ng/ml EGF on Matrigel-coated plates for a week and then the media replaced with NSC medium. These cells are nestin$^+$ and SOX2$^+$ NSCs and can be cultured for long periods of time with proliferation, but without spontaneous differentiation (Wang et al. 2013b). Further differentiation can be performed using DMEM/F12, 1 × N2, 1 × B27, 300 ng/ml cAMP, 200 μM vitamin C, 10 ng/ml BDNF, and 10 ng/ml GDNF with cells plated on poly-*D*-lysine/laminin for 2 weeks, resulting in βIII tubulin$^+$ neuron-like cells. These cells also exhibited action potentials and expressed vesicular glutamate transporter 1 (VGLUT1), vesicular GABA transporter 1 (VGAT), or TH (Wang et al. 2013a, b). Culturing the NSCs in DMEM/F12 and 10% FBS for 2 weeks results in GFAP$^+$ astroglial cells. Culturing the cells on poly-*L*-ornithine in DMEM/F12, 1% N2, 10 ng/ml PDGF-AA, 2 ng/ml NT-3, 2 ng/ml SHH, and 3 nM T3 for 2 weeks and a further week in DMEM/F12, 1% N2, and 3 nM T3 leads to MBP$^+$ oligodendrocytes.

Lee et al. (2015) transduced CD34$^+$ peripheral blood cells with an OCT4 lentivirus in media containing SCF, Fms-related tyrosine kinase 3 (Flt-3), IL-3, and thrombopoietin (TPO), and after 48 h the cells were cultured on Matrigel with 10 ng/ml bFGF added to the media. CD34$^-$ cells were not capable of being induced in this fashion. After 5 days the cells were incubated in DMEM/F12, 1 × N2, 1 × B27, inhibitors of mothers against decapentaplegic (SMAD)-GSK3 such as SB431542 (10 μM), LDN-193189 (100 nM), 3 μM CHIR99021, and 100 ng/ml Noggin. After 14 days the cells were plated on polyornithine-laminin-coated plates, with the addition of 20 ng/ml bFGF and 20 ng/ml EGF to the media. Neurosphere-like structures formed which expressed NSC-associated markers including nestin, PAX6, SOX2, and CD133, as well as the proliferation marker Ki67. These cells could be further differentiated to neurons, astrocytes, and oligodendrocytes in vitro, and differentiation was also observed in vivo. The authors appeared to successfully generate glutamatergic, GABAergic, and DAergic neurons with depolarizing and hyperpolarizing abilities as well as GFAP$^+$ astrocytes and O4$^+$ oligodendrocytes. The cells were transfected with GFP and transplanted into neonatal mouse cortex, and 3 weeks later, GFP$^+$ cells were observed to also express βIII tubulin, MAP 2, or GFAP, but, in contrast to the in vitro data, no oligodendrocytic markers.

The CD34$^+$ hematopoietic population in the peripheral blood can be mobilized by 3 days of granulocyte colony-stimulating factor (G-CSF) treatment and isolated from the blood using an automated cell sorter (Venkatesh et al. 2015). The cells can then be cultured in DMEM and 10% FBS and purified by CD34$^+$ FACS. Culturing the purified cells in DMEM supplemented with 1 μM retinoic acid, 20 ng/ml EGF, 20 ng/ml FGF, and 25 μg/ml insulin for 3 days results in neurosphere formation. A further 3 days in DMEM supplemented with 40 ng/ml thyroxine, 40 ng/ml T3, and 30 nM selenium led to oligodendrocyte precursor cell generation, based on their stellate appearance, Olig2 protein expression and *Olig2*, *CNPase*, *PDGFRα*, and proteolipid protein 1 (*PLP1*) mRNA.

CD133$^+$ early hematopoietic stem cells can also be isolated from peripheral blood by MACS and cultured in RPMI-1640 containing 10% FBS, 100 ng/ml Flt3/fetal liver kinase 2 (Flk2) ligand, and 100 ng/ml IL-6 for 7 days and then transferred to RPMI-1640 and 10% FBS for 4 weeks (Kuci et al. 2008). Culturing in Cambrex NPBM with the addition of EGF, hrbFGF, and neural survivor factor-1 for 14 days led to a neural progenitor-like cell that could be further differentiated by supplementation with 50 ng/ml BDNF and 10 ng/ml GDNF for 7 days. The majority of the neural progenitor-like cells expressed βIII tubulin suggesting they may be neuronal precursors, while around 15% expressed GFAP and MBP, suggesting they were of the astroglial lineage. A proportion of the βIII tubulin-expressing cells also expressed TH and vesicular monoamine transporter 1 (VMAT1) suggesting that they possessed dopaminergic neuron characteristics. A small proportion of the cells also expressed voltage-gated ion channels.

Alternatively the MNC fraction can undergo FACS and a distinct cell population isolated and cultured in DMEM with 10% human AB serum, 10 mM βME, and 500 nM all-trans-retinoic acid for 24 h. Replacement of the media with Neurobasal medium, 2 mM L-glutamine, B27, 25 ng/ml FGF2, 10 ng/ml EGF, and 10% human AB serum can lead to neural differentiation. The optimal cell population, based on nestin levels after 24 h, was observed to be CD133$^+$, chemokine CXC motif receptor 4 (CXCR4)$^+$, and ATP- binding cassette subfamily G member 2 (ABCG2$^+$) (Nichols et al. 2013). In the replacement media, TH$^+$, βIII tubulin$^+$, and NeuN$^+$ cells were observed, but relatively few GFAP$^+$ cells. Coculture of the all-trans-retinoic acid-treated cells with rat or human astrocytes or dopaminergic cells led to high TH expression, though this was reduced in a Transwell setup where no cell-to-cell contact was possible. Coculturing with bone marrow MSCs did not induce expression of neuronal markers. Intracerebroventricular transplantation of CD133$^+$, CXCR4$^+$, and ABCG2$^+$ cells into TBI or uninjured rats was performed, and the cells had been labeled with carboxyfluorescein succinimidyl ester (CFSE) prior to transplant to allow tracking of the cells. After a month in uninjured animals, CXCR4$^+$ cells were observed primarily in or around the ventricles, while within 2 days in TBI rats, the cells were migrating closer to the injury site. Some of these cells were now expressing βIII tubulin and TH, suggesting neuronal differentiation was occurring. After 3 months the majority of surviving cells transplanted into TBI rats were βIII tubulin$^+$ and NeuN$^+$ compared with around 40% in uninjured animals.

Plastic adherent culturing of peripheral blood-derived MNCs in DMEM supplemented with 10% FBS, glutamine, and antibiotics can result in the isolation of a MSC-like cell expressing CD73 and CD90. Culturing the cells in Cambrex's neural progenitor basal medium, supplemented with 2 mM L-glutamine, 10 ng/ml EGF, 10 ng/ml bFGF, and antibiotics for 3 days, leads to the formation of nestin$^+$ neurosphere-like structures that may be NSCs. These can be mechanically disassociated and replated for 1 week in the absence of growth factors to generate NF-M$^+$ or GFAP$^+$ neural cells (Kim et al. 2006). It is unclear whether these neural cells are functional or still precursor cells since no additional study of the cells, e.g., electrophoresis, was performed.

Porat et al. (2006) used a dual density-enrichment step, whereby they subjected peripheral blood to Ficoll centrifugation and the MNCs were then subjected to Percoll or OptiPrep centrifugation. The cells were then cultured in X-Vivo 15 media supplemented with 10% autologous plasma on fibronectin-coated plates. The media were supplemented with 10 ng/ml bFGF, 25 ng/ml BDNF, 50 ng/ml NGF, and 5 U/ml heparin for 8 days. The autologous plasma was replaced with 33% F12, 2% B27, and 20% EGF, and the resulting cells were nestin$^+$, βIII tubulin$^+$, and NeuN$^+$ and responded to glutamate and GABA via voltage-gated calcium channels or were O4$^+$ or GFAP$^+$, suggesting that they were neurons, oligodendrocytes or astrocytes.

An insulin-producing cell that expressed ESC-like transcription factors and CD45$^+$, but was negative for CD34, CD14, CD80, and CD86, markers for hematopoietic and endothelial progenitor cells, was isolated from peripheral blood (Li et al. 2015). The cells were isolated by Ficoll centrifugation and cultured in serum-free media under high CO_2 conditions (8%) for 14 days. After reducing CO_2 to the "normal" culture levels of 5% and adding 10 μM all-trans-retinoic acid for 14 days, the cells morphologically resembled astrocytes and expressed a number of astrocytic markers, such as GFAP, CD44, GLAST, and S100β, but were unable to clear glutamate from the media, suggesting that they were not functioning astrocytes but may be an immature astrocyte cell type.

Another cell type in cord blood has been termed multipotent adult progenitor cells (MAPCs) and can be isolated from the peripheral blood of GFP transgenic swine. These cells were obtained by culturing the MNC fraction in MethoCult H4535 media (Stem Cell Technologies) supplemented with 100 ng/ml LIF, 100 ng/ml bFGF, 100 μM NEAA, 430 μg/ml GlutaMAX, and antibiotics for 14 days (Price et al. 2006). The cells form floating spheroid-like structures that can be replated after disassociation in the above media in a 1:1 ratio with Iscove's minimum essential medium supplemented with 30% FBS, 100 μM βME, 50 ng/ml SCF, 20 ng/ml granulocyte-macrophage colony-stimulating factor (GM-CSF), 20 ng/ml IL-3, 20 ng/ml IL-6, 20 ng/ml G-CSF, 100 ng/ml LIF, 100 ng/ml bFGF, 100 μM NEAA, 430 μg/ml GlutaMAX, and antibiotics for 10 days. These cells were able to proliferate and can be differentiated by incubating in different media. For instance, plating on Matrigel with endothelial cell growth media and DMEM/F12 (1:1) with 10% FBS for 7 days resulted in adherent cells adopting a neurogenic morphology and expressing neural markers such as βIII tubulin, TH, and GFAP (Price et al. 2006). An alternative differentiation media for these cells proposed by the same group is the use of Neurobasal medium containing antibiotics, 100 μM NEAA, 430 μg/ml GlutaMAX, B27, N2, 2.5% Matrigel, 60 ng/ml EGF, 10 ng/ml bFGF, 50 ng/ml SHH, 100 ng/ml FGF8, and 10 μM $GABA_B$ receptor antagonist CGP55845 (or 50 ng/ml bFGF, 25 ng/ml GDNF, 25 ng/ml BDNF, 25 ng/ml NT-3, and 1 × ITS instead of EGF, FGF8, and CBP55845) and plating the cells on poly-*l*-lysine-coated slides (Spitzer et al. 2011). The authors do say that Matrigel alone (or on laminin or fibronectin-coated coverslips) in a basic media without growth factors and cytokines also initiates neural differentiation—though a slightly higher proportion of undifferentiated cells compared to the more complex media mentioned above was observed. The majority of cells using the complete media expressed βIII tubulin and NeuN, while a small number also expressed the presynaptic marker synataxin and synaptic marker synaptophysin, as well as GFAP, acetyl cholinesterase, TH, nestin, O4, MAP 2, and voltage-gated sodium channel proteins. Predifferentiated MAPCs also expressed NeuN, synataxin, βIII tubulin, MAP 2, and voltage-gated sodium channels mRNA suggesting that MAPCs may have a neural predisposition (Price et al. 2006; Spitzer et al. 2011). This predisposition of predifferentiated MAPCs appears to be long lasting since the cells will still differentiate after numerous passages. However once the cells are incubated in differentiation media, they stop proliferating, even after removal of neurogenic signals. Time-lapse microscopy of the process of differentiation demonstrates thin neurite-like projections growing out from the cell body that contain βIII tubulin and end in f-actin-rich growth cones, suggestive of an active dynamic neurite outgrowth process (Spitzer et al. 2011). The differentiated cells also showed depolarizing and hyperpolarizing effects by patch clamping, suggesting that they may be functionally active.

These studies show that a population of NSCs and mature neurons can be obtained by manipulation of culture conditions of the myriad of different cells that can be found within the peripheral blood.

2.2.2 Bone Marrow

Bone marrow contains many different cell types including all of the cells of the hematopoietic system as a mixture of precursor and mature cells and stromal tissue which includes the nonhematopoietic stem cells, such as the BMSCs or MSCs, that provide structural and functional support for the hematopoietic cells.

In 2000, Sanchez-Ramos et al. showed that up to 5% of first or second passage mouse and human BMSCs express the neuronal markers nestin, NeuN, or GFAP following culture in N5 medium with the addition of 5% horse serum, 1% FBS, ITS, 60 μM putrescine, 0.02 μM progesterone, 0.5 μM all-trans-retinoic acid, and 10 ng/ml BDNF (Sanchez-Ramos et al. 2000). This media is similar to that used to differentiate ESCs and NSCs. A similar effect was observed on coculture with rat fetal midbrain cultures. The human or mouse BMSCs were labeled with either 5-chloromethyl fluorescein diacetate or PKH-26 or obtained from transgenic LacZ mice, which would express β-galactosidase, for the coculture studies, so that they could be distinguished from the rat fetal cells. Hematopoietic progenitors were excluded by depletion of $CD34^+$ or stem cell antigen 1^+ ($SCA1^+$) cells and the use of plastic adherence in the culturing conditions. The changes in expression were observed within 5 h of culturing, but it is unclear whether this was a permanent change or whether changing the media again would reverse the expression. In addition only a small proportion of cells were observed to change, suggesting that either there is only a small population that can transform, which could be isolated further or expanded, or that the ability of BMSCs to differentiate is small, or that the culture conditions were not optimal. Later studies suggest that BMSCs are able to differentiate, under the right conditions.

In another study from 2000, rat and human $CD90^+$BMSCs were incubated in media (DMEM) and 20% FBS with antioxidants—1 mM βME for 24 h followed by serum-free DMEM, 2% DMSO, and 200 μM butylated hydroxyanisole (BHA) for up to 6 days or serum-free DMEM, 2% DMSO, 200 μM BHA, 25 mM KCl, 2 mM valproic acid, 10 μM forskolin, 1 μM hydrocortisone, and 5 μg/ml insulin for longer periods (Woodbury et al. 2000). After 5 h, some of the BMSCs adopted a more neuronal phenotype, including rare pyramidal cells and multipolar cells with long processes, and increased their expression of NSE from the minimal basal levels of the normal BMSCs. Many of the changing cells expressed high levels of NF-M, including a few cells where the antigen was localized specifically to the processes, which is a sign of maturing neurons. NeuN-expressing cells were also observed, though no $GFAP^+$ cells were evident, suggesting that the protocol induced a neuronal differentiation rather than astrocytic. More than 50% of the cells appeared to undergo differentiation, and substituting BDNF for the βME increased the neuronal cell numbers to 78%. Nestin was highly expressed at 5 h but had disappeared by 6 days suggesting that the cells may progress to an NSC state before maturing into neurons. In a second study, the authors modified the induction media by replacing the forskolin, hydrocortisone, and insulin with k252a, heparin, N2, bFGF, and PDGF, and at a later time point, the media were exchanged with non-supplemented DMEM to determine if the neuronal induction was reversible. The BMSCs were used from passage 15, and gene analysis prior to induction showed evidence of endodermal, mesodermal, and particularly neuroectodermal gene expression such as NPC transcription factor, neuroD, and GFAP, both of which decreased with time in the media, suggesting the progression of neuronal differentiation. BMSCs cultured for 24 h in the neural induction media, followed by 24 h in unsupplemented DMEM, reversed all the morphological and gene expression changes initiated by neural induction. Unfortunately the authors did not investigate whether the same is true after more long-term differentiation (Woodbury et al. 2002). However, there is some debate about the reliability of this method (see later).

Another study used compounds that would increase intracellular cAMP, such as db-cAMP and IBMX, to induce neural differentiation in BMSCs (Deng et al. 2001). The BMSCs were isolated by Ficoll centrifugation and plastic adherence and were passaged twice in α-MEM, 20% FBS, antibiotics, and 2 mM L-glutamine prior to differentiation. The same media were used, with the addition of 1 mM db-cAMP and 0.5% IBMX for 6 days. Within 2 days, neuronal morphology was evident, and a slight reduction in proliferation was observed. Untreated cells expressed MAP 1B, NSE, βIII tubulin, and vimentin, and both NSE and vimentin increased following induction, but the expression levels of the other proteins or of mature neuron markers such as MAP 2, NF-M, tau, S-100, GFAP, or MBP were not apparent. Of note, removal of IBMX and dbcAMP resulted in the death of the neuron-like cells, and the unchanged cells entered senescence, suggesting that differentiation toward an NSC/NPC phenotype was not reversible, even in cells that did not appear to change.

Tondreau et al. (2004) observed modest levels of early neuroectodermal gene and protein expression, such as nestin and βIII tubulin, and low levels of late neuroectodermal genes and proteins, such as TH, MAP 2, and GFAP in undifferentiated BMSCs. Nestin expression remained stable, while βIII tubulin decreased, and late gene/protein expression increased significantly following culturing in neural induction media. Under basal conditions the cells were cultured with α-MEM supplemented with 15% FBS, 2 mM glutamine, and antibiotics. For differentiation, the cells had been initially seeded by limited dilution to single cells and expanded to 80% confluence before replating and culturing in NPBM, 5 μM cAMP, 5 μM IBMX, 25 ng/ml NGF, and 2.5 μg/ml insulin for 10 days.

Gene expression analysis revealed that while undifferentiated BMSCs may express a low level of a multitude of different genes, this changes once placed under differentiation conditions. For instance, under osteogenic differentiating conditions, expression of neural-related genes such as nestin and βIII tubulin is heavily suppressed, while ECM and bone-related genes increase. Conversely, under neurogenic differentiation (in α-MEM, 15% FBS, antibiotics, and 1 mM βME for 24 h followed by Neurobasal medium, B27 and 20 ng/ml BDNF), the ECM- and bone-related genes are silenced (Egusa et al. 2005). This suggests that BMSCs express a multitude of genes ensuring they have extensive plasticity, until they commit to a phenotype.

Compared to the rapid conversion described above, studies in which NSCs are generated from BMSCs and allowed to proliferate also exist. Human MSCs that are $CD90^+$, $CD105^+$, $CD14^-$, $CD34^-$, and $CD45^-$ were cultured for 2–10 passages and then transferred to P4-8F medium (AthenaES) containing 20 ng/ml EGF and 20 ng/ml FGF-2 under hypoxic conditions (3% O_2). Neurosphere-like structures formed within 10 days, and they were expanded for a further 2–10 weeks (Hermann et al. 2004). For terminal differentiation, the cells were plated on poly-*L*-lysine-coated coverslips in Neurobasal medium with 0.5 μM all-trans-retinoic acid, 1% FBS, 5% horse serum, 1% N2 supplement, and 1% antibiotics. The media were supplemented with 10 ng/ml BDNF for neuronal differentiation or 10 ng/ml PDGF-BB for glial induction. The majority of the cells within the neurospheres expressed nestin, but not MAP 2 or GFAP protein, and expressed neuroectodermal transcripts such as neuroD at significantly higher levels than untreated BMSCs. Following neuronal differentiation, the expression of mature neural cell markers such as GFAP, MBP, βIII tubulin, GalC, and TH was significantly increased, while those for neuroD were reduced. Mature MAP 2^+ neurons and TH^+/DA-producing cells were observed, but the authors stated they could not get patch clamping to work due to a requirement for FBS to be present so as to obtain high seal resistance which caused glial cell overgrowth. Glial differentiation did lead to $GFAP^+$ and $GALC^+$ cells suggesting that astrocytes and oligodendrocytes may have formed. Evidence of a delayed rectifier K^+ channel was also observed from electrophysiological analysis suggesting the presence of developing and adult glial cells.

Using limited dilution to subclone BMSC cultures, Kohyama et al. (2001) obtained cells which they induced to a neuronal phenotype by two different methods. First they cultured the cells with IMDM containing 10% FBS and antibiotics and supplemented the media with a demethylating agent, 10 μM 5-azacytidine, and growth factors 50 ng/ml NGF, 50 ng/ml NT-3, and 50 ng/ml BDNF on fibronectin/poly-*L*-ornithine-coated plates. Four days later, the media were exchanged for DMEM/F12 and B27, 50 ng/ml NGF, 50 ng/ml NT-3, and 50 ng/ml BDNF for a further 26 days. Their second method involved transfection of the cells with the BMP inhibitor Noggin in IMDM and 10% FBS. Once the cells formed neurosphere-like structures (~7–10 days), the cell clusters were transferred to fibronectin/poly-*L*-ornithine-coated plates and incubated with DMEM/F12, 5% FBS, and 20 ng/ml bFGF for 5 days. Subsequent culturing was performed without FBS or bFGF and with the addition of 50 ng/ml NGF, 50 ng/ml NT-3, and 50 ng/ml BDNF for a further 15 days. Prior to differentiation the cells were $CD34^+$, SCA-1^+, $CD140^+$, and $CD29^+$ suggesting that they were more likely to be hematopoietic stem cells, especially since they did not express CD90 or CD105. Using 5-azacytidine, a dramatic change in morphology was observed on day 4 in that 20% of the cells appeared to form neurite-like projections. The cells became positive for NeuN, βIII tubulin, and Hu, suggesting neuronal differentiation, as well as GFAP and GalC, suggesting astrocytic and oligodendritic differentiation. Neuron-specific genes such as *NCAM*, *GAP-43*, and *Trk A-C* were also observed. Since only the neurospheres were plated in the Noggin transfection method, a higher number of cells (50%) exhibited neuron-like characteristics and expressed MAP 2 and βIII tubulin, while a smaller proportion (~5%) were $GFAP^+$ suggesting increased neurogenesis and reduced astrocyte generation compared with the 5-azacytidine method. Cells from both methods also showed a reduced resting membrane potential (−50 mV) and evidence for a voltage-dependent rectifying current. Exposure to a depolarizing stimulus such as high potassium levels or a neurotransmitter such as glutamate resulted in a rapid and reversible calcium uptake by the cells.

Using bone marrow from BMP2 and BMP4 double conditional mice to isolate BM-MSCs, Saxena et al. (2016) determined that knockout of these two proteins for 5 days induced a morphological change toward a neuronal phenotype. By 30 days, axon and dendrite-like projections appeared to be making synaptic-like projections with other cells. A decrease in cell proliferation was also apparent with time. During the initial stage of morphological change, βIII tubulin was detected, but by 10 and 15 days, MAP 2-, NeuN-, and GABA-positive cells were also evident with a concomitant decrease in βIII tubulin. $GFAP^+$ cells were also observed; however these also expressed MAP 2, suggesting GFAP was not a marker of astrocytes in this situation. GFAP expression had ceased by 30 days in culture. The authors also observed that Noggin protein treatment had a similar effect to that of BMP knockout on cell survival and differentiation, supporting that 200 ng/ml Noggin acts via inhibition of BMPs.

Jin et al. (2003) set out to explore the potential influence of a number of different growth factors on the ability of the raw heterogeneous bone marrow cells to differentiate to neuron-like cells. The cells were plated onto poly-*D*-lysine-coated plates and cultured in α-MEM with 1 × GlutaMAX and 20% FBS. The media were then changed to Neurobasal medium containing B27, GlutaMAX, and antibiotics with the addition of one or more growth factors for 7 days. Individually, 20 ng/ml EGF and 10 ng/ml heparin-binding EGF-like growth factor were found to promote the largest increase in nestin expression, while heparin-binding EGF-like growth factor and 40 ng/ml FGF-2 increased immature neuronal markers such as DCX, βIII tubulin, and embryonic nerve cell adhesion molecule (ENCAM). Forty ng/ml NGF and 10 ng/ml SCF appeared to exert their greatest effect on the expression of the neurotransmitters GABA and glutamate, showing that there was little relationship between neurotransmitter expression and degree of differentiation. No GFAP expression was observed. Combining EGF and FGF-2 for 2 weeks led to an increase in most of the factors examined, though there was a tendency for higher expression of the more mature markers. Diffuse and homogenous cytoplasmic labeling of these markers was observed since they were microtubule-associated or cytoplasmic proteins. NeuN expression was not restricted to the nuclear compartment in these cells, in contrast to what you'd expect in mature neurons. Further differentiation for 2 weeks with the addition of 0.5 μM all-trans-retinoic acid and 20 ng/ml NGF to the media led to a more neuronal phenotype and a higher proportion of cells expressing NeuN and MAP 2 and tau, though unlike in mature neurons, the MAP 2 and tau were not preferentially localized to the cell processes. Diffuse expression of synaptophysin and calcium channel subunits was also observed, instead of them being localized to the nerve terminals. Again no GFAP expression was observed. This would suggest that fully functioning mature neurons had not yet been obtained, though it is possible that more long-term culture may have allowed the cells enough time to mature.

Within the MNC fraction of the bone marrow, it is believed that there exists cell populations other than the BMSCs. A cell population known as MAPCs, also found in peripheral blood, has been isolated from both the human and mouse BM-MNC fraction. BM-MNCs are cultured on laminin, fibronectin, collagen type IV, or

Matrigel-coated plates in 60% DMEM, 40% MCDB-201 containing 1 × ITS, 1 × linolenic acid BSA, 1 nM dexamethasone, antibiotics, and 100 μM ascorbic acid 2-phosphate for 3 weeks. The leukocyte common antigen marker, CD45, and erythroid cell marker (52 kD glycophorin A-associated protein; TER119$^+$) cell population were then depleted from the cultured cells by MACS, and the remaining cells plated at low density with the addition of 10 ng/ml EGF, 10 ng/ml PDGF-BB, and 10 ng/ml LIF on fibronectin-coated plates. The cells were observed to be CD13$^+$ and SSEA-1$^+$ and could be expanded many times (Jiang et al. 2002). These cells did not stain for nestin, GFAP, MBP, NF-200, or neurotransmitters. After 50–70 population doublings, the cells were cultured in the above media with 100 ng/ml BDNF instead of EGF, PDGF, and LIF. Within 7 days the majority of the cells expressed nestin protein, and some also expressed NF-200, GFAP, or MBP. Continued incubation in this media led to cell death. However if the BDNF was switched out after 7 days with 100 ng/ml SHH and 10 ng/ml FGF8 for a further 7 days, followed by 7 days in 10 ng/ml BDNF and N2 media, the cells were observed to reduce their nestin expression but increase the expression of neurotransmitter and neuronal cell markers, while astrocytic and oligodendrocytic markers declined. After coculture with fetal brain astrocytes or astrocyte-conditioned media for 9 days, tetrodotoxin-dependent membrane currents were also observed, suggesting the cells maybe maturing to functional neurons (Jiang et al. 2003).

By 2004, several studies had been published, casting doubt on whether neural differentiation of bone marrow-derived cells can occur. Their main causes of concern were that (1) they couldn't replicate the neural differentiation reported by others, (2) a pluripotent subpopulation may exist within the BMSC cultures (3) cell fusion may be occurring when cocultures or transplants were used as evidence, (4) the morphological changes and gene or protein expression changes observed were due to actin cytoskeleton changes and the stress on the cells of being cultured under different conditions rather than the process of transdifferentiation (e.g., see Lu et al. (2004); Neuhuber et al. (2004); and review by Krabbe et al. (2005)). The last point is likely to be valid in those studies where rapid changes were observed. Lu et al. (2004) demonstrated rapid cell shrinkage and neuron-like morphology of BMSCs but also terminally differentiated cells, such as fibroblasts, keratinocytes, and HEK293 cells, following their incubation in induction medias containing βME similar to those described above (especially Woodbury et al. (2000)), or under extremes of pH, or with the use of detergents. These neuron-like morphological changes also occurred following protein inhibition by cycloheximide and so are unlikely to be due to differentiation. Changes in the actin cytoskeleton were also found to cause the rapid cell shrinkage and neuron-like morphology using the Woodbury protocol (Woodbury et al. 2000) according to Neuhuber et al. (2004). Fast changes toward a neuronal phenotype, particularly if no neural stem or progenitor cell-like intermediate state could be identified, would therefore seem unlikely to be true differentiation. Consequently more vigorous testing measures were proposed to help validate the reliability of neural differentiation (Krabbe et al. 2005). This included evidence of long-term expression of the neural biomarkers; electrophysiological characterization of resting membrane potential, action potentials, functional voltage-gated ion channels, and neurotransmitter receptors; and use of a homogenous cell population. Unfortunately many of the prior studies did not (and many newer studies still do not) fulfill all these criteria, with many still using the Woodbury protocol (Woodbury et al. 2000).

To help determine whether neurons can really be transdifferentiated from BMSCs, Song et al. (2007) compared the cells obtained from the differentiation of BMSCs to those from brain-derived cells. They cultured mouse BMSCs in DMEM/F12 and 10% FBS for five passages before replating the cells on adherent plastic dishes using N2 medium, ITS, 60 μM putrescine, 0.02 μM progesterone, 20 ng/ml EGF, and 20 ng/ml bFGF. These cells proliferated, and a high proportion of the cells were nestin$^+$ suggesting the cells may be BM-NSCs. The mouse brain-derived cells were initially cultured in DMEM/F12 and 10% FBS, before transfer to DMEM/F12 containing 20 ng/ml EGF and 20 ng/ml bFGF. These cells formed neurospheres and were also highly nestin$^+$. The nestin$^+$ cells derived from the BMSCs and brain were then replated onto poly-*L*-lysine-coated slides and cultured in N2 medium, ITS, 60 μM putrescine, 0.02 μM progesterone, 0.5 μM all-trans-retinoic acid, 10 ng/ml BDNF, and 4% FBS. After 7 days, both mouse brain- and BMSC-derived cells were observed to express neuronal and glial markers such as nestin, βIII tubulin, NeuN, GFAP, GalC, glutamate decarboxylase (GAD), and TH. Significantly more brain-derived cells expressed nestin, βIII tubulin, NeuN, GFAP, and GAD than BMSC-derived cells. Only 2.5% of the BMSC-derived cells exhibited action potentials following patch clamp studies, compared to 60% of the brain-derived cells, though other characteristics suggested that they were still immature neurons. High GABA uptake was observed by both the brain- and BMSC-derived neurons, and this level was higher in the differentiated cells than the proliferating NSCs. DA uptake was also observed in both cell types, but it was not increased by differentiation.

Neural induction methods include the use of cytokines and growth factors or other chemical reagents. Tao et al. compared the use of a cytokine-based induction method and a chemical reagent induction method on two different isolation methods (plastic adherence and MACS) for obtaining BMSCs (Tao et al. 2005). Plastic adherent BMSCs were isolated by culturing human MNCs in DMEM, 10% FBS, and antibiotics and exchanging the media after 24 h to discard nonadherent cells. The cells were cultured to passage 6 before neuronal induction. The second isolation method utilized negative selection with the CD45$^+$ and glycophorin-A$^+$ cells being removed from the MNCs by MACS. The cells were then cultured in 60% DMEM/40% MCDB-201 with 1 × ITS, 1 × linoleic acid BSA, 1 nM dexamethasone, 100 μM ascorbic acid 2-phosphate, 10% FBS, and antibiotics on fibronectin-coated flasks. A similar number of morphologically and immunophenotypically identical cells were isolated by each protocol expressing low levels of βIII tubulin, NF-M, NSE, and CNPase. Growth factor-driven differentiation was performed in fibronectin-coated flasks using DMEM/F12, ITS, 100 μM putrescine, 0.02 μM progesterone, 10 ng/ml EGF, 10 ng/ml bFGF, and 1 ng/ml PDGF for up to 3 months. Neuron-like morphological changes were observed within 2 weeks and remained throughout the 3-month period. However no proliferation or reversal of differentiation was apparent once differentiation started. Removal of the growth factors essentially terminated the culture after 3 days, highlighting the importance of the presence of the growth factors. Expression levels of βIII tubulin, NF-M, and NSE all increased, while MAP 2, tau, GABA, TH, and serotonin were now also expressed. CNPase levels increased slightly, but no GFAP expression was observed. Chemically induced differentiation was initiated in DMEM, 10% FBS, 1 mM βME, and 10 ng/ml bFGF for 24 h followed by serum-free DMEM, 2% DMSO, and 200 μM BHA for 7 days. The latter media caused neuron-like morphological changes within hrs and a similar change in protein expression to those treated with cytokines. However, no further proliferation was observed, and after 2 days, the cells were nonviable after detaching from the plate, suggesting that this type of differentiation was only transient and is likely to be due to cytoskeletal changes from cellular toxicity as evidenced by other studies (Lu et al. 2004; Neuhuber et al. 2004).

In a study where chemical and biological transdifferentiations were explored, the transient nature of the chemical method was again highlighted using rat BMSCs (Zurita et al. 2008). They used a modified serum-free version of the chemical-induced neuronal differentiation media of Woodbury et al. (2000) and that described above. Initially the BMSCs were cultured in α-MEM and 1 mM βME for 24 h followed by α-MEM, 2% DMSO, 200 μM BHA, 25 mM KCl, 2 mM valproic acid, 10 μM forskolin, 1 μM hydrocortisone, and 5 μg/ml insulin for 72 h. To determine the reversibility of the chemically induced differentiation, α-MEM and 10% FBS were used after 72 h. By 4 h, the cells were expressing nestin and adopting a neuron-like morphology. Expression of βIII tubulin and NF-200 was observed by more than 75% of the BMSCs at 72 h. Four days after the media were returned to α-MEM and FBS, βIII tubulin and NF-200 expression and neuronal morphology were still apparent; however after another week, all the cells had reverted back to a BMSC phenotype that was now negative for neural markers. The biological induction method involved Transwell coculture with Schwann cells using α-MEM and 10% FBS as the media. Nestin expression was shown to increase with 50% of cells expressing nestin at 24 h. However nestin expression had decreased by 72 h, while a neuronal morphology was adopted, and βIII tubulin and NF-200 expression were evident. By 1 week, GFAP expression was also observed, and a similar pattern of βIII tubulin, NF-200, and GFAP expression was observed at 2 weeks. Since no direct contact was made between the BMSCs and Schwann cells in the Transwell cultures, this suggests that the growth factors secreted by the Schwann cells (e.g., BDNF and NGF) were initiating the neural differentiation. After 2 weeks, the removal of the Schwann cells had no effect on the cells showing that while chemical transdifferentiation is rapid but reversible, biological transdifferentiation is slower but permanent.

Further exploring the reversibility of transdifferentiation, Fu et al. (2008) used the above described method of Hermann et al. (2004) to generate neurospheres from BM-MSCs, though they only observed approximately 8% of the MSCs adopting a neuronal phenotype rather than the 60% previously described. The obtaining of NSCs was verified by detection of nestin and Musashi-1 in the population of cells within the neurospheres, and these cells were shown to be able to proliferate and differentiate suggesting that they were NSCs. Plating of the spheres onto poly-*L*-

ornithine-coated coverslips in the presence of 1% FBS, ITS, 20 nM progesterone, and 60 μM putrescine (but absence of EGF and bFGF) led to cells that expressed either GFAP, βIII tubulin, or GalC implying astrocyte, neuron, or oligodendrocyte generation. At this stage, the differentiated cells were unable to transdifferentiate to an alternative lineage (e.g., the mesodermal lineage such as adipocytes, osteoblasts, or chondrocytes). Coculturing of the NSCs (on poly-*D*-lysine cover slips) with astrocytes (on gelatin-coated flasks) in DMEM/F12, ITS, progesterone, and putrescine led to synapse formation (typified by punctuate synapsin-1 staining) within 2 weeks. Patch clamping demonstrated the presence of inducible tetrodotoxin-inhibitable action potentials, suggesting that functional neurons can be differentiated from the NSCs. A small proportion of the NSCs (approximately 4%) could also be reverted back to MSCs if cultured in DMEM, 15% FBS, 2 mM L-glutamine, 1 mM sodium pyruvate, and antibiotics.

While it is important to know whether the generation of NSCs is reversible, it is equally important to know whether the ability of the MSCs to proliferate and differentiate into NSCs is maintained over time and whether it is affected by the age of the donor. Passaging of hBM-MSCs obtained via negative selection for HSCs, Ficoll centrifugation, and plastic adherence was performed by culturing the obtained cells in the media used by Tao et al. (2005) and passaging once 80–90% confluent. The cells were passaged 12 times, and the expansion capabilities were measured in each passage (Khoo et al. 2008). The gene and protein expression and expansion characteristics were similar until the eighth passage, at which point, a reduction in proliferation rate and an increased presence of cell debris and large flat cells were detected within the culture. Viability and, in general, gene expression did not change with passaging, though the proportion of cells expressing the pluripotent marker Oct3/Oct4 and neuroectodermal marker nestin decreased with time. The cells were subjected to DMEM/F12, 1 × N2, antibiotics, 10 ng/ml FGF-2, 10 ng/ml EGF, and 1 ng/ml PDGF on fibronectin-coated plates for 3 weeks at P4–P5 and P11–P12 for comparison. Live cell recording was performed to ensure that the neurite-like processes were originating from the cell body rather than being remnants of cell shrinkage due to cellular stress as proposed for the chemical induction studies (Lu et al. 2004; Neuhuber et al. 2004). While cells were still able to differentiate at P11–P12, there did appear to be a higher proportion of cells that were unresponsive to the differentiation media. Primarily an increase in the expression of early neural markers, nestin and βIII tubulin, rather than in the mature marker, MAP 2, was observed following differentiation at either passage, while there was an increased presence of GFAP at the later passage. This suggests that late passage cells can still generate neural progenitors, but there may be an increased propensity for astrocytes. In another study utilizing commercially available hBM-MSCs (Lonza, Australia), cells were passaged 5, 7, 13, and 20 times (Okolicsanyi et al. 2015). Immunocytochemistry, Western blotting, and FACS demonstrated SOX-2, βIII tubulin, nestin, MAP 2, and GFAP staining in the MSC population at similar levels regardless of the number of passages, suggesting their potential to undergo neural differentiation was not altered. Neurosphere formation was induced only at passage 5, by culturing in low attachment plasticware with knockout DMEM/F12, 20 ng/ml

EGF, 20 ng/ml bFGF, antibiotics, and 10 μg/ml heparin for 10 days. Neurosphere formation was observed within hrs. After neurosphere formation, a higher percentage of cells were positive for nestin and SOX2 than observed in the untreated MSCs. It is unfortunate that the authors did not confirm neural differentiation of passage 20 cells even though the little changed basal neural marker expression by the MSCs suggested their potential ability to differentiate had not changed.

Since one potential use of BM-derived NSCs would be as an autologous cell therapy for age-dependent neurodegenerative disorders, it is important to know whether cells derived from old bone marrow are as plastic as from young patients. Some studies suggest that old BM-MSCs can be readily differentiated to mesodermal cell types such as bone and chondrocytes (Scharstuhl et al. 2007; Hermann et al. 2010). The ability of old (>49) compared to young (<49) cells to differentiate into neural tissue was explored by Hermann et al. (2010), using either a direct neuronal differentiation procedure or a two-step neuronal differentiation via NSC generation. The old BM-MSCs were passaged up to 5X, and they expressed lower levels of nestin, βIII tubulin, GalC, GFAP, and Ki67 than young BM-MSCs. The cells were then plated in P4-8F medium containing 20 ng/ml EGF and 20 ng/ml bFGF under hypoxic (3% O_2) conditions to allow formation of neurospheres and NSC-like cells over a 5-week period. Neurosphere formation was equally present in young and old samples, but the subsequent neuronal differentiation using either Neurobasal medium, with 0.5 μM all-trans-retinoic acid, 1% FBS, 5% horse serum, 1% N2 supplement, 1% antibiotics, and 10 ng/ml BDNF, or Glasgow MEM, with 1% glutamine, 1 mM sodium pyruvate, 0.1 mM nonessential amino acid, 0.1 mM βME, and 10% knockout serum replacement (or N2 after 7 days), did not occur even after addition of astrocyte cocultures. This suggests that the terminal neuroectodermal differentiation process is impaired in old cells, even though osteogenesis is still possible and no alterations in telomerase length are detected meaning that the cells are not senescent.

Determining the right mixture of growth factors for inducing and maintaining NSCs can be difficult, so using a naturally occurring mixture maybe beneficial. The cerebrospinal fluid (CSF) bathes the subventricular zone where neurogenesis naturally occurs, and it therefore is likely to contain growth factors that would optimize the development of newly forming NSCs and maintain their survival. Therefore, Ye et al. (2011) isolated MSCs from human volunteer bone marrow by density-gradient centrifugation and plastic adherence. The cells were passaged 1–3 times and then cultured on poly-*L*-lysine-coated plates for 72 h in DMEM/F12. CSF was obtained from the same healthy volunteers and added to the media in a 1:200 dilution every day for at least 7 days. By comparison an equivalent dilution of media containing 10% FBS, 10 μM EGF, and 10 μM bFGF was performed every day. Neuron-like morphological changes and expression of βIII tubulin and GFAP were observed in both treatment groups, but a higher number of neural-like cells were observed in the CSF-treated group. Long-term culture (more than 54 days) in CSF-supplemented media caused cell death, suggesting long-term culture and expansion in CSF-supplemented media may not be effective. Since, in vivo, newly formed cells migrate away from the SVZ as they mature, meaning that they will be exposed to a different environment and growth factors, this may explain why CSF-supplemented media is not effective long term. In a separate study to compare autologous and healthy donor CSF, a 1:200 dilution of CSF from either the donor of the cells or from a healthy volunteer was added every day for 3 days (Ge et al. 2015). The highest number of neuron-like cells was obtained following the addition of CSF from a healthy volunteer. Unfortunately no data was provided on the age or gender of the patients or healthy volunteers or what the patients were suffering from. It seems likely that the disorder from which the patients are suffering from may be impairing the neural induction as it seems likely that cells would behave more favorably under autologous conditions. Neural differentiation of rat MSCs, previously labeled with BrdU, using human CSF was also effective (based on NSE and GFAP staining) in another study exploring the potential benefits of these cells in a rodent stroke model (Ye et al. 2016). The cells were transplanted after 4 days of differentiation in CSF into Sprague Dawley rats, 7 days after the rats had been subjected to MCAO. Four and 15 days after transplant, the rats given the transformed cells had a significantly reduced modified neurological severity score compared with those transplanted with untreated MSCs and vehicle. At 32 days a higher number of BrdU-labeled cells were observed in the neural-like cell transplant group, and these cells were more likely to be NSE-positive, while untransformed BM-MSCs showed a higher incidence of BrdU/GFAP colocalized cells. In another study, the SCF-differentiated rat cells were transformed by adding 10 μM EGF and 10 μM FGF2 after 72 h in media containing human CSF and being differentiated for a further 7 days (Ye et al. 2018). After transplantation into a rat model of SCI, the NSCs induced from MSCs were more effective than untreated MSCs in improving behavioral tests, increasing neurotrophic factor release into the CSF and neurotransmitter levels within the spinal cord. These studies suggest that using CSF to neurally differentiate BM-MSCs to NSCs is effective.

Genetic manipulation of BM-MSCs could be used to promote neural differentiation. Purchased hBM-MSCs (BioWhittaker or Cambrex) and harvested rat BM-MSCs were initially cultured in α-MEM, FBS, and kanamycin. Cells were transfected with the notch intracellular domain (NICD) and GFP and were then cultured in α-MEM and FBS. The transfected cells were shown to express neural stem cell markers such as glutamate transporter 3 (GLAST3), phosphoglycerate dehydrogenase (3-PGDH), and nestin, while non-transfected MSCs only expressed nestin (Dezawa et al. 2004). The expression of genes relating to neural development were altered by transfection since signal transducer and activator of transcription 1 (STAT1) and 3 were reduced, compared to untransfected MSCs. Endogenous notch expression was also reduced, which may explain the absence of GFAP, since notch is believed to induce glial differentiation (Morrison et al. 2000). However since transfection should increase NICD expression, this seems unlikely to be the explanation unless notch is working through a different pathway. The cells were then cultured in α-MEM containing 10% FBS, 5 μM forskolin, 10 ng/ml bFGF, and 10 ng/ml ciliary neurotrophic factor (CNTF) (TF-MSCs). This media did not induce MAP 2 in non-transfected cells, but did induce MAP 2 expression, as well as NF-M, βIII tubulin, and NeuroD expression in almost all transfected cells (Dezawa et al.

2004). The morphology also changed toward a more neuron-like state. However, these cells did not express functional voltage-gated ion channels, though their resting membrane potential was lower than for untransfected MSCs. The cells were not observed to proliferate suggesting that they were likely to be a late neural progenitor cell or immature neuron rather than an NSC or early progenitor cell. Further neuronal differentiation was achieved using α-MEM containing 10% FBS, 50 ng/ml BDNF, and 50 ng/ml NGF, or α-MEM containing 10% FBS and 50 ng/ml GDNF. The latter media induced TH expression as well as transcription factors relating to dopaminergic cell differentiation (G-MSCs). Functional voltage-gated ion channels were observed 7 days later, suggesting that the cells were maturing toward functional neurons. At 11 days, action potentials were evident in some cells. Following striatal transplantation of G-MSCs, TF-MSCs, or untransfected MSCs into the 6-OHDA rat model of PD, some of the cells were observed to survive for at least 10 weeks, and the G-MSC-treated rats exhibited a significant improvement in rotational behavior, step adjustment, and paw reaching. Transplanted cells positive for TH, DAT, and neurofilament were observed within the striatum with long TH^+ neurites extending beyond the graft.

In a second study, 2 days after NICD transfection, the transfected cells were cultured on low cell-binding dishes in Neurobasal medium with B27, 20 ng/ml bFGF, and 20 ng/ml EGF to generate neurospheres over 7 days (Hayase et al. 2009). Many more and larger neurospheres were formed by the transfected cells compared to nontransfected MSCs. A high percentage of the cells in the neurospheres from transfected MSCs were positive for SOX2, nestin, and NeuroD compared with relatively few cells in nontransfected neurospheres, suggesting NSC formation and a potential proliferating cell line. The cells no longer expressed GFP suggesting that the transfection was transient. The neurospheres were dissociated and plated on laminin-coated slides in Neurobasal medium containing B27, 1% FBS, 10 μM forskolin, 20 ng/ml CNTF, and 20 ng/ml bFGF for 1 week. Morphological changes such as neurite-like projections were observed, and the cells were immunoreactive for βIII tubulin and MAP 2, with a small percentage expressing GFAP. Low levels of neurotransmitter expression were also observed, as well as the ability to secrete DA following a depolarizing stimulus. PCR analysis revealed the presence of *NSE* and voltage-gated sodium channels. Striatal and cortical transplantation of the rat neurosphere-derived MSCs into rats 3 days after MCAo was performed and had no effect on lesion size. However, long-term behavioral recovery was observed at 100 days with the transfected cells compared with nontransfected MSCs. The cells were transfected with GFP by lentivirus prior to transplant to enable identification, and from this it was evident that the transplanted cells had proliferated in vivo, though the absence of Ki67 expression at 100 days suggested that proliferation was only a short-term event. Retrograde tracer was injected into the substantia nigra 7 days before sacrifice. Retrograde tracer was colocalized with GFP in the striatum within neuron-like cells, suggesting that the transplanted cells had sent projections to the substantia nigra. Some of the cells also expressed synaptophysin, showing that they were integrating into the striatum and substantia nigra.

Two further studies were exploring the role of notch in the neuronal differentiation of BMSCs (Yanjie et al. 2007; Jing et al. 2011). Yanjie et al. (2007) transfected a short hairpin RNA for notch-1 into P4 mouse BM-MSCs and induced neuronal differentiation using Neurobasal medium, 5% horse serum, 1% FBS, 2% ITS, 0.5 μM retinoic acid, and 10 ng/ml BDNF for 6 days. Downregulation of notch-1 slightly reduced cell proliferation and increased apoptosis. Marked morphological changes toward a neuronal phenotype were observed following neuronal differentiation of the transfected cells. Nestin expression was high at 24 h, but had declined by day 6, while NSE and neurofilament increased with time. No $GFAP^+$ cells were apparent in the transfected cells after neuronal differentiation. In the second study, Jing et al. (2011) transfected mouse BM-MSCs at P10 with microRNA 9 (miR-9), a small noncoding RNA sequence that can bind to the 3'untranslated region of specific mRNA targets and alter their expression. Target genes of miR-9 include notch and two of its downstream targets Hes1 and Hes5. The cells were differentiated in DMEM, 10% FBS, and 1 mM βME for 24 h and then a further 5 hrs in the absence of FBS. Notch signaling was inhibited by miR-9, which, similar to the shRNA transfection, was reflected by a decreased survival rate. MAP 2 expression was also increased. However, since this differentiation procedure is believed to be due to a change in microtubule stability rather than true differentiation (Lu et al. 2004), it is unclear of the relevance of miR-9 from this study. It is worth noting that miR-9 also alters stathmin, a protein which promotes microtubule instability, which may explain the effects observed in this study. Another target of miR-9 is believed to be zinc finger protein 521 (ZFP521). Mouse BM-MSCs were transfected with miR-9 leading to a decrease in cell survival and upregulation of MAP 2 and NSE. Transfection of an inhibitor of miR-9 decreased MAP 2 and NSE below levels in untransfected cells following a 6-day "differentiation" with βME (Han et al. 2012). ZFP521 levels were lower in miR-9-transfected cells before "differentiation" and decreased further on "differentiation". Dual-Luciferase Reporter Assay confirmed that ZFP521 is a target of miR-9. Again using the βME method of "differentiation", a link between miR-9 and increased autophagy was detected (Zhang et al. 2015). The use of miR-9 in a study using a "proper" neuronal differentiation method for MSCs would help to validate whether these observations are genuine and the potential contribution of miR-9 to neuronal transdifferentiation.

Other miRs may also influence the neural differentiation of MSCs. This was explored by Huat et al. (2015) who differentiated P4 rat BM-MSCs in Neurocult proliferation media supplemented with 10 ng/ml EGF and 10 ng/ml bFGF with, or without, 10 ng/ml IGF-1 for 1 week and performed microarray scanning for miRNAs. The presence of IGF-1 was found to increase cell survival and proliferation and decrease apoptotic and necrotic cell death, as well as result in the largest neurospheres. Forty-six miRs were expressed at two or more time points (1, 3, and 5 days from the start of induction) by the neurally induced and normal BM-MSCs. In the absence of IGF-1, 8 miRs were reduced, and 14 miRs upregulated compared to normal BM-MSCs, but in the presence of IGF-1, 12 miRs were downregulated, and only one upregulated (miR-496). Five miRs were exclusively downregulated in the presence of IGF-1 (miR-22, miR-1224, miR-125a-3p (actually increased in absence), miR-214, and miR-708), and seven were downregulated compared to BM-MSCs in both treatment regimens (let-7b, let-7c, let-7d, let-7e, let-7I, miR-320, and miR-93). GO analysis showed that these miRs were primarily associated with regulating various forms of cell death.

In a different study, the microRNA miR-124 was transfected into rat BM-MSCs, and the cells cultured for 6 days in an unspecified media (Zou et al. 2014). This miR was chosen since GeneChip data showed that miR-124 levels in BM-MSCs were significantly lower than in NSCs and neuronal cells (Zou et al. 2014). The transfected cells showed higher expression of βIII tubulin, MAP 2, and synaptophysin in the soma- and neurite-like structures. The cells were also less susceptible to apoptotic cell death as a consequence of oxygen glucose deprivation, suggesting they may be more likely to survive transplantation. Cells were transplanted into the injured rat spinal cord where the higher number of surviving cells appeared to differentiate into neurofilament positive cells. The microRNA, miR-124, has previously been shown to repress anti-neuronal proteins such as repressor element-1-silencing transcription factor (REST), small c-terminal domain phosphatase 1 (SCP1), and Sox9 (Visvanathan et al. 2007; Cheng et al. 2009). In a study using miR-29a and the βME (potentially artifactual) differentiation method, inhibition of REST was shown to promote "neuronal differentiation" (Duan et al. 2014). The microRNA miR-125b was also shown to promote βME "differentiation" of rat BM-MSCs (Wu et al. 2013). It is unfortunate that many of the miR studies are based on a rapid, but potentially flawed, differentiation technique since the true impact of the miRs is unclear until they are investigated in a study using a "proper" differentiation procedure.

Rat BM-MSCs were transfected with either miR-128 or a miR-128 inhibitor, and the cells were then cultured in DMEM/F12, 2% B27, 20 ng/ml bFGF, 20 ng/ml EGF, and antibiotics for 7 days followed by DMEM/F12, 10% FBS, 2% B27, 10 ng/ml bFGF, 10 ng/ml EGF, and antibiotics for 5 days (Wu et al. 2014). RT-PCR revealed elevated levels of *NSE*, nestin, *GFAP*, neurofilament, *MAP 2*, and βIII tubulin compared to nontransfected, nontreated cells. Overexpression of miR-128 reduced mRNA and protein levels of these neural markers compared to a nontransfected but treated group, while inhibition of miR-128 had the opposite effect. Development of neurites and adoption of a neuron-like cell body occurred in all treated groups, but were more evident in the miR-128-inhibited group, and only a few cells changed in the miR-128-overexpressing group. The overexpression group showed reduced Wnt3a expression, while the inhibited group increased Wnt3a expression suggesting that Wnt3a may be a target of miR-128. This was then confirmed by Dual-Luciferase Reporter Assay showing that miR-128 does bind to Wnt3a. Reduced miR-128 expression will thus increase Wnt3a suggesting it may play a role in neural differentiation.

The contribution of Wnt3a (and other Wnts) to the neuronal differentiation of BM-MSCs was further explored using human cells passaged 3–6 times (Tsai et al. 2014) and cultured in DMEM, 1% FBS, 10 ng/ml BDNF, 20 ng/ml NGF, and 5 μM retinoic acid for 7 days. MAP 2 expression increased during the neuronal differentiation suggesting that the BM-MSCs were transforming into neuron-like cells. The modest expression of the canonical pathway Wnt3a and noncanonical pathway

Wnt5a and low levels of canonical Wnt1a did not change during this process. In contrast, Wnt7a and Wnt7b expression was seen to significantly increase over time during differentiation suggesting an involvement in neuronal differentiation. Addition of human recombinant Wnt1, Wnt3a, Wnt5a, Wnt7a, or LiCl at concentrations ranging from 1–4 mM for 48 h doubled MAP 2 expression. Expression of synapsin-1 (SYN1) was significantly increased by the addition of all Wnts except for Wnt5a. ChAT expression was also increased by all Wnts except for Wnt1, while dopamine beta hydroxylase (DBH) was only increased by Wnt5a and Wnt7a. Since Wnt7a consistently increased neural markers, further studies were performed using Wnt7a and canonical/noncanonical pathway inhibitors to determine how Wnt7a contributed to neuronal differentiation. Expression of nestin, MAP 2, GFAP, MBP, βIII tubulin, and synaptic proteins and neurite formation were increased by Wnt7a using a canonical β-catenin signaling pathway as well as lithium. Conversely, neuron-specific expression, e.g., ChAT and DBH, was inhibited by inhibitors of the noncanonical JNK pathway suggesting that cholinergic and dopaminergic differentiation by Wnt7a was mediated via the noncanonical JNK pathway. The ability of lithium to induce neural RNA/protein expression suggests that there may be a GSK-3β inhibitory component to their expression (since lithium is an inhibitor of GSK-3β).

While Huat et al. (2015) showed that IGF-1 was beneficial in neuronal differentiation, a study by Guan et al. (2014) appeared to show the opposite. Possible explanations for this apparent discrepancy include the use of plastic adherence only to isolate the BM-MSC population, the media being composed of FBS and N2-supplemented DMEM, or that the neurotrophic factors, especially the IGF-1, were at higher levels, compared to Huat et al. (2015) who used density-gradient centrifugation as well as plastic adherence, Neurocult, as the basal media and did not add FBS. Guan et al. (2014) also explored the contribution of NT-3 in the differentiating culture media which consisted of DMEM, 2% FBS, 1% N2, and some combination of the following neurotrophic factors: 50 ng/ml EGF, 10 ng/ml bFGF, 50 ng/ml IGF-1, and 20 ng/ml NT-3. The rat BM-MSCs were plated on poly-*L*-lysine-coated plates and began to exhibit long thin projections within 24 h. Nestin and MAP 2 expression increased, with the largest rise observed in cells exposed to the combination of bFGF, EGF, and NT-3. Addition of IGF-1 to any of the cultures that appeared to be undergoing differentiation reduced the number of differentiating cells, while NT-3 promoted neural differentiation.

Culturing of human BM-MSCs on poly-*D*-lysine-coated plates in the presence of Neurobasal medium, with 0.5% B27, 250 ng/ml SHH, 100 ng/ml FGF8, and 50 ng/ml bFGF for 9 days, resulted in the majority of the cells resembling neurons and expressing TH, NeuN, GFAP, and βIII tubulin (Trzaska et al. 2009). However these cells did not possess the electrophysiological properties of neurons suggesting that they may be DAergic progenitor cells. Consequently additional neurotrophic factors (50 ng/ml of BDNF, NT-3, GDNF, or NGF) were added for a further 3 days to determine if maturation could be induced. An increased expression of TrkB was detected by day 9, suggesting that BDNF was likely to be the most successful candidate as an important neurotrophic factor. Sodium-dependent voltage channels were expressed, and the level of D1 receptors was increased. A calcium-dependent current was observed suggesting functional calcium voltage-gated channels were present. The cells expressed receptors for, and responded electrophysiologically to, the neurotransmitters Ach and GABA, though the GABA response was in an immature excitatory fashion rather than the mature inhibitory action. However, the cells did not respond to glutamate despite expressing glutamate receptors. Calcium depolarization was sufficient to elicit secretion of DA. The other neurotrophic factors (NT-3, GDNF, and NGF) were not able to induce the same responses, suggesting that only BDNF was able to mature the progenitor cells. Of note in this study, the media were not exchanged once culturing for neuronal induction began, as the authors had observed cell death, and differentiation stopped when the media were changed. This would suggest that the differentiating cells may be secreting factors that maintain their survival and differentiation.

Harris et al. (2012) utilized both commercially obtained and donated bone marrow samples to derive BM-MSCs by density-gradient centrifugation and plastic adherence and then subjected them to neural progenitor maintenance media with 20 ng/ml EGF and 120 ng/ml bFGF for 21 days on low-adherence flasks. Neurosphere formation was observed within days, and the cells expressed high levels of nestin and Musashi supporting their putative identity as NSCs. After a week, the cells began to express markers of more mature neural cells such as neurofilament and GFAP and the neural stem cell migration marker CXCR4. The cells could only undergo limited osteogenic and adipogenic differentiation showing that the cells had lost their MSC plasticity. Proliferation of the NSCs was detected. Yang et al. (2008) used a similar media (except DMEM was the base) to generate neurospheres from male and female rat BM-MSCs isolated by plastic adherence only. They further differentiated the cells by plating on poly-*L*-lysine coverslips and adding 10% FBS to the media for 5 days. Nestin levels decreased while GFAP, MAP 2, and NSE increased suggesting the cells maybe differentiating from NSCs in the neurospheres to neuron- and glial-like cells.

Inhibition of cyclin-dependent kinases (CDKs) by either transfection of shRNAs for CDK1, CDK2, or CDK4 or addition of CDK-specific inhibitors to the media for human BM-MSCs and adipose tissue-derived MSCs was investigated to determine how this would affect the cells (Lee et al. 2013). Within 12 h of CDK inhibition, the MSCs began to differentiate into neural-like cells. CDK4 inhibition caused transformation with the highest efficiency. Significantly higher levels of MAP 2, βIII tubulin, GFAP, and neuroligin 3 were observed in CDK4-inhibited cells that were undergoing morphological changes. Cellular proliferation however was suppressed. The presence of CDK4 inhibition also greatly reduced adipogenic and osteogenic differentiation of these cells suggesting that CDK4 inhibition limited the plasticity of these cells to the neural lineage. Removal of CDK4 inhibition restored the plasticity of the cells highlighting the immaturity of the neural-like cells, suggesting that CDK4 inhibition alone is not sufficient to induce generation of neurons, astrocytes, or oligodendrocytes, though a more long-term culture would be necessary to confirm this.

The modified N3 induction medium (Neurobasal medium, 2% B27, 0.5 mM IBMX, 20 ng/ml hEGF, 40 ng/ml bFGF, 10 ng/ml FGF-8, 0.25 mg/ml db-cAMP, 2 mM L-glutamine, and 40 ng/ml NGF) will induce BM-MSCs to adopt a neuron-like phenotype over 12 days so that they express neurofilament, NeuN, MAP 2, Olig2, NSE, and GFAP (Isik et al. 2015). Inhibition of DNA topoisomerase IIβ, an enzyme involved in positive and negative DNA supercoils, by transfection of a short interfering RNA, 1 day prior to induction, reduced the incidence of neuronal differentiation by 50% and also decreased the length of the neurite-like projections by 75%. This suggests that DNA topoisomerase IIβ activity may play an important role in neural differentiation.

These studies suggest that neural differentiation of bone marrow-derived cells to NSCs is possible and they may be capable of maturing into fully functioning neurons, astrocytes, and oligodendrocytes over time under the right conditions.

2.3 Direct Comparison of the Neural Differentiation Capabilities of the Perinatal and Adult Somatic Tissues

In this short section, we will discuss the limited number of studies that directly compare the neural differentiation capabilities of cells isolated from the different perinatal and even adult somatic tissues described in the previous sections. The studies by Azedi et al. (2014, 2017) actually compared MenSCs to BM-MSCs. Their first study (Azedi et al., 2014) demonstrated that both MenSCs and BM-MSCs are equally able to differentiate into glial-like cells, but the differentiated BM-MSCs appeared to be less viable. The second study revealed that the propensities of MenSCs and BM-MSCs to undergo neuronal differentiation in vitro were almost equal (Azedi et al., 2017). In contrast, comparison between MenSCs and UC-MSCs (Chen et al., 2015) showed that UC-MSCs had a higher proliferation capacity, but a lower colony-forming ability. Comparison between WJ-MSC and BM-MSC performed by other groups (e.g. Drela et al. 2016) revealed that WJ-MSCs highly surpasses BM-MSCs in both proliferation and early stemness marker expression like Brachyury, SSEA-4, nestin, and neural markers NF-200 and GFAP. Gene analysis of undifferentiated MenSCs, UC-MSCs/WJ-MSCs, and BM-MSCs performed by the above authors suggested that MenSCs had the highest neural potential based on nestin expression, but unfortunately this did not actually translate to increased neural differentiation. On the other side, Bonaventura et al. (2015) compared the ability of BM-MSCs, hUCB-MSCs, endometrial tissue/MenSCs, and AFSCs to undergo neural differentiation in the presence of enriched, inductive media. Utilizing a culture media of 1 mM dbcAMP, 0.5 mM IBMX, 20 ng/ml EGF, 40 ng/ml bFGF, 10 ng/ml NGF, and 10 ng/ml BDNF, cells from the different sources were neurally induced for 10 days. Based on the protein and mRNA expression of neural markers, the results of this study compared with two others using an unspecified neural inductive media, or not (Kwon et al. 2016 versus Lech et al. 2016), and studying similar sources of MSCs, seem to reveal that BM-MSCs and adipose-derived MSCs had the lowest while hUCB-MSCs/

WJ-MSCs and endometrial-derived cells had the highest neural propensity. Also in comparison to AMSCs and chorion-derived MSCs, the UC-MSCs/WJ-MSCs were shown to have the highest propensity to undergo neural differentiation (Kwon et al. 2016). The proliferative capacity of AMSCs was also observed to decline more rapidly with passaging due to increasing senescence, compared to the other cells. Comparison of BM-MSCs and chorion-derived MSCs revealed an increased nestin expression, as well as other neural markers and even mature neuronal markers, in the chorion-derived cells after retinoic acid supplementation to induce neural differentiation (Ziadlou et al. 2015). In a study comparing neural differentiation of AMSCs, AESCs, and WJ-MSCs, using immunofluorescence cell sorting to detect neural markers, the WJ-MSC-differentiated cells expressed a higher proportion of NGFR and S100 protein and a similar proportion of neurofilament and GFAP to the differentiated AESCs (Sanluis-Verdes et al. 2017). The neural differentiation media used in these studies consisted of DMEM, 10% FBS, 30 μM retinoic acid, 10 ng/ml bFGF, 10 ng/ml EGF, 2 mM l-glutamine, 1% NEAA, 55 mM βME, 1 mM sodium pyruvate, 1% ITS, and 2 mM ascorbic acid, and the cells were cultured for 21 days. While AMSCs showed a lower proportion of neural marker-expressing cells, the ability of these differentiated cells to proliferate was higher compared to those from AESCs and WJ-MSCs, which may suggest that the AMSC-derived cells are of an earlier NSC-type than those from AMSCs and WJ-MSCs, since proliferation capacity tends to decrease as NSCs become NPCs, etc.

2.4 Conclusions

Many studies have suggested that perinatal tissues such as the placenta, amnion, and umbilical cord (blood and Wharton's jelly), together with adult tissues including menstrual and peripheral blood and the bone marrow, can be differentiated to NSCs and ultimately neurons, astrocytes, and oligodendrocytes in vitro. Unfortunately a number of these studies used flawed methods of differentiation, e.g., the use of βME, which has been reported to modify the cytoskeleton rather than cause differentiation or incomplete confirmation of obtaining neural tissue (such as no evidence of action potentials, neurotransmitter release, ion channels, etc.). However, several studies have been more rigorously performed and do appear to confirm that at least some of the cells obtained from these tissues are capable of undergoing differentiation to a neural phenotype. This is based on evidence of action potentials, ion channels, neurotransmitter release, network integration, functionality, etc., though it is unclear what the optimal tissue type is as there are only a few studies that make direct comparisons between the different cell sources.

Several studies have shown that some of the isolated MSCs or other cell populations from the tissues described in this chapter may naturally express relatively low levels of neural markers implying that they have the propensity to undergo neural differentiation.

As mentioned before, WJ as well as the other perinatal tissue-derived MSC populations, often named as preMSCs, possess some unique features making them preferably useful for CNS therapy (Sarnowska and Domańska-Janik 2017). This may be directly related to their exceptional, however, still controversial ontogenic origin responsible for the broadened scope of their multilineage differentiation. The unquestionable pluripotency of the adult-type MSCs is still not commonly accepted and eagerly debated (Langrzyk et al. 2017). The independent discovery of this type of cells in many laboratories and tissues seems to prove the capability of certain MSC subpopulations to differentiate into pluripotent and neuroectodermal lineages (Domanska-Janik et al. 2008; Habich and Domańska-Janik 2011; Qi et al. 2011; Dezawa 2016) and further take part in injured brain functional reconstruction (Kozłowska et al. 2007; Górnicka et al. 2006; Uchida et al. 2017). Recently the presence of such pluripotent stem cells in the human umbilical cord and blood, with characteristics similar to the other undifferentiated, embryonic-like subpopulations present in the different tissue niches, has also been nicely confirmed by Monti's and by Aquino's groups (Monti et al. 2017; Aquino 2017). In contrast to the classical theories attributing the role of all adult stem cells to only renew and repair the tissues of the same lineage, it seems to prove the presence of such pluripotent, embryonic-like cells in various tissue niches. For clarification of this controversy, two advanced hypotheses could be proposed. One of them postulates the existence of a fourth germ layer of neural crest origin in all craniates (Shyamala et al. 2015; Hall 2000). This additional layer would produce, like mesoderm, the exceptional, migratory progeny with preserved early embryonic-like properties which harbor in stem cell niches of developing tissues. The existence of such progenitors with the propensity to undergo neurosphere formation has been well documented recently by Aquino et al. (Aquino 2017). An alternative explanation to the fourth germ layer is that the developmental pedigree of the above "very suspicious," embryonic-like (or preMSC) cells arise from them being an offspring from the first wave of Sox1-expressing neural crest-derived migratory stem cells (Takashima et al. 2007). The other concept concerns the presence of persistent embryonic-like pluripotent cells among the adult tissue-committed stem cells and postulates the existence of the small dormant stem cells (VSELs) closely related to the migratory, pre-gastrulation, primordial germ cells firstly postulated by Melton and Cowan (2004). These cells have been deposited, like the other PGC derivatives into the various stem cell niches (e.g., BMSCs, HSCs), as a specific backup for the adult unipotent populations (Kucia et al. 2006). In addition, while not conclusively shown, there is also some indication that earlier passaged material and cells from earlier sources (e.g., perinatal compared with adult, compared with aged adult) have a higher propensity to undergo neural differentiation suggesting only transient existence and subsequent elimination of the pluripotent cells subpopulation in adult stem cell niches during development. The further discussion of these fascinating theories, however, would be in our opinion beyond the scope of this chapter. Thus, our future efforts should force even better characteristic and mechanical understanding of the phenomenon of perinatal and adult stem cell pluripotency according to the generally accepted scientific criteria.

Disclosures PRS holds a number of patents for the application of umbilical cord-, menstrual and peripheral blood-, and bone marrow-derived stem cells in the treatment of neurodegenerative disorders.

References

Acosta SA, Tajiri N, Shinozuka K et al (2014) Combination therapy of human umbilical cord blood cells and granulocyte colony stimulating factor reduces histopathological and motor impairments in an experimental model of chronic traumatic brain injury. PLoS One 9(3):e90953

Adamzyk C, Kachel P, Hoss M et al (2016) Bone tissue engineering using polyetherketoneketone scaffolds combined with autologous mesenchymal stem cells in a sheep calvarial defect model. J Craniomaxillofac Surg 44(8):985–994

Ali H, Al-Mulla F (2012) Defining umbilical cord blood stem cells. Stem Cell Discovery 2:15–23

Aliaghaei A, Gardaneh M, Maghsoudi N et al (2016) Dopaminergic induction of umbilical cord mesenchymal stem cells by conditioned medium of choroid plexus epithelial cells reduces apomorphine-induced rotation in parkinsonian rats. Arch Iran Med 19(8):561–570

Aquino JB (2017) Uncovering the in vivo source of adult neural crest stem cells. Stem Cells Dev 26(5):303–313

Arnhold S, Post C, Gluer S et al (2008) Neuronal characteristics of amniotic fluid derived cells after adenoviral transformation. Cell Biol Int 32(12):1559–1566

Arnhold S, Gluer S, Hartmann K et al (2011) Amniotic-fluid stem cells: growth dynamics and differentiation potential after a cd-117-based selection procedure. Stem Cells Int 2011:715341

Azedi F, Kazemnejad S, Zarnani AH et al (2014) Differentiation potential of menstrual blood- versus bone marrow-stem cells into glial-like cells. Cell Biol Int 38(5):615–624

Azedi F, Kazemnejad S, Zarnani AH et al (2017) Comparative capability of menstrual blood versus bone marrow derived stem cells in neural differentiation. Mol Biol Rep 44(1):169–182

Baer PC, Griesche N, Luttmann W et al (2010) Human adipose-derived mesenchymal stem cells in vitro: evaluation of an optimal expansion medium preserving stemness. Cytotherapy 12(1): 96–106

Bagher Z, Azami M, Ebrahimi-Barough S (2016) Differentiation of Wharton's jelly-derived mesenchymal stem cells into motor neuron-like cells on three-dimensional collagen-Grafted nanofibers. Mol Neurobiol 53(4):2397–2408

Bonaventura G, Chamayou S, Liprino A et al (2015) Different tissue-derived stem cells: a comparison of neural differentiation capability. PLoS One 10(10):e0140790

Bonilla-Porras AR, Velez-Pardo C, Jimenez-Del-Rio M (2017) Fast transdifferentiation of human Wharton's jelly mesenchymal stem cells into neurospheres and nerve-like cells. J Neurosci Methods 282:52–60

Borlongan CV, Kaneko Y, Maki M et al (2010) Menstrual blood cells display stem cell-like phenotypic markers and exert neuroprotection following transplantation in experimental stroke. Stem Cells Dev 19(4):439–452

Boroujeni ME, Gardaneh M, Shahriari MH (2017) Synergy between choroid plexus epithelial cell-conditioned medium and knockout serum replacement converts human adipose-derived stem cells to dopamine-secreting neurons. Rejuvenation Res 20(4):309–319

Bossolasco P, Montemurro T, Cova L et al (2006) Molecular and phenotypic characterization of human amniotic fluid cells and their differentiation potential. Cell Res 16(4):329–336

Buzanska L, Machaj EK, Zablocka B et al (2001) Human cord blood derived neurons, astrocytes and oligodendrocytes. J. Neurochem 78(Suppl. 1):58

Buzanska L, Machaj EK, Zabłocka B et al (2002) Human cord blood-derived cells attain neuronal and glial features in vitro. J Cell Sci 115(Pt 10):2131–2138

Buzanska L, Habich A, Jurga M et al (2005) Human cord blood – derived neural stem cell line possible implementation in studying neurotoxicity. Toxicology in Vitro 19:991–999
Buzanska L, Jurga M, Stachowiak EK et al (2006) Neural stem-like cell line derived from a nonhematopoietic population of human umbilical cord blood. Stem Cells Dev 15(3):391–406
Buzanska L, Jurga M, Domanska-Janik K (2006a) Neuronal differentiation of human umbilical cord blood neural stem-like cell line. Neurodegener Dis 3(1–2):19–26
Buzanska L, Sypecka J, Nerini-Molteni S et al (2009) A human stem cell-based model for identifying adverse effects of organic and inorganic chemicals on the developing nervous system. Stem Cells 27(10):2591–2601
Calzarossa C, Bossolasco P, Besana A et al (2013) Neurorescue effects and stem properties of chorionic villi and amniotic progenitor cells. Neuroscience 234:158–172
Castillo-Melendez M, Yawno T, Jenkin G (2013) Stem cell therapy to protect and repair the developing brain: a review of mechanisms of action of cord blood and amnion epithelial derived cells. Front Neurosci 7:194
Chen SH, Chang FM, Tsai Y et al (2006) Infusion of human umbilical cord blood cells protect against cerebral ischemia and damage during heatstroke in the rat. Exp Neurol 199(1):67–76
Chen H, Zhang Y, Yang Z (2013) Human umbilical cord Wharton's jelly-derived oligodendrocyte precursor-like cells for axon and myelin sheath regeneration. Neural Regen Res 8(10):890–899
Chen JY, Mou XZ, Du XC et al (2015) Comparative analysis of biological characteristics of adult mesenchymal stem cells with different tissue origins. Asian Pacific J Trop Med 8(9):739–746
Cheng LC, Pastrana E, Tavazoie M et al (2009) Mir-124 regulates adult neurogenesis in the subventricular zone stem cell niche. Nat Neurosci 12(4):399–408
Cipriani S, Bonini D, Marchina E et al (2007) Mesenchymal cells from human amniotic fluid survive and migrate after transplantation into adult rat brain. Cell Biol Int 31(8):845–850
Dabrowska S, Sypecka J, Jablonska A et al (2018) Neuroprotective potential and paracrine activity of stromal vs. culture-expanded hMSC derived from Wharton jelly under co-cultured with hippocampal organotypic slices. Mol Neurobiol 55(7):6021–6036
Datta I, Mishra S, Mohanty L et al (2011) Neuronal plasticity of human Wharton's jelly mesenchymal stromal cells to the dopaminergic cell type compared with human bone marrow mesenchymal stromal cells. Cytotherapy 13(8):918–932
Davies JE, Walker JT, Keating A (2017) Concise review: Wharton's jelly: the rich, but enigmatic, source of mesenchymal stromal cells. Stem Cells Transl Med 6(7):1620–1630
De Coppi P, Bartsch G Jr, Siddiqui MM et al (2007) Isolation of amniotic stem cell lines with potential for therapy. Nat Biotechnol 25(1):100–106
Deng W, Obrocka M, Fischer I et al (2001) In vitro differentiation of human marrow stromal cells into early progenitors of neural cells by conditions that increase intracellular cyclic amp. Biochem Biophys Res Commun 282(1):148–152
Dezawa M (2016) Muse cells provide the pluripotency of mesenchymal stem cells: direct contribution of muse cells to tissue regeneration. Cell Transplant 25(5):849–861
Dezawa M, Kanno H, Hoshino M et al (2004) Specific induction of neuronal cells from bone marrow stromal cells and application for autologous transplantation. J Clin Invest 113(12):1701–1710
Divya MS, Roshin GE, Divya TS et al (2012) Umbilical cord blood-derived mesenchymal stem cells consist of a unique population of progenitors co-expressing mesenchymal stem cell and neuronal markers capable of instantaneous neuronal differentiation. Stem Cell Res Ther 3(6):57
Domanska-Janik K, Buzanska L, Lukomska B (2008) A novel, neural potential of non-hematopoietic human umbilical cord blood stem cells. Int J Dev Biol 52(2–3):237–248
Donaldson AE, Cai J, Yang M et al (2009) Human amniotic fluid stem cells do not differentiate into dopamine neurons in vitro or after transplantation in vivo. Stem Cells Dev 18(7):1003–1012
Drela K, Sarnowska A, Siedlecka P et al (2014) Low oxygen atmosphere facilitates proliferation and maintains undifferentiated state of umbilical cord mesenchymal stem cells in an hypoxia inducible factor-dependent manner. Cytotherapy 16(7):881–892

Drela K, Lech W, Figiel-Dabrowska A et al (2016) Enhanced neuro-therapeutic potential of Wharton's Jelly-derived mesenchymal stem cells in comparison with bone marrow mesenchymal stem cells culture. Cytotherapy 18(4):497–509

Duan P, Sun S, Li B et al (2014) Mir-29a modulates neuronal differentiation through targeting REST in mesenchymal stem cells. PLoS One 9(5):e97684

Egusa H, Schweizer FE, Wang CC et al (2005) Neuronal differentiation of bone marrow-derived stromal stem cells involves suppression of discordant phenotypes through gene silencing. J Biol Chem 280(25):23691–23697

Ende M, Ende N (1972) Hematopoietic transplantation by means of fetal (cord) blood: a new method. Va Med Mon 99(3):276–280

Episkopou V (2005) Sox2 functions in adult neural stem cells. Trends Neurosci 28(5):219–221

Faghihi F, Mirzaei E, Ai J et al (2016) Differentiation potential of human chorion-derived mesenchymal stem cells into motor neuron-like cells in two- and three-dimensional culture systems. Mol Neurobiol 53(3):1862–1872

Fatimah SS, Tan GC, Chua K et al (2013) Stemness and angiogenic gene expression changes of serial-passage human amnion mesenchymal cells. Microvasc Res 86:21–29

Fink KD, Rossignol J, Crane AT et al (2013) Transplantation of umbilical cord-derived mesenchymal stem cells into the striata of R6/2 mice: behavioral and neuropathological analysis. Stem Cell Res Ther 4(5):130

Fisher-Shoval Y, Barhum Y, Sadan O et al (2012) Transplantation of placenta-derived mesenchymal stem cells in the EAE mouse model of MS. J Mol Neurosci 48(1):176–184

Fotia C, Massa A, Boriani F et al (2015) Hypoxia enhances proliferation. Cytotechnology 67(6): 1073–1084

Fu L, Zhu L, Huang Y et al (2008) Derivation of neural stem cells from mesenchymal stem cells: evidence for a bipotential stem cell population. Stem Cells Dev 17(6):1109–1121

Gao Y, Bai C, Zheng D et al (2016) Combination of melatonin and wnt-4 promotes neural cell differentiation in bovine amniotic epithelial cells and recovery from spinal cord injury. J Pineal Res 60(3):303–312

Ge W, Ren C, Duan X et al (2015) Differentiation of mesenchymal stem cells into neural stem cells using cerebrospinal fluid. Cell Biochem Biophys 71(1):449–455

Gluckman E, Rocha V (2009) Cord blood transplantation: state of the art. Haematologica 94(4): 451–454

Gornicka E, Janowski M, Jurga M et al (2006) Human umbilical cord blood-derived neural progenitors (HUCB-NPs) effects on functional recovery from long-lasting motor deficits following ouabain lesion of the rat brain. Behav Pharm 17:555

Gornicka-Pawlak el B, Janowski M, Habich A et al (2011) Systemic treatment of focal brain injury in the rat by human umbilical cord blood cells being at different level of neural commitment. Acta Neurobiol Exp (Wars) 71(1):46–64

Guan M, Xu Y, Wang W et al (2014) Differentiation into neurons of rat bone marrow-derived mesenchymal stem cells. Eur Cytokine Netw 25(3):58–63

Ha Y, Choi JU, Yoon DH, Yeon DS et al (2001) Neural phenotype expression of cultured human cord blood cells in vitro. Neuroreport 12:523–527

Habich A, Domanska-Janik K (2011) Aggregation-promoted expansion of neuraly committed human umbilical cord blood progenitors in vitro. Acta Neurobiol Exp (Wars) 71(1):1–11

Habich A, Jurga M, Markiewicz I et al (2006) Early appearance of stem/progenitor cells with neural-like characteristics in human cord blood mononuclear fraction cultured in vitro. Exp Hematol 34:914–925

Habich A, Szablowska-Gadomska I, Zayat V et al (2013) Epigenetic and molecular signature of human umbilical cord blood-derived neural stem cell (HUCB-NSC) neuronal differentiation. Acta Neurobiol Exp (Wars) 73(1):143–156

Hall BK (2000) The neural crest as a fourth germ layer and vertebrates as quadroblastic not triploblastic. Evol Dev 2(1):3–5

Han R, Kan Q, Sun Y et al (2012) Mir-9 promotes the neural differentiation of mouse bone marrow mesenchymal stem cells via targeting zinc finger protein 521. Neurosci Lett 515(2):147–152
Haris DT, Rogers I (2007) Umbilical cord blood, a unique source of pluripotent stem cells for the regenerative medicine. Curr Stem Cell Res Ther 2:301–309
Harris VK, Faroqui R, Vyshkina T et al (2012) Characterization of autologous mesenchymal stem cell-derived neural progenitors as a feasible source of stem cells for central nervous system applications in multiple sclerosis. Stem Cells Transl Med 1(7):536–547
Hattori H, Suzuki S, Okazaki Y et al (2012) Intracranial transplantation of monocyte-derived multipotential cells enhances recovery after ischemic stroke in rats. J Neurosci Res 90(2): 479–488
Hayakawa K, Esposito E, Wang X et al (2016) Transfer of mitochondria from astrocytes to neurons after stroke. Nature 535(7613):551–555
Hayase M, Kitada M, Wakao S et al (2009) Committed neural progenitor cells derived from genetically modified bone marrow stromal cells ameliorate deficits in a rat model of stroke. J Cereb Blood Flow Metab 29(8):1409–1420
Hayati AR, Nur Fariha MM, Tan GC et al (2011) Potential of human decidua stem cells for angiogenesis and neurogenesis. Arch Med Res 42(4):291–300
Hermann A, Gastl R, Liebau S et al (2004) Efficient generation of neural stem cell-like cells from adult human bone marrow stromal cells. J Cell Sci 117(Pt 19):4411–4422
Hermann A, List C, Habisch HJ et al (2010) Age-dependent neuroectodermal differentiation capacity of human mesenchymal stromal cells: limitations for autologous cell replacement strategies. Cytotherapy 12(1):17–30
Hida N, Nishiyama N, Miyoshi S et al (2008) Novel cardiac precursor-like cells from human menstrual blood-derived mesenchymal cells. Stem Cells 26(7):1695–1704
Hoehn H, Bryant EM, Karp LE et al (1974) Cultivated cells from diagnostic amniocentesis in second trimester pregnancies. I. Clonal morphology and growth potential. Pediatr Res 8(8): 746–754
Hoehn H, Bryant EM, Karp LE et al (1975) Cultivated cells from diagnostic amniocentesis in second trimester pregnancies. II. Cytogenetic parameters as functions of clonal type and preparative technique. Clin Genet 7(1):29–36
Hosseini SM, Vasaghi A, Nakhlparvar N (2015) Differentiation of Wharton's jelly mesenchymal stem cells into neurons in alginate scaffold. Neural Regen Res 10(8):1312–1316
Hsu YC, Wu YT, Yu TH et al (2016) Mitochondria in mesenchymal stem cell biology and cell therapy: from cellular differentiation to mitochondrial transfer. Semin Cell Dev Biol 52: 119–131
Hu W, Guan FX, Li Y et al (2013) New methods for inducing the differentiation of amniotic-derived mesenchymal stem cells into motor neuron precursor cells. Tissue Cell 45(5):295–305
Huang YC, Yang ZM, Chen XH et al (2009) Isolation of mesenchymal stem cells from human placental decidua basalis and resistance to hypoxia and serum deprivation. Stem Cell Rev 5(3):247–255
Huang YC, Yang ZM, Jiang NG et al (2010) Characterization of MSCs from human placental decidua basalis in hypoxia and serum deprivation. Cell Biol Int 34(3):237–243
Huat TJ, Khan AA, Abdullah JM et al (2015) MicroRNA expression profile of neural progenitor-like cells derived from rat bone marrow mesenchymal stem cells under the influence of IGF-1, bFGF and EGF. Int J Mol Sci 16(5):9693–9718
Igura K, Zhang X, Takahashi K et al (2004) Isolation and characterization of mesenchymal progenitor cells from chorionic villi of human placenta. Cytotherapy 6(6):543–553
Ilancheran S, Michalska A, Peh G et al (2007) Stem cells derived from human fetal membranes display multilineage differentiation potential. Biol Reprod 77(3):577–588
In 't Anker PS, Scherjon SA, Kleijburg-van der Keur C et al (2003) Amniotic fluid as a novel source of mesenchymal stem cells for therapeutic transplantation. Blood 102(4):1548–1549
In 't Anker PS, Scherjon SA, Kleijburg-van der Keur C et al (2004) Isolation of mesenchymal stem cells of fetal or maternal origin from human placenta. Stem Cells 22(7):1338–1345

Jablonska A, Kozlowska H, Markiewicz I et al (2010) Transplantation of neural stem cells derived from human cord blood to the brain of adult and neonatal rats. Acta Neurobiol Exp (Wars) 70(4):337–350. Erratum in: Acta Neurobiol Exp (Wars) 71(2):300–302.
Janowski M, Lukomska B, Kozlowska H et al (2004) Intra-carotic administration of neural progenitors derived from human umbilical cord blood mononuclear fraction gives robust behavioral recovery in ouabain model of stroke. Exp Neurol 187:210
Janowski M, Walczak P, Kropiwnicki T et al (2014) Long-term MRI cell tracking after intraventricular delivery in a patient with global cerebral ischemia and prospects for magnetic navigation of stem cells within the CSF. PLoS One 9(2):e97631
Jeong JA, Gang EJ, Hong SH et al (2004) Rapid neural differentiation of human cord blood-derived mesenchymal stem cells. Neuroreport 15(11):1731–1734
Jezierski A, Gruslin A, Tremblay R et al (2010) Probing stemness and neural commitment in human amniotic fluid cells. Stem Cell Rev 6(2):199–214
Jiang Y, Jahagirdar BN, Reinhardt RL et al (2002) Pluripotency of mesenchymal stem cells derived from adult marrow. Nature 418(6893):41–49
Jiang Y, Henderson D, Blackstad M et al (2003) Neuroectodermal differentiation from mouse multipotent adult progenitor cells. Proc Natl Acad Sci USA 100(Suppl 1):11854–11860
Jiang G, Di Bernardo J, Maiden MM et al (2014) Human transgene-free amniotic-fluid-derived induced pluripotent stem cells for autologous cell therapy. Stem Cells Dev 23(21):2613–2625
Jin K, Mao XO, Batteur S et al (2003) Induction of neuronal markers in bone marrow cells: differential effects of growth factors and patterns of intracellular expression. Exp Neurol 184(1): 78–89
Jing L, Jia Y, Lu J et al (2011) Microrna-9 promotes differentiation of mouse bone mesenchymal stem cells into neurons by notch signaling. Neuroreport 22(5):206–211
Jones GN, Moschidou D, Puga-Iglesias TI et al (2012) Ontological differences in first compared to third trimester human fetal placental chorionic stem cells. PLoS One 7(9):e43395
Jozwiak S, Habich A, Kotulska K et al (2010) Intracerebroventricular transplantation of cord blood-derived neural progenitors in a child with severe global brain ischemic injury. Cell Med 1(2): 71–80
Jurga M, Lipkowski AW, Lukomska B et al (2009) Generation of functional neural artificial tissue from human umbilical cord blood stem cells. Tissue Eng Part C Methods 15(3):365–372
Jurga M, Dainiak MB, Sarnowska A (2011) The performance of laminin-containing cryogel scaffolds in neural tissue regeneration. Biomaterials 32(13):3423–3434
Kakishita K, Elwan MA, Nakao N et al (2000) Human amniotic epithelial cells produce dopamine and survive after implantation into the striatum of a rat model of Parkinson's disease: a potential source of donor for transplantation therapy. Exp Neurol 165(1):27–34
Kaneko Y, Hayashi T, Yu S et al (2011) Human amniotic epithelial cells express melatonin receptor mt1, but not melatonin receptor mt2: a new perspective to neuroprotection. J Pineal Res 50(3):272–280
Karahuseyinoglu S, Cinar O, Kilic E et al (2007) Biology of stem cells in human umbilical cord stroma: in situ and in vitro surveys. Stem Cells 25(2):319–331
Kheirandish M, Gavgani SP, Samiee S (2017) The effect of hypoxia preconditioning on the neural and stemness genes expression profiling in human umbilical cord blood mesenchymal stem cells. Transfus Apher Sci 56(3):392–399
Kholodenko IV, Yarygin KN, Gubsky LV et al (2012) Intravenous xenotransplantation of human placental mesenchymal stem cells to rats: comparative analysis of homing in rat brain in two models of experimental ischemic stroke. Bull Exp Biol Med 154(1):118–123
Khoo ML, Shen B, Tao H et al (2008) Long-term serial passage and neuronal differentiation capability of human bone marrow mesenchymal stem cells. Stem Cells Dev 17(5):883–896
Kim S, Honmou O, Kato K et al (2006) Neural differentiation potential of peripheral blood- and bone-marrow-derived precursor cells. Brain Res 1123(1):27–33
Kodama H, Inoue T, Watanabe R et al (2006) Neurogenic potential of progenitors derived from human circulating cd14+ monocytes. Immunol Cell Biol 84(2):209–217

Kogler G, Sensken S, Airey JA et al (2004) A new human somatic stem cell from placental cord blood with intrinsic pluripotent differentiation potential. J Exp Med 200:123–135
Kohyama J, Abe H, Shimazaki T et al (2001) Brain from bone: efficient "Meta-differentiation" of marrow stroma-derived mature osteoblasts to neurons with noggin or a demethylating agent. Differentiation 68(4–5):235–244
Koning JJ, Konijn T, Lakeman KA et al (2016) Nestin-expressing precursors give rise to both endothelial as well as nonendothelial lymph node stromal cells. J Immunol 197(7):2686–2694
Kozlowska H, Jablonka J, Janowski M et al (2007) Transplantation of a novel human cord blood-derived neural-like stem cell line in a rat model of cortical infarct. Stem Cells Dev 16(3): 481–488
Krabbe C, Zimmer J, Meyer M (2005) Neural transdifferentiation of mesenchymal stem cells--a critical review. APMIS 113(11–12):831–844
Kuci S, Kuci Z, Schmid S et al (2008) Efficient in vitro generation of adult multipotent cells from mobilized peripheral blood cd133+ cells. Cell Prolif 41(1):12–27
Kucia M, Machalinski B, Ratajczak MZ (2006) The developmental deposition of epiblast/germ cell-line derived cells in various organs as a hypothetical explanation of stem cell plasticity? Acta Neurobiol Exp (Wars) 66(4):331–341
Kucia M, Halasa M, Wysoczynski M et al (2007) Morphological and molecular characterization of novel population of CXCR4+ SSEA-4+ Oct-4+ very small embryonic-like cells purified from human cord blood: preliminary report. Leukemia 21:297–303
Kuh SU, Cho YE, Yoon DH et al (2005) Functional recovery after human umbilical cord blood cells transplantation with brain derived neurotrophic factor into the spinal cord injured rat. Acta Neurochir (Wien) 147:985–992
Kuwana M, Okazaki Y, Kodama H et al (2003) Human circulating cd14+ monocytes as a source of progenitors that exhibit mesenchymal cell differentiation. J Leukoc Biol 74(5):833–845
Kwon A, Kim Y, Kim M et al (2016) Tissue-specific differentiation potency of mesenchymal stromal cells from perinatal tissues. Sci Rep 6:23544
Langrzyk A, Nowak WN, Stępniewski J (2018) Critical view on mesenchymal stromal cells in regenerative medicine. Antioxid Redox Signal 29(2):169–190
Lech W, Figiel-Dabrowska A, Sarnowska A et al (2016) Phenotypic, functional, and safety control at preimplantation phase of MSC-based therapy. Stem Cells Int 2016:2514917
Lee J, Baek JH, Choi KS et al (2013) Cyclin-dependent kinase 4 signaling acts as a molecular switch between syngenic differentiation and neural transdifferentiation in human mesenchymal stem cells. Cell Cycle 12(3):442–451
Lee JH, Mitchell RR, McNicol JD et al (2015) Single transcription factor conversion of human blood fate to npcs with cns and pns developmental capacity. Cell Rep 11(9):1367–1376
Li F, Miao ZN, Xu YY et al (2012) Transplantation of human amniotic mesenchymal stem cells in the treatment of focal cerebral ischemia. Mol Med Rep 6(3):625–630
Li W, Zhu B, Strakova Z et al (2014) Two-way regulation between cells and aligned collagen fibrils: local 3D matrix formation and accelerated neural differentiation of human decidua parietalis placental stem cells. Biochem Biophys Res Commun 450(4):1377–1382
Li H, Li J, Sheng W et al (2015) Astrocyte-like cells differentiated from a novel population of CD45-positive cells in adult human peripheral blood. Cell Biol Int 39(1):84–93
Liao W, Huang N, Yu J et al (2015) Direct conversion of cord blood CD34+ cells Into neural stem cells by OCT4. Stem Cells Transl Med 4(7):755–763
Liu X, Li D, Jiang D, Fang Y (2013) Acetylcholine secretion by motor neuron-like cells from umbilical cord mesenchymal stem cells. Neural Regen Res 8(22):2086–2092
Locher H, Frijns JH, Huisman MA et al (2014) Tubb3: neuronal marker or melanocyte mimic? Cell Transplant 23(11):1471–1473
Lu P, Blesch A, Tuszynski MH (2004) Induction of bone marrow stromal cells to neurons: differentiation, transdifferentiation, or artifact? J Neurosci Res 77(2):174–191

Lu GH, Yong WS, Xu ZM et al (2012) Human placental decidua basalis-derived mesenchymal stem cells differentiate into dopamine neuron-like cells with no response to long-term culture in vitro. Neuroreport 23(8):513–518

Mahrouf-Yorgov M, Augeul L, Da Silva CC et al (2017) Mesenchymal stem cells sense mitochondria released from damaged cells as danger signals to activate their rescue properties. Cell Death Differ 24(7):1224–1238

Maraldi T, Bertoni L, Riccio M et al (2014) Human amniotic fluid stem cells: neural differentiation in vitro and in vivo. Cell Tissue Res 357(1):1–13

Marcus AJ, Coyne TM, Rauch J et al (2008) Isolation, characterization, and differentiation of stem cells derived from the rat amniotic membrane. Differentiation 76(2):130–144

Mareschi K, Rustichelli D, Comunanza V et al (2009) Multipotent mesenchymal stem cells from amniotic fluid originate neural precursors with functional voltage-gated sodium channels. Cytotherapy 11(5):534–547

McGuckin CP, Forraz N, Allouard Q et al (2004) Umbilical cord blood stem cells can expand hematopoietic and neuroglial progenitors in vitro. Exp Cell Res 295:350

McGuckin CP, Forraz N, Baradez MO (2005) Production of stem cells with embryonic characteristics from human umbilical cord blood. Cell Prolif 38:245–255

Melton DA, Cowan C (2004) "Stemness": definitions, criteria, and standards. In: Lanza R et al (eds) Handbook of stem cells. Elsevier, Amsterdam, pp 25–31

Meng X, Ichim TE, Zhong J et al (2007) Endometrial regenerative cells: a novel stem cell population. J Transl Med 5:57

Miki T, Lehmann T, Cai H et al (2005) Stem cell characteristics of amniotic epithelial cells. Stem Cells 23(10):1549–1559

Mitchell KE, Weiss ML, Mitchell BM (2003) Matrix cells from Wharton's jelly form neurons and glia. Stem Cells 21(1):50–60

Monti M, Imberti B, Bianchi N et al (2017) Novel method for isolation of pluripotent stem cells from human umbilical cord blood. Stem Cells Dev 6(17):1258–1269

Morrison SJ, Perez SE, Qiao Z et al (2000) Transient notch activation initiates an irreversible switch from neurogenesis to gliogenesis by neural crest stem cells. Cell 101(5):499–510

Nan Z, Grande A, Sanberg CD (2005) Infusion of human umbilical cord blood ameliorates neurologic deficits in rats with hemorrhagic brain injury. Ann NY Acad Sci 1049:84–96

Neuhuber B, Gallo G, Howard L et al (2004) Reevaluation of in vitro differentiation protocols for bone marrow stromal cells: disruption of actin cytoskeleton induces rapid morphological changes and mimics neuronal phenotype. J Neurosci Res 77(2):192–204

Nichols JE, Niles JA, DeWitt D et al (2013) Neurogenic and neuro-protective potential of a novel subpopulation of peripheral blood-derived CD133+ ABCG2+CXCR4+ mesenchymal stem cells: development of autologous cell-based therapeutics for traumatic brain injury. Stem Cell Res Ther 4(1):3

Niknejad H, Peirovi H, Ahmadiani A et al (2010) Differentiation factors that influence neuronal markers expression in vitro from human amniotic epithelial cells. Eur Cell Mater 19:22–29

Obtulowicz P, Lech W, Strojek L (2016) Induction of endothelial phenotype from Wharton's jelly-derived MSCs and comparison of their vasoprotective and neuroprotective potential with primary WJ-MSCs in CA1 hippocampal region ex vivo. Cell Transplant 25:715–727

Okawa H, Okuda O, Arai H et al (2001) Amniotic epithelial cells transform into neuron-like cells in the ischemic brain. Neuroreport 12(18):4003–4007

Okolicsanyi RK, Camilleri ET, Oikari LE et al (2015) Human mesenchymal stem cells retain multilineage differentiation capacity including neural marker expression after extended in vitro expansion. PLoS One 10(9):e0137255

Paldino E, Cenciarelli C, Giampaolo A (2014) Induction of dopaminergic neurons from human Wharton's jelly mesenchymal stem cell by forskolin. J Cell Physiol 229(2):232–244

Park S, Koh SE, Maeng S et al (2011) Neural progenitors generated from the mesenchymal stem cells of first-trimester human placenta matured in the hypoxic-ischemic rat brain and mediated restoration of locomotor activity. Placenta 32(3):269–276

Park S, Kim E, Koh SE et al (2012) Dopaminergic differentiation of neural progenitors derived from placental mesenchymal stem cells in the brains of Parkinson's disease model rats and alleviation of asymmetric rotational behavior. Brain Res 1466:158–166
Park S, Koh SE, Hur CY et al (2013) Comparison of human first and third trimester placental mesenchymal stem cell. Cell Biol Int 37(3):242–249
Patel AN, Park E, Kuzman M et al (2008) Multipotent menstrual blood stromal stem cells: Isolation, characterization, and differentiation. Cell Transplant 17(3):303–311
Perrier AL, Tabar V, Barberi T et al (2004) Derivation of midbrain dopamine neurons from human embryonic stem cells. Proc Natl Acad Sci USA 101(34):12543–12548
Peterson DA (2004) Umbilical cord blood cells and brain stroke injury: bringing in fresh blood to address an old problem. J Clin Invest 114:312–314
Poloni A, Rosini V, Mondini E et al (2008) Characterization and expansion of mesenchymal progenitor cells from first-trimester chorionic villi of human placenta. Cytotherapy 10(7):690–697
Porat Y, Porozov S, Belkin D et al (2006) Isolation of an adult blood-derived progenitor cell population capable of differentiation into angiogenic, myocardial and neural lineages. Br J Haematol 135(5):703–714
Portmann-Lanz CB, Schoeberlein A, Huber A et al (2006) Placental mesenchymal stem cells as potential autologous graft for pre- and perinatal neuroregeneration. Am J Obstet Gynecol 194(3):664–673
Portmann-Lanz CB, Schoeberlein A, Portmann R et al (2010) Turning placenta into brain: placental mesenchymal stem cells differentiate into neurons and oligodendrocytes. Am J Obstet Gynecol 202(3):294.e1–294.e11
Price EM, Prather RS, Foley CM (2006) Multipotent adult progenitor cell lines originating from the peripheral blood of green fluorescent protein transgenic swine. Stem Cells Dev 15(4):507–522
Prusa AR, Marton E, Rosner M et al (2004) Neurogenic cells in human amniotic fluid. Am J Obstet Gynecol 191(1):309–314
Qi H, Yang YL, Xie HB (2011) Teratoma derived from bone marrow mesenchymal stem cells in a diabetic microenvironment. Pancreas 40(6):981–983
Qiao C, Xu W, Zhu W (2008) Human mesenchymal stem cells isolated from the umbilical cord. Cell Biol Int 32(1):8–15
Rafieemehr H, Kheirandish M, Soleimani M (2015) Improving the neuronal differentiation efficiency of umbilical cord blood-derived mesenchymal stem cells cultivated under appropriate conditions. Iran J Basic Med Sci 18(11):1100–1106
Resca E, Zavatti M, Bertoni L et al (2013) Enrichment in c-kit+ enhances mesodermal and neural differentiation of human chorionic placental cells. Placenta 34(7):526–535
Rodrigues MC, Lippert T, Nguyen H et al (2016) Menstrual blood-derived stem cells: in vitro and in vivo characterization of functional effects. Adv Exp Med Biol 951:111–121
Sabapathy V, Ravi S, Srivastava V et al (2012) Long-term cultured human term placenta-derived mesenchymal stem cells of maternal origin displays plasticity. Stem Cells Int 2012:174328
Sakuragawa N, Thangavel R, Mizuguchi M et al (1996) Expression of markers for both neuronal and glial cells in human amniotic epithelial cells. Neurosci Lett 209(1):9–12
Sakuragawa N, Misawa H, Ohsugi K et al (1997) Evidence for active acetylcholine metabolism in human amniotic epithelial cells: applicable to intracerebral allografting for neurologic disease. Neurosci Lett 232(1):53–56
Sakuragawa N, Kakinuma K, Kikuchi A et al (2004) Human amnion mesenchyme cells express phenotypes of neuroglial progenitor cells. J Neurosci Res 78(2):208–214
Sanchez-Ramos J, Song S, Cardozo-Pelaez F et al (2000) Adult bone marrow stromal cells differentiate into neural cells in vitro. Exp Neurol 164(2):247–256
Sanchez-Ramos JR, Song S, Kamath SG et al (2001) Expression of neural markers in human umbilical cord blood. Exp Neurol 171:109–115
Sandvig I, Gadjanski I, Vlaski-Lafarge M et al (2017) Strategies to enhance implantation and survival of stem cells after their injection in ischemic neural tissue. Stem Cells Dev 26(8): 554–565

Sanluis-Verdes A, Sanluis-Verdes N, Manso-Revilla MJ et al (2017) Tissue engineering for neurodegenerative diseases using human amniotic membrane and umbilical cord. Cell Tissue Bank 18:1–15

Sarnowska A, Domańska-Janik K (2017) The basis to application of cell therapy in neurological disorders. Neurologia Praktyczna (in polish) 6(99):7–14

Sarnowska A, Jablonska A, Jurga M et al (2013) Encapsulation of mesenchymal stem cells by bioscaffolds protects cell survival and attenuates neuroinflammatory reaction in injured brain tissue after transplantation. Cell Transplant 22(Suppl 1):S67–S82

Sarugaser R, Lickorish D, Baksh D (2005) Human umbilical cord perivascular (HUCPV) cells: a source of mesenchymal progenitors. Stem Cells 23(2):220–229

Saxena M, Prashar P, Yadav PS et al (2016) Mouse bone marrow stromal cells differentiate to neuron-like cells upon inhibition of BMP signaling. Differentiation 92(1–2):1–9

Scharstuhl A, Schewe B, Benz K et al (2007) Chondrogenic potential of human adult mesenchymal stem cells is independent of age or osteoarthritis etiology. Stem Cells 25(12):3244–3251

Secco M, Zucconi E, Vieira NM et al (2008) Multipotent stem cells from umbilical cord: cord is richer than blood! Stem Cells 26(1):146–150

Semenov OV, Koestenbauer S, Riegel M et al (2010) Multipotent mesenchymal stem cells from human placenta: critical parameters for isolation and maintenance of stemness after isolation. Am J Obstet Gynecol 202(2):193.e1–193.e13

Shyamala K, Yanduri S, Girish HC (2015) Neural crest: the fourth germ layer. J Oral Maxillofac Pathol 19(2):221–229

Song S, Zhang H, Cuevas J et al (2007) Comparison of neuron-like cells derived from bone marrow stem cells to those differentiated from adult brain neural stem cells. Stem Cells Dev 16(5): 747–756

Spitzer N, Sammons GS, Butts HM et al (2011) Multipotent progenitor cells derived from adult peripheral blood of swine have high neurogenic potential in vitro. J Cell Physiol 226(12):3156–3168

Sridharan I, Kim T, Strakova Z et al (2013) Matrix-specified differentiation of human decidua parietalis placental stem cells. Biochem Biophys Res Commun 437(3):489–495

Strakova Z, Livak M, Krezalek M et al (2008) Multipotent properties of myofibroblast cells derived from human placenta. Cell Tissue Res 332(3):479–488

Sun W, Buzanska L, Domanska-Janik K (2005) Voltage-sensitive and ligand-gated channels in differentiating neural stem-like cells derived from the nonhematopoietic fraction of human umbilical cord blood. Stem Cells 23(7):931–945

Sun H, Hou Z, Yang H et al (2014) Multiple systemic transplantations of human amniotic mesenchymal stem cells exert therapeutic effects in an als mouse model. Cell Tissue Res 357(3): 571–582

Taguchi A, Soma T, Tanaka H (2004) Administration of CD34+ cells after stroke enhances neurogenesis via angiogenesis in a mouse model. J Clin Invest 114(3):330–338

Takashima Y, Era T, Nakao K, Kondo S et al (2007) Neuroepithelial cells supply an initial transient wave of MSC differentiation. Cell 129(7):1377–1388

Talwadekar M, Fernandes S, Kale V et al (2016) Valproic acid enhances the neural differentiation of human placenta derived-mesenchymal stem cells in vitro. J Tissue Eng Regen Med 11(11): 3111–3123

Tang X, Wang S, Bai Y et al (2016) Conversion of adult human peripheral blood mononuclear cells into induced neural stem cell by using episomal vectors. Stem Cell Res 16(2):236–242

Tao H, Rao R, Ma DD (2005) Cytokine-induced stable neuronal differentiation of human bone marrow mesenchymal stem cells in a serum/feeder cell-free condition. Dev Growth Differ 47(6): 423–433

Taylor HS (2004) Endometrial cells derived from donor stem cells in bone marrow transplant recipients. JAMA 292(1):81–85

Tio M, Tan KH, Lee W et al (2010) Roles of db-cAMP, IBMX and RA in aspects of neural differentiation of cord blood derived mesenchymal-like stem cells. PLoS One 5(2):e9398

Tondreau T, Lagneaux L, Dejeneffe M et al (2004) Bone marrow-derived mesenchymal stem cells already express specific neural proteins before any differentiation. Differentiation 72(7): 319–326
Toselli M, Cerbai E, Rossi F et al (2008) Do amniotic fluid-derived stem cells differentiate into neurons in vitro? Nat Biotechnol 26(3):269–270 author reply 270–271
Tracy ET, Zhang CY, Gentry T (2011) Isolation and expansion of oligodendrocyte progenitor cells from cryopreserved human umbilical cord blood. Cytotherapy 13(6):722–729
Troyer DL, Weiss ML (2008) Wharton's jelly-derived cells are a primitive stromal cell population. Stem Cells 26(3):591–599
Trzaska KA, King CC, Li KY et al (2009) Brain-derived neurotrophic factor facilitates maturation of mesenchymal stem cell-derived dopamine progenitors to functional neurons. J Neurochem 110(3):1058–1069
Tsai HL, Deng WP, Lai WF et al (2014) Wnts enhance neurotrophin-induced neuronal differentiation in adult bone-marrow-derived mesenchymal stem cells via canonical and noncanonical signaling pathways. PLoS One 9(8):e104937
Tsai PH, Chang YC, Lee YY et al (2015) Differentiation of blood T cells: reprogramming human induced pluripotent stem cells into neuronal cells. J Chin Med Assoc 78(6):353–359
Uchida H, Niizuma K, Kushida Y et al (2017) Human muse cells reconstruct neuronal circuitry in subacute lacunar stroke model. Stroke 48(2):428–435
Venkatesh K, Srikanth L, Vengamma B et al (2015) In vitro transdifferentiation of human cultured cd34+ stem cells into oligodendrocyte precursors using thyroid hormones. Neurosci Lett 588: 36–41
Visvanathan J, Lee S, Lee B et al (2007) The microrna mir-124 antagonizes the anti-neural rest/scp1 pathway during embryonic cns development. Genes Dev 21(7):744–749
Wang HS, Hung SC, Peng ST et al (2004) Mesenchymal stem cells in the Wharton's jelly of the human umbilical cord. Stem Cells 22(7):1330–1337
Wang H, Jin P, Sabatino M et al (2012) Comparison of endometrial regenerative cells and bone marrow stromal cells. J Transl Med 10(2012):207
Wang CH, Wu CC, Hsu SH et al (2013a) The role of rhoa kinase inhibition in human placenta-derived multipotent cells on neural phenotype and cell survival. Biomaterials 34(13):3223–3330
Wang T, Choi E, Monaco MC et al (2013b) Derivation of neural stem cells from human adult peripheral cd34+ cells for an autologous model of neuroinflammation. PLoS One 8(11):e81720
Wang T, Choi E, Monaco MC et al (2015) Direct induction of human neural stem cells from peripheral blood hematopoietic progenitor cells. J Vis Exp 95:52298
Wei PC, Chao A, Peng HH et al (2014) Sox9 as a predictor for neurogenesis potentiality of amniotic fluid stem cells. Stem Cells Transl Med 3(10):1138–1147
Wharton T (1656) Adenographia: sive glandularum totius corporis descriptio. Wharton, London
Willing AE, Lixian J, Milliken M et al (2003) Intravenous versus intrastriatal cord blood administration in a rodent model of stroke. J Neurosci Res 73(3):296–307
Wolff EF, Gao XB, Yao KV et al (2011) Endometrial stem cell transplantation restores dopamine production in a Parkinson's disease model. J Cell Mol Med 15(4):747–755
Woodbury D, Schwarz EJ, Prockop DJ et al (2000) Adult rat and human bone marrow stromal cells differentiate into neurons. J Neurosci Res 61(4):364–370
Woodbury D, Reynolds K, Black IB (2002) Adult bone marrow stromal stem cells express germline, ectodermal, endodermal, and mesodermal genes prior to neurogenesis. J Neurosci Res 69(6):908–917
Wu R, Wang N, Li M et al (2013) Experimental study on the facilitative effects of mir-125b on the differentiation of rat bone marrow mesenchymal stem cells into neuron-like cells. Cell Biol Int 37(8):812–819
Wu R, Tang Y, Zang W et al (2014) Microrna-128 regulates the differentiation of rat bone mesenchymal stem cells into neuron-like cells by wnt signaling. Mol Cell Biochem 387(1–2):151–158
Wu CG, Zhang JC, Xie CQ et al (2015a) In vivo tracking of human placenta derived mesenchymal stem cells in nude mice via (1)(4)c-tdr labeling. BMC Biotechnol 15:55

Wu KJ, Yu SJ, Chiang CW et al (2015b) Transplantation of human placenta-derived multipotent stem cells reduces ischemic brain injury in adult rats. Cell Transplant 24(3):459–470

Yan ZJ, Zhang P, Hu YQ et al (2013) Neural stem-like cells derived from human amnion tissue are effective in treating traumatic brain injury in rat. Neurochem Res 38(5):1022–1033

Yang Q, Mu J, Li Q et al (2008) A simple and efficient method for deriving neurospheres from bone marrow stromal cells. Biochem Biophys Res Commun 372(4):520–524

Yang P, Li Y, Zhang JT et al (2017) Application of human umbilical cord mesenchymal stem cells in spinal cord injury. J Biom Tissue Eng 7(5):393–400

Yanjie J, Jiping S, Yan Z et al (2007) Effects of notch-1 signalling pathway on differentiation of marrow mesenchymal stem cells into neurons in vitro. Neuroreport 18(14):1443–1447

Yarygin KN, Kholodenko IV, Konieva AA et al (2009) Mechanisms of positive effects of transplantation of human placental mesenchymal stem cells on recovery of rats after experimental ischemic stroke. Bull Exp Biol Med 148(6):862–868

Ye Y, Zeng YM, Wan MR et al (2011) Induction of human bone marrow mesenchymal stem cells differentiation into neural-like cells using cerebrospinal fluid. Cell Biochem Biophys 59(3): 179–184

Ye Y, Peng YR, Hu SQ et al (2016) In vitro differentiation of bone marrow mesenchymal stem cells into neuron like cells by cerebrospinal fluid improves motor function of middle cerebral artery occlusion rats. Front Neurol 7:183

Ye Y, Feng TT, Peng YR et al (2018) The treatment of spinal cord injury in rats using bone marrow-derived neural-like cells induced by cerebrospinal fluid. Neurosci Lett 666:85–91

Yen BL, Huang HI, Chien CC et al (2005) Isolation of multipotent cells from human term placenta. Stem Cells 23(1):3–9

Yen BL, Chien CC, Chen YC et al (2008) Placenta-derived multipotent cells differentiate into neuronal and glial cells in vitro. Tissue Eng Part A 14(1):9–17

Zeng G, Wang G, Guan F et al (2011) Human amniotic membrane-derived mesenchymal stem cells labeled with superparamagnetic iron oxide nanoparticles: the effect on neuron-like differentiation in vitro. Mol Cell Biochem 357(1–2):331–341

Zhang L, Zhang HT, Hong SQ (2009) Cografted Wharton's jelly cells-derived neurospheres and BDNF promote functional recovery after rat spinal cord transection. Neurochem Res 34(11): 2030–2039

Zhang GY, Wang J, Jia YJ et al (2015) Microrna-9 promotes the neuronal differentiation of rat bone marrow mesenchymal stem cells by activating autophagy. Neural Regen Res 10(2):314–320

Zhou H, Zhang H, Yan Z et al (2016) Transplantation of human amniotic mesenchymal stem cells promotes neurological recovery in an intracerebral hemorrhage rat model. Biochem Biophys Res Commun 475(2):202–208

Zhu M, Feng Q, Sun Y et al (2017) Effect of cartilaginous matrix components on the chondrogenesis and hypertrophy of mesenchymal stem cells in hyaluronic acid hydrogels. J Biomed Mater Res B Appl Biomater 105(8):2292–2300

Ziadlou R, Shahhoseini M, Safari F et al (2015) Comparative analysis of neural differentiation potential in human mesenchymal stem cells derived from chorion and adult bone marrow. Cell Tissue Res 362(2):367–377

Zigova T, Willing A, Cardozo-Pelaez F et al (2001) Expression of neural markers in human umbilical cord blood. Exp Neurol 171:109–115

Zigova T, Song S, Willing AE et al (2002) Human umbilical cord blood cells express neural antigens after transplantation into the developing rat brain. Cell Transplant 11(3):265–274

Zou D, Chen Y, Han Y et al (2014) Overexpression of microRNA-124 promotes the neuronal differentiation of bone marrow-derived mesenchymal stem cells. Neural Regen Res 9(12):1241–1248

Zurita M, Bonilla C, Otero L et al (2008) Neural transdifferentiation of bone marrow stromal cells obtained by chemical agents is a short-time reversible phenomenon. Neurosci Res 60(3): 275–280

Chapter 3
Neural Stem Cells Derived from Human-Induced Pluripotent Stem Cells and Their Use in Models of CNS Injury

Pavla Jendelova, Eva Sykova, and Slaven Erceg

Abstract Induced pluripotent stem (iPS) cells are derived from differentiated cells by different reprogramming techniques, by introducing specific transcription factors responsible for pluripotency. Induced pluripotent stem cells can serve as an excellent source for differentiated neural stem/progenitor cells (NSCs/NPs). Several methods and protocols are utilized to create a robust number of NSCs/NPs without jeopardizing the safety issues required for in vivo applications. A variety of disease-specific iPS cells have been used to study nervous system diseases. In this chapter, we will focus on some of the derivation and differentiation approaches and the application of iPS-NPs in the treatment of spinal cord injury and stroke.

Keywords Induced pluripotent stem cells · Neural stem cells · Neuronal differentiation · Stroke · Spinal cord injury

3.1 Introduction

Neural stem cells (NSCs) are multipotent cells with the ability to differentiate into neurons, oligodendrocytes, and astrocytes. They are one of the most promising cell sources that have been studied so far in the treatment of neural disorders (spinal cord injury, stroke, neurodegenerative diseases). Their derivatives are able to replace lost neurons and/or remyelinate axons and provide neuroprotection and local trophic support. These cells are present in the adult and developing CNS and can be isolated and expanded in vitro. However, fetal or adult neural tissue is difficult to obtain in

---

P. Jendelova (✉)
Institute of Experimental Medicine ASCR, Prague, Czech Republic
e-mail: pavla.jendelova@iem.cas.cz

E. Sykova
2nd Faculty of Medicine, Charles University, Prague, Czech Republic

S. Erceg
Institute of Experimental Medicine ASCR, Prague, Czech Republic

Research Center Principe Felipe, Valencia, Spain

L. Buzanska (ed.), *Human Neural Stem Cells*, Results and Problems in Cell Differentiation 66, https://doi.org/10.1007/978-3-319-93485-3_3

the required quantity and quality. Therefore, NSC can be prepared as differentiated derivatives from embryonic (Salewski et al. 2015b; Yang et al. 2013) or induced pluripotent stem cells (Romanyuk et al. 2015; Sareen et al. 2014). Induced pluripotent stem (iPS) cells are derived from differentiated cells by different reprogramming techniques by introducing specific transcription factors responsible for pluripotency. The first iPS cells were created by retrovirus-mediated transfer of Oct3/4, Sox2, Klf4, and c-Myc genes into mouse fibroblast cells (Takahashi and Yamanaka 2006). Currently, a variety of methods, integrative or non-integrative, have been developed to derive iPS cells (Wang et al. 2013). Similarly, several protocols exist to prepare NSCs or neural progenitor cells (NPs) such as dual SMAD inhibition (Chambers et al. 2013) or embryoid body formation, followed by differentiation into neural rosettes (Muratore et al. 2014; Polentes et al. 2012; described below).

3.1.1 iPS-NPs Derivation and Differentiation

There are two basic methods for generating neurons from iPS cells. The first one uses embryoid body (EB) or neurospheres (Koch et al. 2009; Zhang 2006; Zhang et al. 2001), and the second includes monolayer or adherent culture conditions (Chambers et al. 2012; Lukovic et al. 2017).

The most well-studied iPS cell differentiation system involves the formation of three-dimensional structures called EBs. These structures appear when clumps of iPS cells aggregate in culture dishes that do not favor cell adhesion or attachment. These structures faithfully mimic three embryonic layers: endoderm, mesoderm, and ectoderm. Further neural specification is reached by generating the neuroectodermal cell population, isolating the rosette type of NPCs. Neural rosettes are radial arrangements of columnar cells resembling neuroepithelial cells of the neural tube during developmental stage. However, spontaneous differentiation of EBs yields only a small fraction of cells with neural lineages. Therefore, to induce neural differentiation, EBs are treated with different morphogens and growth factors (Erceg et al. 2009). In some procedures, in order to enrich and expand NPCs, neural rosettes are manually isolated and cultured in suspension, forming neurospheres, dynamic 3D physiological microincubators of NPCs. These neurospheres can be continuously propagated as a highly homogenous population (Koch et al. 2009) without losing the expression of neural progenitor markers (Jensen and Parmar 2006). The NPCs can be easily identified by expressing Pax6, Sox2, Sox1, and Nestin. These cells have shown stable neuro- and gliogenic potential. In neural transplantation, neurospheres are the most commonly used neural progenitors that are injected into the brain, due to their easy delivery and ability to rapidly migrate to the neurogenic areas of the brain (Koch et al. 2009). The heterogeneous nature of embryoid body differentiation toward cells of ectoderm, endoderm, and mesoderm and the poor yield of protocols based on the selective survival of neural progeny are

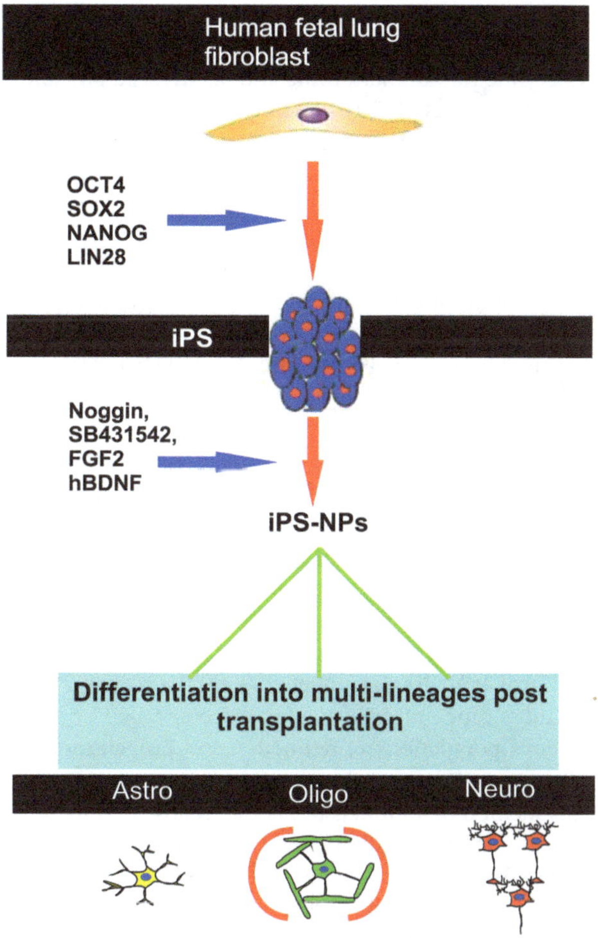

Fig. 3.1 Derivation and differentiation of iPS-NPs from human fetal lung fibroblasts iMR90 (Polentes et al. 2012). These cells, when grafted into a model of stroke or spinal cord injury, differentiated mainly into neuronal phenotype, less into astrocytes, and very rarely into oligodendrocytes

the main disadvantages of these procedures. It is more convenient to have the direct differentiation in monolayer, adherent culture conditions.

In order to derive the human iPS line, a lentivirus-mediated combination of OCT4, SOX2, NANOG, and LIN28 human cDNA (Takahashi and Yamanaka 2006) was used for transduction of female human fetal lung fibroblasts. Early neural precursors were produced in low-attachment cultures in the presence of Noggin, the transforming growth factor-b pathway inhibitor SB431542, bFGF, and hBDNF. NPs were produced from early neural precursors in the presence of 20 ng/ml hBDNF for 7 days (Polentes et al. 2012) (Fig. 3.1). These cells were tested in models of acute and chronic spinal cord injury (SCI) and stroke (Amemori et al. 2015; Polentes et al. 2012; Romanyuk et al. 2015; Ruzicka et al. 2017). A novel technique with automated tissue chopper was used to transform adherent iPS cells into free-floating spheres that were easy to maintain, were expandable, and can be differentiated toward NP spheres with a spinal cord phenotype, using a combination of all-trans retinoic acid (RA) and epidermal growth factor (EGF) and fibroblast growth factor-2 (FGF-2) mitogens. These cells had an in vitro as well in vivo (grafted into spinal cord of athymic rat) similar profile and behavior to NP spheres derived from human fetal tissue (Sareen et al. 2014).

To avoid the EB step in neural differentiation, other approaches are used such as simple medium conditions or recombination factors treatment of undifferentiated iPS cells, to directly generate a pure neuroepithelial cell population. The most used adherent method was developed by Studer Group (Chambers et al. 2009) using dual inhibition of SMAD signaling, promoting efficient neuronal differentiation. Dual SMAD inhibition rapidly differentiates a confluent, feeder-free culture of human iPS cells into early neuroectoderm. This rapid differentiation is caused by blocking the two signaling pathways that utilize SMADs for transduction: BMP and TGF-β. The neuroepithelial cells can be further committed to specific neural cell types such as cortical neurons.

Other studies have used undefined factors such as an animal extracellular matrix, to form neural rosettes, followed by retinoic acid (RA) exposure (Lee et al. 2007; Shin et al. 2006). The majority of these cell lines are differentiated in the presence of animal feeder cell lines or animal components, which bear the risk of xenogenetic pathogen cross-transfer, and as such they are unsuitable for medical applications. Additionally these protocols include recombinant factors increasing the expenses of the neural differentiation procedure. In spite of recent advances in xeno-free protocols (Aoi et al. 2008; Chin et al. 2009; Kim et al. 2009; Nguyen et al. 2014), to date, the most efficient protocol for the controlled conversion of iPS cells into homogeneous populations of defined neural progenitors, avoiding the formation of EBs under animal-free conditions, was developed by Erceg et al. (Lukovic et al. 2017). This method is minimalistic, is less costly, and includes the initial differentiation of iPS cells in chemically defined and animal-free medium and adherent human substrate. EB step is avoided, and differentiation process results in formation of rosettes and the neural tubelike structures previously identified as typical neural progenitor cells (Li et al. 2005; Shin et al. 2006). In addition, the defined media is combined with the defined human surface components such as "CELLstart (defined human matrix from Thermo Fisher) and insulin-transferrin-sodium selenite medium, which all together create defined conditions for neural differentiation. The yield of obtained neural progenitors was higher than in previously published protocols, where chemically defined medium and adherent conditions were used (Chambers et al. 2009; Joannides et al. 2007; Koch et al. 2009; Nat et al. 2007; Roese-Koerner et al. 2013; Shin et al. 2006). This strategy could represent a standard differentiation procedure suitable for clinical applications, including neurodegenerative diseases and spinal cord injury. Taking into account that these neural progenitors can be derived from patient-specific iPS cells, this could provide an attractive human in vitro cellular tool for disease modeling and pharmacological screening.

NPs can also be generated directly from somatic cells without an intermediate pluripotent state. The Wernig Lab initially directly reprogrammed NPs by using Sox2, FoxG1, and Brn2 reprogramming factors (Lujan et al. 2012). Several other combinations of factors that can directly reprogram somatic cells to NPs have subsequently been discovered (Han et al. 2012; Ring et al. 2012; Zou et al. 2014).

3.1.2 iPS-NPs in the Treatment of Injured CNS

NSCs and NPs can have dual importance for neuroscience research. Apart from application in cell therapy, they can be used for modeling neurodegenerative diseases (Okano and Yamanaka 2014). A variety of disease-specific iPS cells have been used to study nervous system diseases, including amyotrophic lateral sclerosis (Dimos et al. 2008; Egawa et al. 2012), spinal muscular atrophy (Ebert et al. 2009), spinobulbar muscular atrophy (Nihei et al. 2013), Friedreich's ataxia (Ku et al. 2010), Alzheimer's disease (Israel et al. 2012; Yagi et al. 2011), Parkinson's disease (Devine et al. 2011), Huntington's disease (Zhang et al. 2010), Machado-Joseph disease (Koch et al. 2011), fragile X-syndrome (Urbach et al. 2010), Rett's syndrome (Marchetto et al. 2010), schizophrenia (Brennand et al. 2011), and Dravet's syndrome of intractable epilepsy (Horiuchi et al. 2013). However, in this chapter, we will focus on the possible use and the mechanisms of action in the treatment of spinal cord injury and stroke.

3.1.3 iPS-NPs in the Treatment of SCI

Human SCIs are very heterogeneous. Generally, traumatic injury to the spinal cord is defined by two broad events: a primary event (also called primary SCI), attributable to the mechanical impact and shear forces themselves, and a secondary component (secondary SCI) that consists of a series of systemic and local neurochemical changes that occur in the nervous tissue after the initial traumatic shock (Klussmann and Martin-Villalba 2005). Any mechanical deformation of the spinal cord leads to the rupture of neuronal cell membranes with the release of the intracellular contents, localized edema, breakdown of the blood–brain barrier, etc. All of these processes trigger a chain of events that are accompanied by an inflammatory reaction leading to secondary necrotic cell death at the core of the injury site and apoptotic cell death in the surrounding areas. The acute and subacute phase turns into the chronic stage, key features of which are cavity formation and astroglial scar as well as the activation of inhibitory molecules at the injury site. Currently, research is focused on (1) the application of stem cells to preserve spared tissue or replace lost tissue after injury to the host cells; this treatment is effective in the acute and subacute phase of SCI; (2) the reactivation of neuroplasticity with digesting glial scar with chondroitinase ABC or blocking inhibitory molecules with anti-NOGO antibody; and (3) filling the cavity with different types of biomaterials in order to provide a scaffold for cell growth and axonal outgrowth. The last two approaches can be part of multifaceted treatment of the chronic stage of SCI and can be supplemented by the use of stem cells.

3.1.4 iPS-NPs in the Treatment of Acute SCI

Neural stem and progenitor cells have been shown to be particularly useful for transplantation therapy for SCI due to their ability to provide an unlimited source of nerve cells for cell replacement (Cummings et al. 2005; Hooshmand et al. 2009; Volarevic et al. 2013; Yan et al. 2007), as well as trophic support for endogenous neuroregeneration (Hsu et al. 2007; Sharp et al. 2010; Yan et al. 2004). The first studies using iPS-NPs were published in 2010 and were focused mainly on safety issues. iPS-derived neurospheres, which had been pre-evaluated as non-tumorigenic, were transplanted into the spinal cord 9 days after contusive injury. Grafted cells differentiated into all three neural lineages without forming teratomas or other tumors. However, the transplantation of iPS-derived neurospheres pre-evaluated as "unsafe" showed robust teratoma formation and locomotor functional loss in the SCI model (Tsuji et al. 2010). The effect of iPS-NPs in animal models of SCI was studied worldwide with similar results. Okano's group transplanted human iPSC-NPCs into SCI model of NOD/SCID mice (Nori et al. 2011) and later into SCI models of common marmosets (Kobayashi et al. 2012) and showed that the transplanted cells mainly differentiated into neurons which made a synaptic connection with host axons. Grafted cells enhanced axonal regrowth, angiogenesis, and preservation of the whole spinal cord and white matter area. These beneficial effects contributed to locomotor functional recovery. Importantly, the NPs did not make tumors in mice or in marmosets. Tuszynski's group induced a cervical hemisection model of immunodeficient rats and transplanted the human iPS-NPs in fibrin matrices containing a growth factor cocktail for better survival at the lesion site. The grafted cells showed neurite elongation within very long distances 12 weeks after injury, with synapse formation between donor and host neurons (Lu et al. 2014). As an interesting source of cells for iPS derivation, a dissected disc was used, which can be acquired during SCI stabilizing surgery (Oh et al. 2015). The disc-derived iPS-NPs mainly differentiated into neurons and contributed to motor functional recovery in a subacute SCI mouse model.

Our group compared the efficacy of human iPS-NPs transplantation into SCI with other types of cells such as human bone marrow-derived mesenchymal stem cells and human spinal fetal cell-derived NPs (Ruzicka et al. 2017). Among these cells, the iPS-NPs provided the most beneficial effect to preserve host tissue, reduce glial scar, increase axonal sprouting, and promote motor functional recovery. These cells robustly survived for at least 4 months and slowly matured. Eight weeks after implantation, grafted iPS-NPs were still immature as they expressed early neural markers such as doublecortin, MAP2, and βIII-tubulin and the astroglial marker GFAP. A small number of cells were positive for oligodendrocytical marker CNPase. Four months after grafting, the cells differentiated mainly into neuronal phenotype; individual cells in the graft were also positive for NF70 and NF200—an advanced marker of neuronal differentiation. Transplanted cells were also able to differentiate toward interneurons (calbindin$^+$), dopaminergic neurons (tyrosine hydroxylase$^+$), serotoninergic neurons (serotonin$^+$), and newly forming motor neurons that were found in the ventral part of the spinal cord (choline acetyltransferase$^+$) (Romanyuk et al. 2015). Grafted cells also differentiate into astrocytes (GFAP$^+$); however, the differentiation was not so robust as described with other sources of NSCs (Amemori et al. 2013). The effect of grafted cells on functional outcome was much faster than the differentiation and maturation. Already 3 weeks after grafting the animals with iPS-NP, their locomotor skills significantly improved, as was detected in the BBB test and walking across the flat beam (Romanyuk et al. 2015; Ruzicka et al. 2017). Therefore, it is evident that strong paracrine effect underlines the locomotor recovery. Therefore, we tested whether these cells can be applied only intrathecally, since the intrathecal application has some advantages over injection into the spinal cord tissue. It eliminates the risk of direct surgical implantation with no need of deep analgesia and anesthesia of the animal and yet still guarantees a wide dissemination of cells through the subarachnoid space and around the lesion site. The cells were injected via lumbar puncture into the subarachnoid space of rats with a balloon-induced spinal cord compression lesion. Applied iPS-NPs did not survive in the spinal cord canal for longer than 2 weeks; however, we observed a moderate effect on locomotor recovery, spared white matter in the lesion center, and increased axonal sprouting (Amemori et al. 2015). Nevertheless, pure paracrine action without direct cell contact is not sufficient to exert a long-term sustainable effect on damaged spinal cord tissue, leading to a robust increase in axonal sprouting, the upregulation of neurotrophic genes, and glial scar reduction. Nakashima's group administered diphtheria toxin to ablate the human-grafted iPS-NPs 7 weeks after injury and observed a loss of improved locomotor function. These results support the role of the transplanted cells in neuronal activity associated with spinal cord tissue regeneration (Fujimoto et al. 2012).

Studies using iPS-NPs in the SCI treatment mentioned above show that grafted cells predominantly differentiate into neurons, less into astrocytes, and rarely into oligodendrocytes. However, remyelination is an important feature in function restoration after SCI. Oligodendrocytes die during secondary injury due to their susceptibility to oxidative stress and glutamate excitotoxicity, leaving demyelinated axons vulnerable to further damage and loss of function. Therefore several protocols predifferentiating iPS cells into oligodendrocyte precursors (OPCs) have been developed. PiggyBac transposon system (https://en.wikipedia.org/wiki/PiggyBac_Transposon_System) used for reprogrammed iPS cells (Woltjen et al. 2009) was directed to NSCs using neurosphere expansion methods (Salewski et al. 2013; Smukler et al. 2006). NSCs derived from wildtype (wt) and nonmyelinating shiverer iPS cell lines were intraspinally grafted into thoracic SCI. Both iPS-NSC lines successfully integrated into the injured spinal cord and predominantly differentiated into oligodendrocytes, but only the wt-iPS-NSC treatment led to remyelination of the axons and resulted in functional improvement (Salewski et al. 2015a). Okano's group used the pre-evaluated safe line described earlier (Nori et al. 2011) and induced their differentiation into oligodendrocyte precursor cell-enriched NPs. These cells can in vitro produce mature MBP^{+} oligodendrocytes and produce growth factors VEGF and PDGF-AA. About 35% of grafted cells differentiated in vivo into oligodendrocytes, which migrated into host white matter and contributed to remyelination of demyelinated axons (Kawabata et al. 2016).

The majority of studies using iPS-NPs for SCI treatment have focused on predifferentiation into neurons or oligodendrocytes. However, astrocytes are also an important part of the nervous tissue maintaining ion homeostasis and governing levels of glutamate via glutamate transporters. Astrocytes, differentiated from iPS and transduced with lentivirus to express glutamate transporter GLT1, were injected into mice and rats with cervical contusion. GLT1-overexpressing astrocytes reduced lesion size within the injured cervical spinal cord and reduced morphological and functional diaphragm denervation (Li et al. 2015).

3.1.5 iPS-NPs in the Treatment of Chronic SCI

At the chronic stage, which can last for years, substantial tissue loss leads to cavity formation at the site of injury. In addition, glial scar formation, which mainly consists of chondroitin sulfate proteoglycans (CSPGs), prevents axonal growth. These conditions generate an unfavorable environment for neural cell survival, resulting in the failure of spinal cord regeneration. Therefore, to enhance the potential of transplanted NPs and recover the locomotor function, it is important to modulate the microenvironment at the chronic phase by reducing the glial scar and/or bridging the cavity with the scaffolds, which serve as cell carriers. Chondroitinase ABC (ChABC) is a bacterially derived enzyme degrading CSPGs and has been successfully used for treatment in animal models of SCI (Bradbury et al. 2002). Seven weeks after cervical spinal cord injury, Suzuki et al. applied via osmotic pump ChABC to create a more permissive environment for transplanted iPS-NPs (Suzuki et al. 2017). Cells were grafted into the spinal cord rostrally and caudally from the lesion center 1 week later. ChABC administration reduced the scar and resulted in significantly improved iPS-NP survival, with clear differentiation into all three neuroglial lineages. Neurons derived from transplanted cells also formed functional synapses with host circuits. Furthermore, the combined treatment based on application of ChABC and iPS-NP transplantation led to recovery in forelimb grip strength and locomotion assessed by catwalk.

To bridge the pseudocyst cavities, which develop during the chronic stage of the SCI, different biomaterials can be used as cell carriers. We have assessed the use of laminin-coated hydrogel with dual porosity, seeded with iPS-NPs, in the treatment of chronic SCI. The iPS-NPs were cultured for 3 weeks in hydrogel in vitro prior to transplantation and were positive for nestin, GFAP, and MAP2. These cell-polymer constructs were implanted into the balloon compression lesion 5 weeks after lesion induction. Spinal cord tissue was immunohistochemically analyzed 4 months later. The implanted iPS-NPs survived in the scaffold for the entire experimental period.

Host axons, astrocytes, and blood vessels grew into the implant, and an increased sprouting of host TH^+ fibers was observed in the lesion vicinity. However, to significantly improve the behavioral recovery of chronically injured animals, further co-therapies, such as infusion of ChABC, might be necessary to further modify the nonpermissive environment of the chronic lesion.

3.1.6 iPS-NPs in the Treatment of Stroke

Of all stroke cases, 87% are ischemic in nature, and the rest are hemorrhagic. In ischemic stroke, a clot occludes a brain vessel (most commonly the middle cerebral artery or its branches), and blood flow to the brain region supplied by that vessel is ceased, causing a cascade of pathological events associated with energy failure, acidosis, excessive glutamate release, elevated intracellular Ca^{2+} levels, generation of free radicals (especially after reperfusion), blood-brain-barrier disruption, inflammation, and eventually massive excitotoxic cell death composed of necrosis, apoptosis, and autophagy (Wei et al. 2017). Hemorrhagic stroke, on the other hand, occurs when a blood vessel ruptures in the brain leading to intracranial hemorrhage. Recent research has focused on developing strategies that facilitate neuroplasticity to maximize functional outcome post stroke. The benefits of exogenous stem cell-based strategies include their potential to rescue damaged or replace dead neurons and to reconstruct injured circuitry with neural progenitors derived from stem cells. Transplanted cells could also act in synergy with endogenous stem cells to exert immunomodulation, neuroprotection, and the stimulation of angiogenesis. One of the first studies used iPS cells mixed in the fibrin glue and grafted subdurally into rats with middle cerebral artery occlusion. Cells significantly decreased the infarct size and improved the functional outcome in rotarod and grasping tasks. Analysis of cytokine expression in cell-treated ischemic brains revealed a significant reduction of pro-inflammatory cytokines and an increase of anti-inflammatory cytokines (Chen et al. 2010). Our study with iPS-NPs proved the dual effect of grafted cells, neuroprotective effects, and reconstruction of impaired pathways. Grafting reversed stroke-induced somatosensory and motor deficits and protected the host substantia nigra (SN) from the atrophy. These effects, together with graft innervation by tyrosine hydroxylase fibers, were early events, occurring weeks after lesion induction. In the long term, grafted cells formed a mixed tissue with host-derived blood vessels and astrocytes. GABAergic striatal neurons formed clusters of fully differentiated hDARPP-32-positive neurons and subpopulations of calretinin-positive interneurons and neurons with type 2 dopaminergic receptors. Moreover, after 2 months, a dense graft-derived axonal network was observed in SN *pars reticulata*, in close association to TH-positive dendrites (Polentes et al. 2012).

For possible transfer of iPS-based cell therapy for stroke patients into clinic, different protocols aiming at safety were tested. Serum-free medium and retinoic acid (Yuan et al. 2013) or vector-free and transgene-free hiPS cells (Mohamad et al. 2013) were used in rodent models of stroke. iPS-NPs expressed mature neuronal markers in vivo, migrated into the lesion area, restored neurovascular coupling, and promoted behavioral recovery after stroke. To eliminate the c/myc and klf4 oncogenes, 6 h after transfection with Sox2 and Oct4 plasmids, mouse embryonic fibroblasts were repeatedly pretreated with hypoxia (3% O_2 for 24 h). iPS cells were differentiated into NPs and grafted into an ischemic stroke mouse model, where differentiation into neurons and astrocytes together with functional improvement was observed (Liu et al. 2014). To increase the cell survival and differentiation after transplantation into ischemic lesion, iPS-NPs were delivered encapsulated in a hyaluronic acid based hydrogel. The hydrogel did not promote cell survival; however, it did increase differentiation into neuroblasts (Lam et al. 2014).

3.1.7 iPS-NPs and Their Translation to Clinical Medicine

Several experimental studies using rodent models of SCI and stroke have shown a positive outcome. However, these results are difficult to simply transfer into clinical settings. We should take into consideration better animal models with a chronic stage of the disease, aged animals, or animals with comorbidities, such as atherosclerosis, diabetes, and hypertension. Regarding the cell preparation, many of the described protocols are too expensive and time-consuming to be tailored for individual patients in the acute or subacute phase. It will take months to years to perform derivation and differentiation and to pass all safety and quality control tests. Therefore allogenic transplantations with a banking system will be required. Kyoto University's Center for iPS Cell Research and Application (CiRA) has produced iPS in clinical-grade quality for a cell bank, and these cells are currently under quality control tests (Okano and Yamanaka, 2014). These iPS cell stocks are homozygous at the three major human leukocyte antigen (HLA) gene loci and match the patient's HLA type (Turner et al. 2013). Based on the progress of basic and preclinical research with iPS-NPs, Okano and Yamanaka laboratories are currently planning iPS-based cell therapy for SCI patients in the subacute phase using clinical-grade integration-free human iPS cell lines. As a further step, iPS cell-based therapy for stroke patients is planned (Okano and Yamanaka 2014).

3.2 Conclusions

Though there are still many unanswered questions regarding the derivation, differentiation, application, safety, and quality control of the iPS-NPs in the clinically relevant models of acute and chronic stroke and SCI, iPS-NPs have to be considered as potentially strong players in the field of regenerative medicine, which are slowly finding their way into clinical medicine.

Acknowledgment This study was supported by the Czech Science Foundation GACR P304/12/G069, Operational Programme Research, Development and Education in the framework of the project "Centre of Reconstructive Neuroscience," registration number CZ.02.1.01/0.0./0.0/15_003/0000419, from National Sustainability Program LO1309 and from CZ.1.05/1.1.00/02.0109 BIOCEV.

References

Amemori T, Romanyuk N, Jendelova P et al (2013) Human conditionally immortalized neural stem cells improve locomotor function after spinal cord injury in the rat. Stem Cell Res Ther 4:68

Amemori T, Ruzicka J, Romanyuk N, Jhanwar-Uniyal M et al (2015) Comparison of intraspinal and intrathecal implantation of induced pluripotent stem cell-derived neural precursors for the treatment of spinal cord injury in rats. Stem Cell Res Ther 6:257

Aoi T, Yae K, Nakagawa M et al (2008) Generation of pluripotent stem cells from adult mouse liver and stomach cells. Science 321:699–702

Bradbury EJ, Moon LD, Popat RJ et al (2002) Chondroitinase ABC promotes functional recovery after spinal cord injury. Nature 416:636–640

Brennand KJ, Simone A, Jou J et al (2011) Modelling schizophrenia using human induced pluripotent stem cells. Nature 473:221–225

Chambers SM, Fasano CA, Papapetrou EP et al (2009) Highly efficient neural conversion of human ES and iPS cells by dual inhibition of SMAD signaling. Nature Biotechnol 27:275–280

Chambers SM, Qi Y, Mica Y et al (2012) Combined small-molecule inhibition accelerates developmental timing and converts human pluripotent stem cells into nociceptors. Nat Biotechnol 30:715–720

Chambers SM, Mica Y, Lee G et al (2013) Dual SMAD inhibition/WNT activation-based methods to induce neural crest and derivatives from human pluripotent stem cells. In: Turksen K (ed) Human embryonic stem cell protocols. Springer, New York, pp 329–343

Chen SJ, Chang CM, Tsai SK et al (2010) Functional improvement of focal cerebral ischemia injury by subdural transplantation of induced pluripotent stem cells with fibrin glue. Stem Cells and Dev 19:1757–1767

Chin MH, Mason MJ, Xie W et al (2009) Induced pluripotent stem cells and embryonic stem cells are distinguished by gene expression signatures. Cell Stem Cell 5:111–123

Cummings BJ, Uchida N, Tamaki SJ et al (2005) Human neural stem cells differentiate and promote locomotor recovery in spinal cord-injured mice. Proc Natl Acad Sci USA 102:14069–14074

Devine MJ, Ryten M, Vodicka P et al (2011) Parkinson's disease induced pluripotent stem cells with triplication of the alpha-synuclein locus. Nature Commun 2:440

Dimos JT, Rodolfa KT, Niakan KK et al (2008) Induced pluripotent stem cells generated from patients with ALS can be differentiated into motor neurons. Science 321:1218–1221

Ebert AD, Yu J, Rose FF Jr et al (2009) Induced pluripotent stem cells from a spinal muscular atrophy patient. Nature 457:277–280

Egawa N, Kitaoka S, Tsukita K et al (2012) Drug screening for ALS using patient-specific induced pluripotent stem cells. Science Transl Med 4:145ra104

Erceg S, Ronaghi M, Stojkovic M (2009) Human embryonic stem cell differentiation toward regional specific neural precursors. Stem Cells 27:78–87

Fujimoto Y, Abematsu M, Falk A et al (2012) Treatment of a mouse model of spinal cord injury by transplantation of human induced pluripotent stem cell-derived long-term self-renewing neuroepithelial-like stem cells. Stem Cells 30:1163–1173

Han DW, Tapia N, Hermann A et al (2012) Direct reprogramming of fibroblasts into neural stem cells by defined factors. Cell Stem Cell 10:465–472

Hooshmand MJ, Sontag CJ, Uchida N et al (2009) Analysis of host-mediated repair mechanisms after human CNS-stem cell transplantation for spinal cord injury: correlation of engraftment with recovery. PloS One 4:e5871

Horiuchi Y, Kano S, Ishizuka K et al (2013) Olfactory cells via nasal biopsy reflect the developing brain in gene expression profiles: utility and limitation of the surrogate tissues in research for brain disorders. Neurosci Res 77:247–250

Hsu YC, Lee DC, Chiu IM (2007) Neural stem cells, neural progenitors, and neurotrophic factors. Cell Transpl 16:133–150

Israel MA, Yuan SH, Bardy C et al (2012) Probing sporadic and familial Alzheimer's disease using induced pluripotent stem cells. Nature 482:216–220

Jensen JB, Parmar M (2006) Strengths and limitations of the neurosphere culture system. Mol Neurobiol 34:153–161

Joannides AJ, Fiore-Heriche C, Battersby AA et al (2007) A scaleable and defined system for generating neural stem cells from human embryonic stem cells. Stem Cells 25:731–737

Kawabata S, Takano M, Numasawa-Kuroiwa Y et al (2016) Grafted human iPS cell-derived oligodendrocyte precursor cells contribute to robust remyelination of demyelinated axons after spinal cord injury. Stem Cell Rep 6:1–8

Kim JB, Sebastiano V, Wu G et al (2009) Oct4-induced pluripotency in adult neural stem cells. Cell 136:411–419

Klussmann S, Martin-Villalba A (2005) Molecular targets in spinal cord injury. J Mol Med (Berl) 83:657–671

Kobayashi Y, Okada Y, Itakura G et al (2012) Pre-evaluated safe human iPSC-derived neural stem cells promote functional recovery after spinal cord injury in common marmoset without tumorigenicity. PloS One 7:e52787

Koch P, Opitz T, Steinbeck JA et al (2009) A rosette-type, self-renewing human ES cell-derived neural stem cell with potential for in vitro instruction and synaptic integration. Proc Natl Acad Sci USA 106:3225–3230

Koch P, Breuer P, Peitz M et al (2011) Excitation-induced ataxin-3 aggregation in neurons from patients with Machado-Joseph disease. Nature 480:543–546

Ku S, Soragni E, Campau E et al (2010) Friedreich's ataxia induced pluripotent stem cells model intergenerational GAATTC triplet repeat instability. Cell Stem Cell 7:631–637

Lam J, Lowry WE, Carmichael ST et al (2014) Delivery of iPS-NPCs to the stroke cavity within a hyaluronic acid matrix promotes the differentiation of transplanted cells. Adv Funct Mater 24: 7053–7062

Lee H, Shamy GA, Elkabetz Y et al (2007) Directed differentiation and transplantation of human embryonic stem cell-derived motoneurons. Stem Cells 25:1931–1939

Li XJ, Du ZW, Zarnowska ED et al (2005) Specification of motoneurons from human embryonic stem cells. Nat Biotechnol 23:215–221

Li K, Javed E, Scura D et al (2015) Human iPS cell-derived astrocyte transplants preserve respiratory function after spinal cord injury. Exp Neurol 271:479–492

Liu SP, Fu RH, Wu DC et al (2014) Mouse-induced pluripotent stem cells generated under hypoxic conditions in the absence of viral infection and oncogenic factors and used for ischemic stroke therapy. Stem Cells and Dev 23:421–433

Lu P, Woodruff G, Wang Y et al (2014) Long-distance axonal growth from human induced pluripotent stem cells after spinal cord injury. Neuron 83:789–796

Lujan E, Chanda S, Ahlenius H et al (2012) Direct conversion of mouse fibroblasts to self-renewing, tripotent neural precursor cells. Proc Natl Acad Sci USA 109:2527–2532

Lukovic D, Diez Lloret A, Stojkovic P et al (2017) Highly efficient neural conversion of human pluripotent stem cells in adherent and animal-free conditions. Stem Cells Transl Med 6:1217–1226

Marchetto MC, Carromeu C, Acab A et al (2010) A model for neural development and treatment of Rett syndrome using human induced pluripotent stem cells. Cell 143:527–539

Mohamad O, Drury-Stewart D, Song M et al (2013) Vector-free and transgene-free human iPS cells differentiate into functional neurons and enhance functional recovery after ischemic stroke in mice. PloS One 8:e64160

Muratore CR, Srikanth P, Callahan DG et al (2014) Comparison and optimization of hiPSC forebrain cortical differentiation protocols. PloS One 9:e105807

Nat R, Nilbratt M, Narkilahti S et al (2007) Neurogenic neuroepithelial and radial glial cells generated from six human embryonic stem cell lines in serum-free suspension and adherent cultures. Glia 55:385–399

Nguyen HX, Nekanti U, Haus DL et al (2014) Induction of early neural precursors and derivation of tripotent neural stem cells from human pluripotent stem cells under xeno-free conditions. J Comp Neurol 522:2767–2783

Nihei Y, Ito D, Okada Y et al (2013) Enhanced aggregation of androgen receptor in induced pluripotent stem cell-derived neurons from spinal and bulbar muscular atrophy. J Biol Chem 288: 8043–8052

Nori S, Okada Y, Yasuda A et al (2011) Grafted human-induced pluripotent stem-cell-derived neurospheres promote motor functional recovery after spinal cord injury in mice. Proc Natl Acad Sci USA 108:16825–16830

Oh J, Lee KI, Kim HT et al (2015) Human-induced pluripotent stem cells generated from intervertebral disc cells improve neurologic functions in spinal cord injury. Stem Cell Res Ther 6:125

Okano H, Yamanaka S (2014) iPS cell technologies: significance and applications to CNS regeneration and disease. Mol Brain 7:22

Polentes J, Jendelova P, Cailleret M et al (2012) Human induced pluripotent stem cells improve stroke outcome and reduce secondary degeneration in the recipient brain. Cell Transpl 21: 2587–2602

Ring KL, Tong LM, Balestra ME et al (2012) Direct reprogramming of mouse and human fibroblasts into multipotent neural stem cells with a single factor. Cell Stem Cell 11:100–109

Roese-Koerner B, Stappert L, Koch P et al (2013) Pluripotent stem cell-derived somatic stem cells as tool to study the role of microRNAs in early human neural development. Curr Mol Med 13: 707–722

Romanyuk N, Amemori T, Turnovcova K et al (2015) Beneficial effect of human induced pluripotent stem cell-derived neural precursors in spinal cord injury repair. Cell Transplant 24: 1781–1797

Ruzicka J, Machova-Urdzikova L, Gillick J et al (2017) A comparative study of three different types of stem cells for treatment of rat spinal cord injury. Cell Transplant 26:585–603

Salewski RP, Buttigieg J, Mitchell RA et al (2013) The generation of definitive neural stem cells from PiggyBac transposon-induced pluripotent stem cells can be enhanced by induction of the NOTCH signaling pathway. Stem Cells and Dev 22:383–396

Salewski RP, Mitchell RA, Li L et al (2015a) Transplantation of induced pluripotent stem cell-derived neural stem cells mediate functional recovery following thoracic spinal cord injury through remyelination of axons. Stem Cells Transl Med 4:743–754

Salewski RP, Mitchell RA, Shen C et al (2015b) Transplantation of neural stem cells clonally derived from embryonic stem cells promotes recovery after murine spinal cord injury. Stem Cells and Dev 24:36–50

Sareen D, Gowing G, Sahabian A et al (2014) Human induced pluripotent stem cells are a novel source of neural progenitor cells (iNPCs) that migrate and integrate in the rodent spinal cord. J Comp Neurol 522:2707–2728

Sharp J, Frame J, Siegenthaler M et al (2010) Human embryonic stem cell-derived oligodendrocyte progenitor cell transplants improve recovery after cervical spinal cord injury. Stem Cells 28: 152–163

Shin S, Mitalipova M, Noggle S et al (2006) Long-term proliferation of human embryonic stem cell-derived neuroepithelial cells using defined adherent culture conditions. Stem Cells 24: 125–138

Smukler SR, Runciman SB, Xu S et al (2006) Embryonic stem cells assume a primitive neural stem cell fate in the absence of extrinsic influences. J Cell Biol 172:79–90
Suzuki H, Ahuja CS, Salewski RP et al (2017) Neural stem cell mediated recovery is enhanced by Chondroitinase ABC pretreatment in chronic cervical spinal cord injury. PloS One 12:e0182339
Takahashi K, Yamanaka S (2006) Induction of pluripotent stem cells from mouse embryonic and adult fibroblast cultures by defined factors. Cell 126:663–676
Tsuji O, Miura K, Okada Y et al (2010) Therapeutic potential of appropriately evaluated safe-induced pluripotent stem cells for spinal cord injury. Proc Natl Acad Sci USA 107: 12704–12709
Turner M, Leslie S, Martin NG et al (2013) Toward the development of a global induced pluripotent stem cell library. Cell Stem Cell 13:382–384
Urbach A, Bar-Nur O, Daley GQ et al (2010) Differential modeling of fragile X syndrome by human embryonic stem cells and induced pluripotent stem cells. Cell Stem Cell 6:407–411
Volarevic V, Erceg S, Bhattacharya SS et al (2013) Stem cell-based therapy for spinal cord injury. Cell Transpl 22:1309–1323
Wang H, Fang H, Dai J et al (2013) Induced pluripotent stem cells for spinal cord injury therapy: current status and perspective. Neurol Sci 34:11–17
Wei L, Wei ZZ, Jiang MQ et al (2017) Stem cell transplantation therapy for multifaceted therapeutic benefits after stroke. Prog Neurobiol 157:49–78
Woltjen K, Michael IP, Mohseni P et al (2009) piggyBac transposition reprograms fibroblasts to induced pluripotent stem cells. Nature 458:766–770
Yagi T, Ito D, Okada Y et al (2011) Modeling familial Alzheimer's disease with induced pluripotent stem cells. Human Mol Gen 20:4530–4539
Yan J, Welsh AM, Bora SH et al (2004) Differentiation and tropic/trophic effects of exogenous neural precursors in the adult spinal cord. J Comp Neurol 480:101–114
Yan J, Xu L, Welsh AM, Hatfield G et al (2007) Extensive neuronal differentiation of human neural stem cell grafts in adult rat spinal cord. PLoS Med 4:e39
Yang JR, Liao CH, Pang CY et al (2013) Transplantation of porcine embryonic stem cells and their derived neuronal progenitors in a spinal cord injury rat model. Cytotherapy 15:201–208
Yuan T, Liao W, Feng NH et al (2013) Human induced pluripotent stem cell-derived neural stem cells survive, migrate, differentiate, and improve neurologic function in a rat model of middle cerebral artery occlusion. Stem Cell Res Ther 4:73
Zhang SC (2006) Neural subtype specification from embryonic stem cells. Brain Pathol 16: 132–142
Zhang SC, Wernig M, Duncan ID et al (2001) In vitro differentiation of transplantable neural precursors from human embryonic stem cells. Nat Biotechnol 19:1129–1133
Zhang N, An MC, Montoro D et al (2010) Characterization of human Huntington's disease cell model from induced pluripotent stem cells. PLoS Curr 2:RRN1193
Zou Q, Yan Q, Zhong J et al (2014) Direct conversion of human fibroblasts into neuronal restricted progenitors. J Biol Chem 289:5250–5260

Chapter 4
Generation of Human Neural Stem Cells by Direct Phenotypic Conversion

Daekee Kwon, Hee-Jin Ahn, and Kyung-Sun Kang

Abstract Human neural stem cells (hNSC) are multipotent adult stem cells. Various studies are underway worldwide to identify new methods for treatment of neurological diseases using hNSC. This chapter summarizes the latest research trends in and fields for application of patient-specific hNSC using direct phenotypic conversion technology. The aim of the study was to analyze the advantages and disadvantages of current technology and to suggest relevant directions for future hNSC research.

Keywords Cell replacement therapy · Direct phenotypic conversion · Disease modeling · Drug screening · Induced neural stem cells · Reprogramming

4.1 Introduction

4.1.1 Neural Stem Cells

Neural stem cells (NSC) are adult stem cells that are present in the nervous system (Ma et al. 2009). These cells have self-renewal capacity for symmetric division into undifferentiated states (Homem et al. 2015). However, unlike pluripotent embryonic stem cells (ESC) or induced pluripotent stem cells (iPSC), they exhibit limited self-renewal potential (Calzolari et al. 2015). In addition, NSC are multipotent and can differentiate into neurons and glia, the cells that constitute the nervous system.

D. Kwon · H.-J. Ahn
Stem Cells and Regenerative Bioengineering Institute in Kangstem Biotech, Seoul National University, Seoul, South Korea

K.-S. Kang (✉)
Stem Cells and Regenerative Bioengineering Institute in Kangstem Biotech, Seoul National University, Seoul, South Korea

Adult Stem Cell Research Center, College of Veterinary Medicine, Seoul National University, Seoul, South Korea
e-mail: kangpub@snu.ac.kr

L. Buzanska (ed.), *Human Neural Stem Cells*, Results and Problems in Cell Differentiation 66, https://doi.org/10.1007/978-3-319-93485-3_4

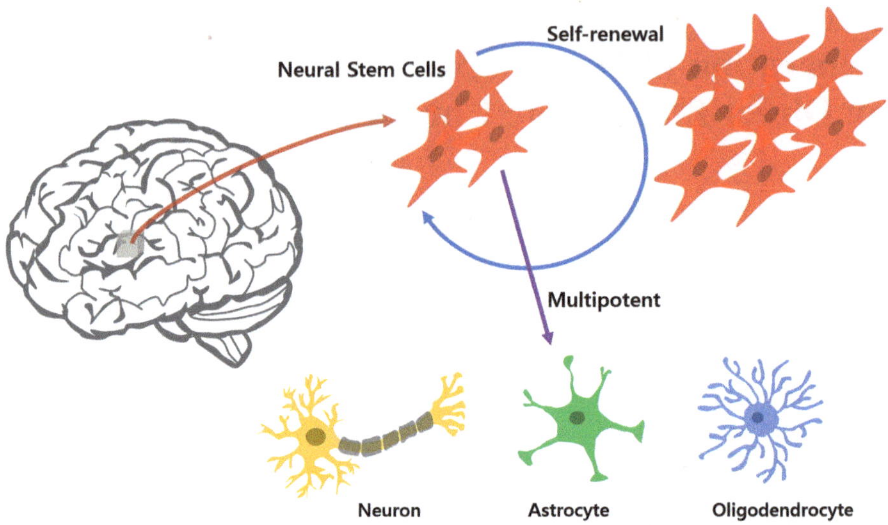

Fig. 4.1 Neural stem cells (NSC). NSC was first obtained by in vitro isolated culture of the subventricular zone of the fetal brain. These cells are characterized by self-renewality and multipotency. Given proper stimulation, NSC can differentiate into neurons and glia (astrocyte and oligodendrocyte)

Traditionally, NSC have been isolated from fetal brain tissue for in vitro culture. In 1989, multipotent stem cells with the capacity for self-renewal were identified in the subventricular zone (SVZ) of the mouse brain (Temple 1989). In 1992, NSC were isolated and cultured from the striatum of the mouse brain for the first time (Reynolds and Weiss 1992). Subsequently, NSC have been isolated and cultured not only from the brain tissue of various species, including humans, but also from the spinal cord (Taupin and Gage 2002).

The key features of NSC are their capacity for self-renewal and their multipotency (Fig. 4.1). In general, NSC undergo symmetric division, with a population doubling time of approximately 24–48 h per cell line (Kim et al. 2008a, b). Under appropriate induction conditions, they can also differentiate into neurons, oligodendrocytes, and astrocytes, which are typical cells that constitute the nervous system (Gage 2000). Neurons are cells that carry electrical signals from the distal to the central nervous system. Oligodendrocyte acts as an insulator when electrical signals are transmitted through neuron. Astrocytes play a role in supporting the survival and functioning of neurons and oligodendrocytes. Oligodendrocytes and astrocytes are collectively referred to as glia because they support the function of neurons. NSC can be cultured as adherent cell cultures. Interestingly, unlike other types of mammalian cells, NSC can be also cultured as a spherical cell mass in suspension, called a neurosphere culture (Reynolds and Rietze 2005).

Following the successful isolation and culture of NSC from the fetal brain, they have been studied intensively for use in cell replacement therapy for treatment of neurological diseases. For example, NSC have been studied for treatment of spinal cord injuries and Parkinson's disease (Han et al. 2015; Lee et al. 2009; Mothe and Tator 2013; Pardal and Lopez-Barneo 2012). However, the therapeutic effect of the cells was not sufficient, and the reason for their ineffectiveness remains unclear. However, the cells used in the experiment were isolated and cultured from fetal brain tissue, and issues such as immune rejection after transplantation may have affected the results. Therefore, the acquisition of patient-specific NSC for use in cell replacement therapy for neurological diseases remains an important issue.

4.1.2 Reprogramming Technology

Among the reprogramming technologies to secure patient-specific NSC, the technique that has been studied for the longest is somatic cell nuclear transfer (SCNT) (Gurdon 1960, 1962; Gurdon et al. 1958; Wilmut et al. 1997). SCNT technology can be classified into two types, depending on the target: reproductive cloning and therapeutic cloning (Yang et al. 2007). The purpose of reproductive cloning is to produce clones replicated through SCNT. The purpose of therapeutic cloning is to produce patient-specific pluripotent ESC through SCNT. Pluripotent ESC were first isolated and cultured from embryos derived from mouse and human in vitro fertilization (IVF) (Evans and Kaufman 1981; Thomson et al. 1998). This ESC isolation/culture technique using IVF-derived embryo was used to establish patient-specific ESC via fusion with SCNT technology (Chung et al. 2014; Tachibana et al. 2013; Wakayama et al. 2001). The convergence of SCNT and ESC technologies solved the post-transplantation immune rejection problem, which was a major disadvantage of ESC from IVF embryos. Since SCNT-ESC are pluripotent, they have the potential to differentiate into all cell types, including NSC (Fig. 4.2).

Successful differentiation of IVF-ESC into NSC has been reported (Reubinoff et al. 2001; Zhang et al. 2001). In order to differentiate ESC into NSC, the ESC colonies from adherent culture were detached, and embryoid bodies were formed in a suspension culture. Then, an appropriate growth factor (Vitamin A etc.) was added to induce differentiation into NSC. NSC produced using this process have been confirmed to have NSC-specific self-renewal capacity and multipotency. The protocol used to differentiate IVF-ESC into NSC has been applied to SCNT-ESC for differentiation into NSC. NSC generation using SCNT technology has the advantage of not requiring extra genetic material, but the ethical barrier remains high, because use of human oocytes is required, greatly decreasing its potential for use in the field of regenerative medicine (Kfoury 2007; Table 4.1).

SCNT technology is difficult to apply to regenerative medicine, due to the ethical considerations surrounding use of human oocytes and the technical difficulty. iPSC technology using a defined transcription factor was developed to overcome the ethical issues of SCNT technology (Takahashi et al. 2007; Takahashi and Yamanaka 2006; Yu et al. 2007; Fig. 4.2 and Table 4.1). It was confirmed that somatic cells were reprogrammed to iPSC almost similar to ESC when four transcription factors of

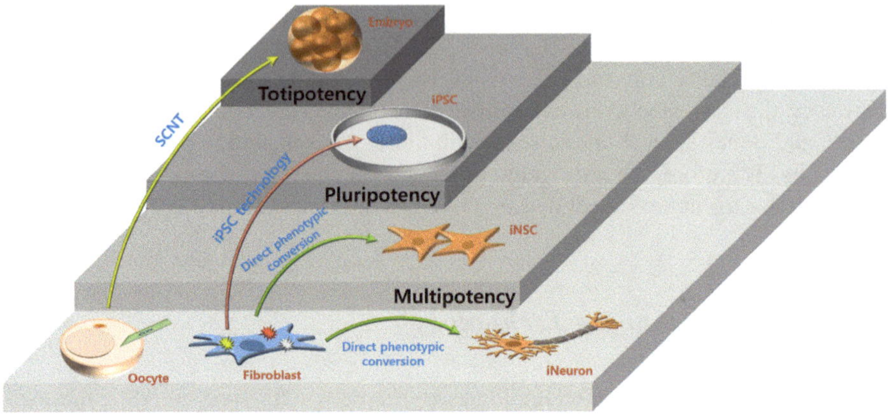

Fig. 4.2 Diverse reprogramming technologies. Somatic cell nuclear transfer (SCNT) is characterized by the introduction of somatic cell nuclei into enucleated oocytes, and somatic cell nuclei are reprogrammed to the totipotency stage. Induced pluripotent stem cell (iPSC) technology is characterized by introducing a defined transcription factor that induces reprogramming in somatic cells, and somatic nuclei are reprogrammed to the pluripotency stage. Direct phenotypic conversion, like iPSC technology, is characterized by the introduction of defined transcription factors into somatic cells, and conversion is possible to various lineages of cells depending on the type of transcription factor introduced

Table 4.1 Comparison of SCNT, iPSC technology, and direct phenotypic conversion

	SCNT	iPSC technology	Direct phenotypic conversion	References
Major pros	– No immune rejection	– No ethical problems – Use of defined factors – No immune rejection	– No ethical problems – Use of defined factors – No immune rejection – No tumorigenesis	Cheng et al. (2014), Ring et al. (2012), Tachibana et al. (2013), Takahashi et al. (2007), Yu et al. (2015)
Major cons	– Ethical problems – Use of undefined factors – Tumorigenesis	– Tumorigenesis	– Insufficient understanding of molecular mechanism	

OCT4, *KLF4*, *SOX2*, and *c-MYC* (*OKSM*) were introduced into somatic cells using retrovirus (Takahashi et al. 2007; Takahashi and Yamanaka 2006). Then, in order to overcome the problem that the *OKSM* gene used for reprogramming is integrated into the genome and threatens genetic integrity of the cells, techniques for producing iPSC using mRNA, miRNA, protein, etc., which are not integrated in the genome, have been developed in succession (Anokye-Danso et al. 2011; Kim et al. 2009; Nemes et al. 2014; Warren et al. 2010; Zhou et al. 2009). The iPSC technology has

overcome the ethical problems of SCNT and has become a universal reprogramming technique because its technical difficulty is relatively low; it has therefore been reproduced in many research groups around the world. Dr. Shinya Yamanaka was awarded the Nobel Prize in 2012 for his efforts in developing iPSC generation technology.

Like ESC, successful differentiation of iPSC to NSC was reported (Yuan et al. 2013). Because iPSC has the same basic cell character as ESC according to morphology, marker expression, global gene expression profile, self-renewality and pluripotency. So the protocol used to differentiate ESC to NSC could be applied almost identically in case of iPSC. Also because iPSC technology does not use human oocyte, the biggest problem of SCNT, the ethical problem, is completely overcome. However, when reprogramming into the iPSC stage, the cells undergo a pluripotency stage, and there is always the possibility that some cells may remain undifferentiated without reacting to signals when differentiated into NSC. Therefore, there is still a possibility that teratoma may occur when iPSC-derived NSC are transplanted in vivo, so it is still a concern for use in clinical trials (Gao et al. 2016).

Development of a technique for introducing a defined transcription factor into somatic cells, producing iPSC with characteristics similar to ESC, has stimulated development of direct phenotypic conversion technology. In fact, the first direct phenotypic conversion technology was developed in the 1980s, long before iPSC reprogramming technology was developed (Davis et al. 1987; Tapscott et al. 1988). In this case, the *myoD* gene was introduced into somatic cells, and the skin cells were successfully converted into muscle cells by direct phenotypic conversion. However, gene cloning was difficult at the time, and few researchers were working on direct phenotypic conversion technology. Following establishment of iPSC technology using defined transcription factors, direct phenotypic conversion to neurons was reported by introducing *Neuro D* into somatic cells, and direct phenotypic conversion to various cell types has subsequently been reported (Chanda et al. 2014; Jiang et al. 2015; Pang et al. 2011; Pfisterer et al. 2011; Yamamoto et al. 2015; Fig. 4.2). Direct phenotypic conversion bypasses the pluripotent state, unlike SCNT or iPSC technology, and therefore has the advantage that there is little possibility of teratoma formation following in vivo transplantation (Kelaini et al. 2014).

Using this direct phenotypic conversion technique, induced neural stem cells (iNSC) were produced from human somatic cells (Ring et al. 2012). iNSC were obtained by introducing *SOX2* into somatic cells using a viral vector and culturing the cells in NSC culture conditions. The iNSC produced had the capacity for self-renewal and were multipotent, and neurons differentiated from these iNSC were shown to be functional neurons in electrophysiological experiments. Interestingly, since the pluripotent state was bypassed in the generation of iNSC, the cells did not form a teratoma when transplanted in vivo. For iNSC, there is also no immune rejection in vivo transplantation, and the ethical problem is completely ruled out, which means it is the most suitable cell replacement therapy for neurological diseases (Table 4.1).

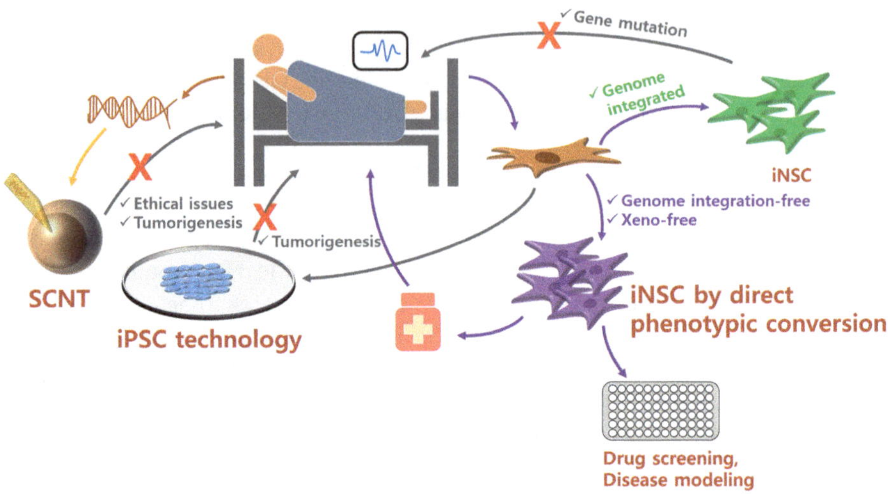

Scheme 4.1 Summary of this chapter. This chapter reviewed the various reprogramming techniques (SCNT, iPSC technology, and direct phenotypic conversion) and patient-specific NSC generation using this. Among them, status of development of patient-specific iNSC using direct phenotypic conversion technology, application fields, advantages and shortcomings, and future directions of development were reviewed in detail

4.1.3 Purpose

As described above, direct phenotypic conversion is considered to be the most suitable reprogramming technology for regenerative medicine of nervous system diseases, beyond SCNT and iPSC technology. This chapter will summarize the latest trends and fields for application of patient-specific iNSC research using the direct phenotypic conversion technology (Scheme 4.1). It will present the advantages and disadvantages of the current direct phenotypic conversion technology and potential future directions for human iNSC generation research (Fig. 4.3).

4.2 Generation of iNSC

4.2.1 Generation of Genome-Integrated iNSC

Studies of direct phenotypic conversion from somatic cells to iNSC were first initiated using somatic cells from animal models such as mice. In the early days, as in the case of iPSC technology, studies were mainly conducted using mouse embryonic fibroblasts (MEF) and a genome-integrating viral vector system (Han et al. 2012; Kim et al. 2011, 2014; Thier et al. 2012). The viral vector system allowed easy overexpression of the gene and enabled efficient reprogramming (Rao and Malik 2012). In general, direct phenotypic conversion into iNSC is achieved by

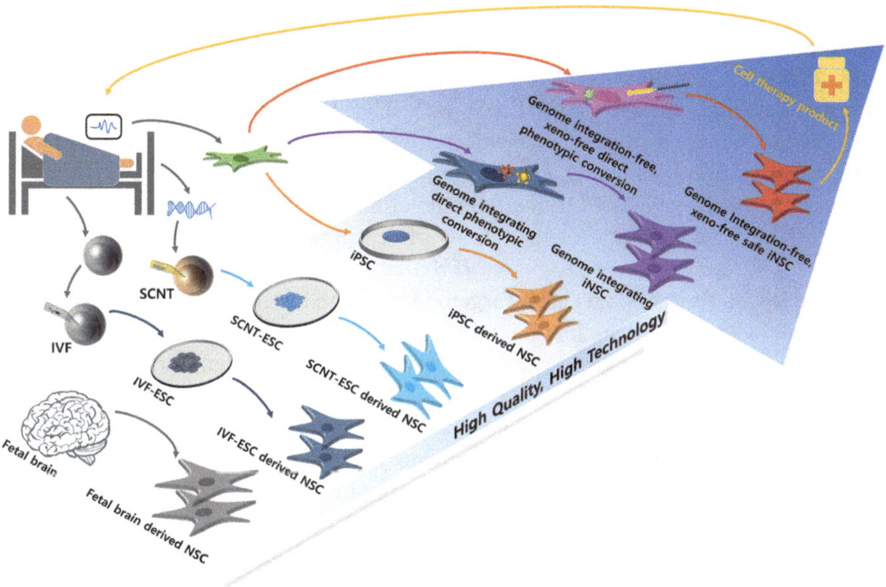

Fig. 4.3 Development direction of patient-specific NSC generation method. The first NSC was obtained by in vitro culture of the SVZ of the fetal brain. Subsequently, based on in vitro fertilization (IVF) embryo-derived ESC-based protocol to differentiate into NSC, patient-specific NSC derived from SCNT and iPSC were produced. Recently, a method for producing patient-specific iNSC that have been directly phenotypically converted from somatic cells using genome-integrating retrovirus or lentivirus has been developed. Currently, studies are underway to produce safe genome integration-free iNSC using small molecule, mRNA, episomal vector, and genome integration-free virus. Genome integration-free and xeno-free system development will accelerate the application of regenerative medicine for neurological diseases of iNSC derived from direct phenotypic conversion

introducing *OKSM*, which was used in the initial iPSC reprogramming study in MEFs, into cells using a viral vector system and culturing the transduced cells in NSC culture conditions (Kim et al. 2011). Interestingly, the iNSC produced by the introduction of the *OKSM* gene did not form teratoma after in vivo transplantation unlike iPSC. This may indicate that the pluripotent state was bypassed, resulting in direct phenotypic conversion into the targeted iNSC (Kelaini et al. 2014). Early studies using *OKSM* have progressed in the direction of decreasing the number of reprogramming factors, and direct phenotypic conversion from MEF to iNSC has been successful with the introduction of only one gene—*SOX2* (Ring et al. 2012). It has also been reported that iNSC can be produced using somatic cells from other animals as rats in addition to mice (Xi et al. 2013). Early studies on the generation of iNSC using animal somatic cells and genome-integrating viral vector system formed a strong foundation for the stable application of direct phenotypic conversion technology to human somatic cells.

Direct phenotypic conversion of human somatic cells to iNSC has been initiated based on the results of a study that showed that direct phenotypic conversion to iNSC could be achieved by gene transfer using a genome-integrating viral vector system in animal somatic cells (Table 4.2). The direct phenotypic conversion to iNSC using human somatic cells was achieved by introducing *SOX2* gene into human dermal fibroblast using a retroviral vector system (Ring et al. 2012). Various studies have been carried out to increase the efficiency of direct phenotypic conversion to iNSC using various combinations of reprogramming factors in addition to *SOX2*. As a representative example, *HMGA2* can be used, which expression is inhibited by let-7b in human cells. Targeted inhibition of let-7b resulted in increased expression of *HMGA2*, which is linked to increased efficiency and kinetics of iNSC direct phenotypic conversion. Therefore, the efficiency and kinetics of direct phenotypic conversion to iNSC can be dramatically improved by simultaneously introducing *SOX2* and *HMGA2* retroviral vector system in human somatic cells (Yu et al. 2015). The study of iNSC generation using the genome-integrating viral vector system has the purpose of introducing the minimum number of genes into the cell to obtain maximum reprogramming efficiency and kinetics. However, the resulting genome-integrating iNSC clearly has limitations that it is difficult to use the cells for clinical trial purposes because of safety issues.

4.2.2 Generation of Genome Integration-Free iNSC

When iNSC is produced using a genome-integrating viral system, viral vectors are inevitably inserted into the genome (Hindmarsh and Leis 1999; Marini et al. 2015). If a viral vector is inserted into tumor suppressor such as *p53* or proto-oncogene such as *MYC*, which regulates cellular homeostasis, the resulting mutations may make the iNSC unfit for use (Lane et al. 2010). This is an important factor that threatens the safety of clinical trials using iNSC. Therefore, in existing iPSC technology, although the initial proof-of-principle experiment used the genome-integrating viral system, the trend is shifting toward the generation of genome integration-free iPSC using mRNA, miRNA, and protein. Recent studies on the generation of iNSC using MEF have also demonstrated that direct treatment of nine small molecules, including signal inhibitors with target various cellular mechanisms, can lead to direct phenotypic conversion to iNSC (Zhang et al. 2016). In this study, genome integration-free iNSC was derived by treating MEF with glycogen synthase kinase (GSK) inhibitor, CHIR99021; BMP type 1 receptor ALK2/3 inhibitor, LDN193189; transforming growth factor-b (TGF-b) type 1 receptor ALK 4/5/7 inhibitor, A83-01; DNA methyltransferase inhibitor, RG108; and lysine-specific demethylase 1, monoamine oxidase inhibitor, parnate, retinoic acid, and bFGF. This study provided valuable information on the mechanism of direct phenotypic conversion and suggested the possibility of efficient generation of genome integration-free iNSC using human somatic cells.

Table 4.2 List of human iNSC generation

Starting cells	Reprogramming factors	Delivery system	Genome integration	Efficiency	References
Adult fibroblasts UCBMSC Blood CD34+ cells	*SOX2* and *HMGA2*	Retrovirus	Yes	~0.6%	Yu et al. (2015)
Fetal fibroblasts	*SOX2*	Retrovirus	Yes	N/A	Ring et al. (2012)
Adult fibroblasts	*OCT4, KLF4, SOX2,* and *c-MYC*	Retrovirus	Yes	~0.004%	Matsui et al. (2012)
BJ fibroblasts	*OCT4, KLF4, SOX2,* and *ZIC3*	Retrovirus	Yes	~0.035%	Kumar et al. (2012)
Neonatal fibroblasts Adult ASC	*OCT4, KLF4, SOX2,* and *c-MYC*	Lentivirus	Yes	~0.07%	Cairns et al. (2016)
Fibroblasts	*SOX2*	Lentivirus	Yes	N/A	Bagó et al. (2017)
BJ fibroblasts	*OCT4, KLF4, SOX2, L-MYC,* and *NANOG*	Lentivirus	Yes	N/A	Miura et al. (2015)
Fibroblast	*OCT4, KLF4, SOX2, L-MYC, LIN28* with small hairpin directed against *P53*	Epstein-Barr virus (EBV)-derived oriP/EBNA1	No	~0.04%	Capetian et al. (2016)
Urine cells	*OCT4, KLF4, SOX2, SV40LT,* and microRNA cluster *MIR302-367*	oriP/EBNA episomal vectors	No	~0.2%	Wang et al. (2013)
Adult peripheral blood mononuclear cells	*OCT4, KLF4, SOX2, c-MYC, NANOG,* and *LIN28*	oriP/EBNA1-based episomal vector	No	~0.00015%	Tang et al. (2016)
Adult peripheral blood hematopoietic progenitor cells	*OCT4, KLF4, SOX2,* and *c-MYC*	Sendai virus	No	N/A	Wang et al. (2015)
Urinary cells	Small molecule cocktail (GSK-3 kinase inhibitor, HDAC inhibitor, and TGF-b inhibitor)	Medium additive	No	N/A	Cheng et al. (2014)

ASC adipose-derived stem cells, *GSK-3* glycogen synthase kinase-3, *HDAC* histone deacetylase, *HMGA2* high-mobility group AT-hook 2, *N/A* not available, *TGF-b* transforming growth factor-b, *UCBMSC* umbilical cord blood-derived mesenchymal stem cells, *ZIC3* zinc-finger transcription factor in cerebellum-3

A Chinese research team has also generated genome integration-free iNSC using human somatic cells (Cheng et al. 2014; Table 4.2). In this study, direct phenotypic conversion to iNSC was achieved by treatment of human urine cells with cocktail of three signal inhibitors (histone deacetylase (HDAC) inhibitor, GSK-3 kinase inhibitor, and TGF-b inhibitor). Interestingly, in normoxia conditions, direct phenotypic conversion efficiency was very low, whereas in hypoxia conditions, the efficiency was dramatically increased. This is consistent with a previous iPSC reprogramming study that reported the efficiency of iPSC generation was significantly higher in hypoxia conditions than in normoxia conditions (Yoshida et al. 2009). Genome integration-free iNSC generation research possesses a great advantage in obtaining a safe iNSC, although the efficiency is lower than that of genome-integrating iNSC using existing viruses.

4.2.3 Rationale of Direct Phenotypic Conversion

Biochemical or biophysical microenvironments have been used as a tool to mimic gene expression of NSC through direct phenotypic conversion. Typical examples of a biochemical microenvironment are transcription factors and signal inhibitors. A typical example of a biophysical microenvironment is the extracellular matrix.

The transcription factor binds to the promoter region of a gene and recruits several proteins necessary for transcription by RNA polymerase II to the promoter region, to enhance transcription (Dignam et al. 1983; Orphanides et al. 1996). The principle of general direct phenotypic conversion is that an exogenous master transcription factor binds to a specific promoter on the genome and recruits transcriptional co-regulators to induce conversion to the target cells. The *SOX2* transcription factor is one of three members of the *SOXB1* subgroup (Boyer et al. 2005). *SOX2* plays a role in maintaining pluripotency with *OCT4* and occasionally plays a role in germ layer fate selection (Thomson et al. 2011). In mouse ESC, *SOX2* inhibits mesendodermal differentiation and induces neuroectodermal differentiation. When *SOX2* gene is introduced into ESC and overexpressed, consequently overexpression of *SOX21* is induced within 3 h. Then, *SOX21* specifically induces expression of the neuroectodermal lineage genes *TAPA1*, *ATBF1*, *NEUROD1*, *MASH1*, *HES1*, *HES6*, and *ID2* and specifically decreases *NANOG* and *SALL4* (Maucksch et al. 2013). Interestingly, *OCT4* plays the opposite role to *SOX2*. Thus, it can be said that *SOX2* is the master regulatory gene located at the top of the signal cascade of NSC. So, in general, genes that can code for the *SOX2* transcription factor are commonly used to induce an endogenous neural program in somatic cells (Maucksch et al. 2013).

Various signal inhibitor cocktails are used to mimic NSC transcriptome and proteome through direct phenotypic conversion process. The most commonly used signal inhibitors are the TGF-b inhibitor, the GSK-3 kinase inhibitor, and the HDAC inhibitor. TGF-b inhibitors induce cells into neural ectoderm lineage by inhibiting mesoderm and endoderm specification (Chambers et al. 2009; Smith and Harland 1992). TGF-b signal inhibitors include Repsox, SB431542, and Tranilast. The GSK-3 kinase inhibitor plays a role in favoring neural development with bFGF (Li et al. 2011a, b). Examples of GSK-3 kinase signal inhibitors include CHIR99021, lithium carbonate, and lithium chloride. HDAC inhibitor loosen the DNA-histone complex to facilitate DNA access to the reprogramming factor. Examples of HDAC inhibitors include sodium butyrate, trichostatin A, and valproic acid (VPA). In summary, the signal inhibitors used to induce direct phenotypic conversion to iNSC have the role of specifying the fate of cells to ectoderm, and maintaining an advantageous form of the DNA-histone complex, facilitating expression of reprogramming-related gene. In fact, when the signal inhibitor cocktail was processed for direct phenotypic conversion to iNSC, the intermediate genes *Elk1* and *Gli2*, which determine the neural identity under *Sox2*, were activated, and it was confirmed that many downstream neural linage genes were activated (Zhang et al. 2016).

In addition to the biochemical microenvironment, which acts within the cell to determine the direct phenotypic conversion of cells, there is also a biophysical microenvironment that acts physically outside the cell and affects direct phenotypic conversion. The extracellular matrix (ECM), which is in direct contact with the exterior of cells, can provide biophysical stimulation to the cell through the cell membrane proteins, e.g., integrins, which have a decisive influence on cell adhesion, proliferation, and gene expression (Choi et al. 1992; Roskelley et al. 1995). Generally, when establishing iNSC, the proteins laminin or fibronectin (ligands of integrin a6b1, which is expressed in NSC) are used to coat the surface of culture dishes, to increase cell adhesion (Campos 2005; Flanagan et al. 2006; Hall et al. 2006; Prowse et al. 2011). Recently, it has been reported that reprogramming efficiency is significantly improved by using a nanogrooved substrate pattern for direct phenotypic conversion to dopaminergic neuron (Yoo et al. 2015). Nanogrooved substrate pattern used in direct phenotypic conversion increases the level of histone H3 tri-methylation at lysin 4 (H3K4me3), which is known to be involved in transcription activation by biophysical stimulation of cells; it is considered that the reprogramming efficiency was increased by inducing mesenchymal to epithelial transition (Li et al. 2011a, b). In another study, in direct phenotypic conversion to neuron, reprogramming efficiency was increased by culturing cells on a soft substrate (www.bmes.org/userfiles/uploads/Song_Li.pdf). The increase of direct phenotypic conversion efficiency by stiffness seems to involve various signaling pathways such as cytoskeleton and nuclear matrix. Direct phenotypic conversion of cells by biophysical stimulation is in a very early stage of research that has not yet been reported in the field of iNSC generation, and much further study is needed.

In summary, biochemical/biophysical microenvironments have different starting points, but ultimately achieve NSC-specific gene expression, so the final goal is the same. Direct phenotypic conversion into iNSC can be achieved using various initial triggers (transcription factor, signal inhibitor, extracellular matrix, etc.), and it can be summarized as the process of inducing neural identity in somatic cells by inducing cascading expression of genes that determine downstream neural identity through the activation of endogenous *SOX2*, the master neural regulator (Fig. 4.4).

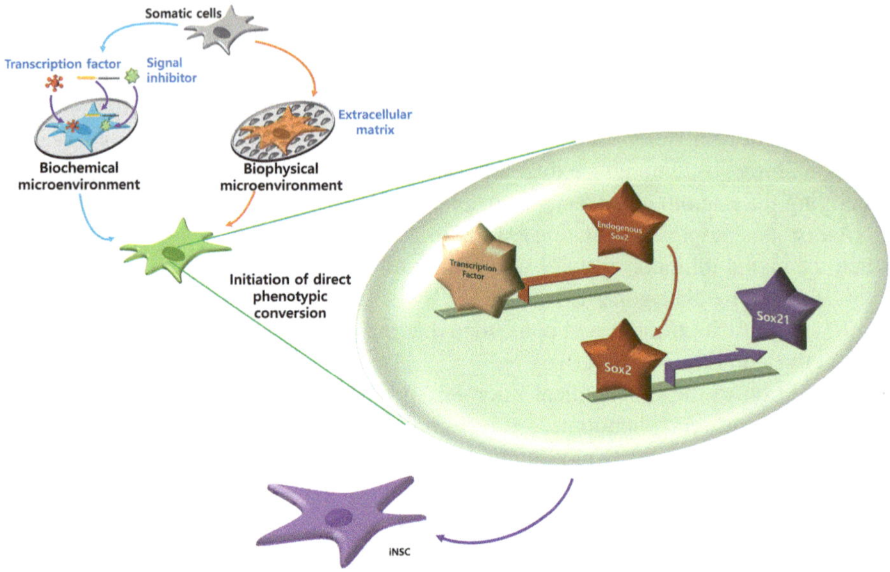

Fig. 4.4 Various factors affecting direct phenotypic conversion. Factors that affect direct phenotypic conversion from somatic cells to iNSC include biochemical microenvironments such as transcription factors and signal inhibitors and biophysical microenvironments such as extracellular matrix. The important point is that the endogenous *SOX2* gene of the NSC master regulator gene is overexpressed by the biochemical/biophysical microenvironment and that the NSC-specific genes are sequentially overexpressed, thereby giving neural identity

4.3 Application Fields of iNSC

4.3.1 Application Fields of Genome-Integrating iNSC

Genome-integrating iNSC can be used in vitro for drug screening for neurological diseases. In general, when administering a medicament for treating a neurological disease to a patient, it is often dependent on a physician's knowledge and experience. However, this approach is unlikely to deliver the best drug with the least side effects while giving the best effect to the patient. This is because individual patients have polymorphisms that are microscopic differences on the genome, and their reactivity to drugs may be different (Baudin 2000; Weinshilboum 2003; Wilkinson 2005). If variety of drugs that can be administered to patient-specific iNSC or differentiated neurons or glia can be tested in vitro in advance, the possibility of selecting the optimal drug for the patient can be greatly increased. For example, the NSC of Niemann-Pick type C (NPC) disease mice, which have mutations in the intracellular *NPC1* gene causing problems with cholesterol metabolism, have defects in cholesterol homeostasis and neural differentiation. However, it has been shown that NSC derived from NPC mice can be treated with HDAC inhibitor VPA in vitro to alter intracellular cholesterol homeostasis and neural differentiation levels close to normal (Kim et al. 2007).

Genome-integrating iNSC can also be used for in vitro disease modeling of neurological diseases. Frequently, direct phenotypic conversion of easy accessible cells from patients with mutations in specific genes allows to observe disease phenotypes in the reprogrammed cells (Hu et al. 2015). The study of these cells allows disease modeling that examines the mechanisms involved in initiation and progression of disease in vitro. For example, in vitro study of NSC derived from mice model of NPC disease has been reported (Kim et al. 2008a, b). NSC from NPC mice showed very high levels of nitric oxide (NO), which was found to be associated with reduced self-renewal of NSC. This could provide a clue that treatment of NPC disease may be possible through control of NO production levels. Therefore, this approach allows iNSC technology to be useful in identifying the disease mechanism of patients with neurological diseases at the cellular level.

4.3.2 Application Fields of Genome Integration-Free iNSC

Since long ago, there have been studies to treat neurological diseases using NSC of various sources (De Feo et al. 2012; Kim et al. 2002; Lu et al. 2003; Wu et al. 2008). In general, major neurological disease targets have been found to be highly treatable with in vivo transplantation of a single type of cell. Examples include spinal cord injury, which is likely to be treated when an oligodendrocyte is transplanted into the spinal cord, and Parkinson's disease, which is more likely to be treated when dopaminergic neurons are implanted in the brain (Li and Leung 2015; Vernier et al. 2004). However, after implantation of NSC in vivo, problems arose such as the occurrence of teratoma, and treatment effects are often difficult to see (Brederlau et al. 2006; Salewski et al. 2015; Sonntag et al. 2007; Tropepe et al. 2001). In addition, even when therapeutic effects were confirmed, it was often mainly confined to a rat or mouse, which are small animal models, and the results are difficult to apply to humans (Kim et al. 2002; Zhang et al. 2014). Therefore, a safe genome integration-free iNSC derived from direct phenotypic conversion with little potential for teratoma formation in vivo is likely to become a highly feasible therapeutic agent for neurological diseases (Capetian et al. 2016; Cheng et al. 2014; Tang et al. 2016; Wang et al. 2013, 2015).

4.4 Pros and Cons of iNSC

Direct phenotypic conversion has several advantages over other reprogramming techniques (Table 4.3). First, direct phenotypic conversion has no ethical problems. SCNT has to use human eggs, so there is a big ethical problem, but direct phenotypic conversion does not have such concern. Second, unlike SCNT and iPSC technology, cells produced by direct phenotypic conversion are significantly less likely to

Table 4.3 Pros and cons of human iNSC

Pros	Cons	Overcoming strategy
– No ethical problems – No tumorigenesis – No immune rejection – Produced in a short time	– Insufficient understanding of molecular mechanisms – Use of some xenogeneic compound	– Complete understanding of the molecular mechanism of direct phenotypic conversion

develop teratoma after in vivo transplantation. There have been no reported cases of teratoma occurring from iNSC (Ring et al. 2012). Third, cells produced by direct phenotypic conversion have no potential to develop immune rejection in vivo if the donor and recipient of transplant are the same. Fourth, iNSC can be produced in a short time. SCNT has to be transplanted into enucleated oocytes and cultured to the early embryonic stage, followed by separate culture of ESC and further differentiation into NSC. Although iPSC technology is comparatively easy compared to SCNT, it also requires time-consuming steps of reprograming and differentiation. However, direct phenotypic conversion is advantageous in that the targeted cells can be obtained very quickly because unnecessary processes are omitted and reprogramming of the targeted NSC is performed directly.

However, iNSC research through direct phenotypic conversion is only about 5 years old, and it is at the initial stage of research all over the world. Therefore, the short study period is not sufficient for various studies in various perspectives (Table 4.3). There are few reports on the generation of genome integration-free iNSC for in vivo therapy (Capetian et al. 2016; Cheng et al. 2014; Tang et al. 2016; Wang et al. 2013, 2015). For this reason, the in vivo therapeutic effect of iNSC is lacking verification. In addition, understanding of the molecular mechanism of direct phenotypic conversion needs to be improved (Xu et al. 2015). Increased understanding of the molecular mechanism of direct phenotypic conversion is essential because it can lead to an increase in reprogramming efficiency. Finally, in the current direct phenotypic conversion, some xenogenic compounds derived from porcine or bovine have been used, which should be converted to a xeno-free system for future clinical applications.

Active research from various angles, high efficiency of direct phenotypic conversion, and implementation of a xeno-free system can all be achieved if a complete understanding of the molecular mechanism of direct phenotypic conversion can be achieved (Table 4.3). A variety of reprogramming enhancers validated in the existing SCNT and iPSC technology studies may be a good tool for understanding the molecular mechanism of direct phenotypic conversion (Kwon et al. 2017). Proliferation modulator, epigenetic modulator, and ROS modulator have greatly improved the reprogramming efficiency in SCNT and iPSC technology research. By using various materials with these functions in direct phenotypic conversion studies, it will be possible to increase the potential for identifying the molecular mechanisms that have yet to be clarified.

4.5 Conclusions and Future Perspectives

In the last about 5 years of research, direct phenotypic conversion technology has been used to establish both genome-integrating human iNSC and genome integration-free human iNSC. iNSC can be useful in fields such as in vitro drug screening, in vitro disease modeling, and in vivo cell replacement therapy. iNSC produced through direct phenotypic conversion has advantages that it is free from ethical issues, has no possibility of teratoma formation in vivo, has no possibility of in vivo immune rejection, and can be manufactured in a relatively short time. However, research into this technology is still in the early stages, and research from various angles is lacking, and the molecular mechanism of direct phenotypic conversion remains to be fully clarified. Also, in order to use iNSC for cell replacement therapy, a xeno-free system must also be developed. In summary, patient-specific iNSC generation through direct phenotypic conversion is a cutting-edge technology that goes beyond conventional SCNT and iPSC technologies in many ways. In the future, the technology for generating patient-specific iNSC through direct phenotypic conversion will play a pivotal role in regenerative medicine for treatment of neurological diseases.

Acknowledgments We thank Min-Jun Jee for his contribution in illustration. This work was supported by the National Research Foundation of Korea (NRF) grant funded by the Korea government (MSIT) (2016R1C1B1010742 and 2017K1A3A1A61091969).

Compliance with Ethical Standards

Disclosures The authors indicate no potential conflicts of interest.

References

Anokye-Danso F, Trivedi CM, Juhr D et al (2011) Highly efficient miRNA-mediated reprogramming of mouse and human somatic cells to pluripotency. Cell Stem Cell 8:376–388

Bagó JR, Okolie O, Dumitru R et al (2017) Tumor-homing cytotoxic human induced neural stem cells for cancer therapy. Sci Transl Med 9:eaah6510

Baudin B (2000) Angiotensin I-converting enzyme gene polymorphism and drug response. Clin Chem Lab Med 38:853–856

Boyer LA, Lee TI, Cole MF et al (2005) Core transcriptional regulatory circuitry in human embryonic stem cells. Cell 122:947–956

Brederlau A, Correia AS, Anisimov SV et al (2006) Transplantation of human embryonic stem cell-derived cells to a rat model of Parkinson's disease: effect of in vitro differentiation on graft survival and teratoma formation. Stem Cells 24:1433–1440

Cairns DM, Chwalek K, Moore YE et al (2016) Expandable and rapidly differentiating human induced neural stem cell lines for multiple tissue engineering applications. Stem Cell Reports 7:557–570

Calzolari F, Michel J, Baumgart EV et al (2015) Fast clonal expansion and limited neural stem cell self-renewal in the adult subependymal zone. Nat Neurosci 18:490–492

Campos LS (2005) Beta1 integrins and neural stem cells: making sense of the extracellular environment. Bioessays 27:698–707
Capetian P, Azmitia L, Pauly MG et al (2016) Plasmid-based generation of induced neural stem cells from adult human fibroblasts. Front Cell Neurosci 10:245
Chambers SM, Fasano CA, Papapetrou EP et al (2009) Highly efficient neural conversion of human ES and iPS cells by dual inhibition of SMAD signaling. Nat Biotechnol 27:275–280
Chanda S, Ang CE, Davila J et al (2014) Generation of induced neuronal cells by the single reprogramming factor ASCL1. Stem Cell Reports 3:282–296
Cheng L, Hu W, Qiu B et al (2014) Generation of neural progenitor cells by chemical cocktails and hypoxia. Cell Res. 24:665–679
Choi AM, Olsen DR, Cook KG et al (1992) Differential extracellular matrix gene expression by fibroblasts during their proliferative life span in vitro and at senescence. J Cell Physiol 151: 147–155
Chung YG, Eum JH, Lee JE et al (2014) Human somatic cell nuclear transfer using adult cells. Cell Stem Cell 14:777–780
Davis RL, Weintraub H, Lassar AB (1987) Expression of a single transfected cDNA converts fibroblasts to myoblasts. Cell 51:987–1000
De Feo D, Merlini A, Laterza C et al (2012) Neural stem cell transplantation in central nervous system disorders: from cell replacement to neuroprotection. Curr Opin Neurol 25:322–333
Dignam JD, Lebovitz RM, Roeder RG (1983) Accurate transcription initiation by RNA polymerase II in a soluble extract from isolated mammalian nuclei. Nucleic Acids Res 11:1475–1489
Evans MJ, Kaufman MH (1981) Establishment in culture of pluripotential cells from mouse embryos. Nature 292:154–156
Flanagan LA, Rebaza LM, Derzic S et al (2006) Regulation of human neural precursor cells by laminin and integrins. J Neurosci Res 83:845–856
Gage FH (2000) Mammalian neural stem cells. Science 287:1433–1438
Gao M, Yao H, Dong Q et al (2016) Tumourigenicity and immunogenicity of induced neural stem cell grafts versus induced pluripotent stem cell grafts in syngeneic mouse brain. Sci Rep 6:29955
Gurdon JB (1960) The developmental capacity of nuclei taken from differentiating endoderm cells of Xenopus laevis. J Embryol Exp Morphol 8:505–526
Gurdon JB (1962) The developmental capacity of nuclei taken from intestinal epithelium cells of feeding tadpoles. J Embryol Exp Morphol 10:622–640
Gurdon JB, Elsdale TR, Fischberg M (1958) Sexually mature individuals of Xenopus laevis from the transplantation of single somatic nuclei. Nature 182:64–65
Hall PE, Lathia JD, Miller NG et al (2006) Integrins are markers of human neural stem cells. Stem Cells 24:2078–2084
Han DW, Tapia N, Hermann A et al (2012) Direct reprogramming of fibroblasts into neural stem cells by defined factors. Cell Stem Cell 10:465–472
Han F, Baremberg D, Gao J et al (2015) Development of stem cell-based therapy for Parkinson's disease. Transl Neurodegener 4:16
Homem CC, Repic M, Knoblich JA (2015) Proliferation control in neural stem and progenitor cells. Nat Rev Neurosci 16:647–659
Hu W, Qiu B, Guan W et al (2015) Direct conversion of normal and Alzheimer's disease human fibroblasts into neuronal cells by small molecules. Cell Stem Cell 17:204–212
Hindmarsh P, Leis J (1999) Retroviral DNA integration. Microbiol Mol Biol Rev 63:836–843
Jiang H, Xu Z, Zhong P et al (2015) Cell cycle and p53 gate the direct conversion of human fibroblasts to dopaminergic neurons. Nat Commun 6:10100
Kelaini S, Cochrane A, Margariti A (2014) Direct reprogramming of adult cells: avoiding the pluripotent state. Stem Cells Cloning 7:19–29
Kfoury C (2007) Therapeutic cloning: promises and issues. Mcgill J Med 10:112–120
Kim JH, Auerbach JM, Rodriguez-Gomez JA et al (2002) Dopamine neurons derived from embryonic stem cells function in an animal model of Parkinson's disease. Nature 418:50–56

Kim SJ, Lee BH, Lee YS et al (2007) Defective cholesterol traffic and neuronal differentiation in neural stem cells of Niemann-Pick type C disease improved by valproic acid, a histone deacetylase inhibitor. Biochem Biophys Res Commun 360:593–599
Kim SU, Lee HJ, Park IH et al (2008a) Human neural stem cells for brain repair. Int J Stem Cells 1:27–35
Kim SJ, Lim MS, Kang SK et al (2008b) Impaired functions of neural stem cells by abnormal nitric oxide-mediated signaling in an in vitro model of Niemann-Pick type C disease. Cell Res 18:686–694
Kim D, Kim CH, Moon JI et al (2009) Generation of human induced pluripotent stem cells by direct delivery of reprogramming proteins. Cell Stem Cell 4:472–476
Kim J, Efe JA, Zhu S et al (2011) Direct reprogramming of mouse fibroblasts to neural progenitors. Proc Natl Acad Sci USA 108:7838–7843
Kim SM, Flasskamp H, Hermann A et al (2014) Direct conversion of mouse fibroblasts into induced neural stem cells. Nat Protoc 9:871–881
Kumar A, Declercq J, Eggermont K et al (2012) Zic3 induces conversion of human fibroblasts to stable neural progenitor-like cells. J Mol Cell Biol 4:252–255
Kwon D, Ji M, Lee S et al (2017) Reprogramming enhancers in somatic cell nuclear transfer, iPSC technology, and direct conversion. Stem Cell Rev 13:24–34
Lane DP, Cheok CF, Lain S (2010) p53-based cancer therapy. Cold Spring Harb Perspect Biol 2(9): a001222
Lee SH, Chung YN, Kim YH et al (2009) Effects of human neural stem cell transplantation in canine spinal cord hemisection. Neurol Res 31:996–1002
Li N, Leung GK (2015) Oligodendrocyte precursor cells in spinal cord injury: a review and update. Biomed Res Int 2015:235195
Li Y, Chu JS, Kurpinski K et al (2011a) Biophysical regulation of histone acetylation in mesenchymal stem cells. Biophys J 100:1902–1909
Li W, Sun W, Zhang Y et al (2011b) Rapid induction and long-term self-renewal of primitive neural precursors from human embryonic stem cells by small molecule inhibitors. Proc Natl Acad Sci USA 108:8299–8304
Lu P, Jones LL, Snyder EY et al (2003) Neural stem cells constitutively secrete neurotrophic factors and promote extensive host axonal growth after spinal cord injury. Exp Neurol 181:115–129
Ma DK, Bonaguidi MA, Ming GL et al (2009) Adult neural stem cells in the mammalian central nervous system. Cell Res 19:672–682
Marini B, Kertesz-Farkas A, Ali H et al (2015) Nuclear architecture dictates HIV-1 integration site selection. Nature 521:227–231
Matsui T, Takano M, Yoshida K et al (2012) Neural stem cells directly differentiated from partially reprogrammed fibroblasts rapidly acquire gliogenic competency. Stem Cells 30:1109–1119
Maucksch C, Jones KS, Connor B (2013) Concise review: the involvement of SOX2 in direct reprogramming of induced neural stem/precursor cells. Stem Cells Transl Med 2:579–583
Miura T, Sugawara T, Fukuda A et al (2015) Generation of primitive neural stem cells from human fibroblasts using a defined set of factors. Biol Open 4:1595–1607
Mothe AJ, Tator CH (2013) Review of transplantation of neural stem/progenitor cells for spinal cord injury. Int J Dev Neurosci 31:701–713
Nemes C, Varga E, Polgar Z et al (2014) Generation of mouse induced pluripotent stem cells by protein transduction. Tissue Eng Part C Methods 20:383–392
Orphanides G, Lagrange T, Reinberg D (1996) The general transcription factors of RNA polymerase II. Genes Dev 10:2657–2683
Pang ZP, Yang N, Vierbuchen T et al (2011) Induction of human neuronal cells by defined transcription factors. Nature 476:220–223
Pardal R, Lopez-Barneo J (2012) Neural stem cells and transplantation studies in Parkinson's disease. Adv Exp Med Biol 741:206–216
Pfisterer U, Kirkeby A, Torper O et al (2011) Direct conversion of human fibroblasts to dopaminergic neurons. Proc Natl Acad Sci USA 108:10343–10348

Prowse AB, Chong F, Gray PP et al (2011) Stem cell integrins: implications for ex-vivo culture and cellular therapies. Stem Cell Res 6:1–12
Rao MS, Malik N (2012) Assessing iPSC reprogramming methods for their suitability in translational medicine. J Cell Biochem 113:3061–3068
Reubinoff BE, Itsykson P, Turetsky T et al (2001) Neural progenitors from human embryonic stem cells. Nat Biotechnol 19:1134–1140
Reynolds BA, Rietze RL (2005) Neural stem cells and neurospheres--re-evaluating the relationship. Nat Methods 2:333–336
Reynolds BA, Weiss S (1992) Generation of neurons and astrocytes from isolated cells of the adult mammalian central nervous system. Science 255:1707–1710
Ring KL, Tong LM, Balestra ME et al (2012) Direct reprogramming of mouse and human fibroblasts into multipotent neural stem cells with a single factor. Cell Stem Cell 11:100–109
Roskelley CD, Srebrow A, Bissell MJ (1995) A hierarchy of ECM-mediated signalling regulates tissue-specific gene expression. Curr Opin Cell Biol 7:736–747
Salewski RP, Mitchell RA, Shen C et al (2015) Transplantation of neural stem cells clonally derived from embryonic stem cells promotes recovery after murine spinal cord injury. Stem Cells Dev 24:36–50
Smith WC, Harland RM (1992) Expression cloning of noggin, a new dorsalizing factor localized to the Spemann organizer in Xenopus embryos. Cell 70:829–840
Sonntag KC, Pruszak J, Yoshizaki T et al (2007) Enhanced yield of neuroepithelial precursors and midbrain-like dopaminergic neurons from human embryonic stem cells using the bone morphogenic protein antagonist noggin. Stem Cells 25:411–418
Tachibana M, Amato P, Sparman M et al (2013) Human embryonic stem cells derived by somatic cell nuclear transfer. Cell 153:1228–1238
Takahashi K, Yamanaka S (2006) Induction of pluripotent stem cells from mouse embryonic and adult fibroblast cultures by defined factors. Cell 126:663–676
Takahashi K, Tanabe K, Ohnuki M et al (2007) Induction of pluripotent stem cells from adult human fibroblasts by defined factors. Cell 131:861–872
Tang X, Wang S, Bai Y et al (2016) Conversion of adult human peripheral blood mononuclear cells into induced neural stem cell by using episomal vectors. Stem Cell Res 16:236–242
Tapscott SJ, Davis RL, Thayer MJ et al (1988) MyoD1: a nuclear phosphoprotein requiring a Myc homology region to convert fibroblasts to myoblasts. Science 242:405–411
Taupin P, Gage FH (2002) Adult neurogenesis and neural stem cells of the central nervous system in mammals. J Neurosci Res 69:745–749
Temple S (1989) Division and differentiation of isolated CNS blast cells in microculture. Nature 340:471–473
Thier M, Worsdorfer P, Lakes YB et al (2012) Direct conversion of fibroblasts into stably expandable neural stem cells. Cell Stem Cell 10:473–479
Thomson JA, Itskovitz-Eldor J, Shapiro SS et al (1998) Embryonic stem cell lines derived from human blastocysts. Science 282:1145–1147
Thomson M, Liu SJ, Zou LN et al (2011) Pluripotency factors in embryonic stem cells regulate differentiation into germ layers. Cell 145:875–889
Tropepe V, Hitoshi S, Sirard C et al (2001) Direct neural fate specification from embryonic stem cells: a primitive mammalian neural stem cell stage acquired through a default mechanism. Neuron 30:65–78
Vernier P, Moret F, Callier S et al (2004) The degeneration of dopamine neurons in Parkinson's disease: insights from embryology and evolution of the mesostriatocortical system. Ann NY Acad Sci 1035:231–249
Wakayama T, Tabar V, Rodriguez I et al (2001) Differentiation of embryonic stem cell lines generated from adult somatic cells by nuclear transfer. Science 292:740–743
Wang L, Wang L, Huang W et al (2013) Generation of integration-free neural progenitor cells from cells in human urine. Nat Methods 10:84–89

Wang T, Choi E, Monaco MC et al (2015) Direct induction of human neural stem cells from peripheral blood hematopoietic progenitor cells. J Vis Exp 2015:52298

Warren L, Manos PD, Ahfeldt T et al (2010) Highly efficient reprogramming to pluripotency and directed differentiation of human cells with synthetic modified mRNA. Cell Stem Cell 7: 618–630

Weinshilboum R (2003) Inheritance and drug response. N Engl J Med 348:529–537

Wilkinson GR (2005) Drug metabolism and variability among patients in drug response. N Engl J Med 352:2211–2221

Wilmut I, Schnieke AE, McWhir J et al (1997) Viable offspring derived from fetal and adult mammalian cells. Nature 385:810–813

Wu S, Sasaki A, Yoshimoto R et al (2008) Neural stem cells improve learning and memory in rats with Alzheimer's disease. Pathobiology 75:186–194

Xi G, Hu P, Qu C et al (2013) Induced neural stem cells generated from rat fibroblasts. Genomics Proteomics Bioinformatics 11:312–319

Xu J, Du Y, Deng H (2015) Direct lineage reprogramming: strategies, mechanisms, and applications. Cell Stem Cell 16:119–134

Yamamoto K, Kishida T, Sato Y et al (2015) Direct conversion of human fibroblasts into functional osteoblasts by defined factors. Proc Natl Acad Sci USA 112:6152–6157

Yang X, Smith SL, Tian XC et al (2007) Nuclear reprogramming of cloned embryos and its implications for therapeutic cloning. Nat Genet 39:295–302

Yoo J, Noh M, Kim H et al (2015) Nanogrooved substrate promotes direct lineage reprogramming of fibroblasts to functional induced dopaminergic neurons. Biomaterials 45:36–45

Yoshida Y, Takahashi K, Okita K et al (2009) Hypoxia enhances the generation of induced pluripotent stem cells. Cell Stem Cell 5:237–241

Yu J, Vodyanik MA, Smuga-Otto K et al (2007) Induced pluripotent stem cell lines derived from human somatic cells. Science 318:1917–1920

Yu KR, Shin JH, Kim JJ et al (2015) Rapid and efficient direct conversion of human adult somatic cells into neural stem cells by HMGA2/let-7b. Cell Rep 10:441–452

Yuan T, Liao W, Feng NH et al (2013) Human induced pluripotent stem cell-derived neural stem cells survive, migrate, differentiate, and improve neurologic function in a rat model of middle cerebral artery occlusion. Stem Cell Res Ther 4:73

Zhang SC, Wernig M, Duncan ID et al (2001) In vitro differentiation of transplantable neural precursors from human embryonic stem cells. Nat Biotechnol 19:1129–1133

Zhang N, Fang M, Chen H et al (2014) Evaluation of spinal cord injury animal models. Neural Regen Res 9:2008–2012

Zhang M, Lin YH, Sun YJ et al (2016) Pharmacological reprogramming of fibroblasts into neural stem cells by signaling-directed transcriptional activation. Cell Stem Cell 18:653–667

Zhou H, Wu S, Joo JY et al (2009) Generation of induced pluripotent stem cells using recombinant proteins. Cell Stem Cell 4:381–384

Part II
Differentiation and Underlying Mechanisms

Chapter 5
Epigenetic Regulation of Human Neural Stem Cell Differentiation

Mizuki Honda, Kinichi Nakashima, and Sayako Katada

Abstract Emerging evidence has demonstrated that epigenetic programs influence many aspects of neural stem cell (NSC) behavior, including proliferation and differentiation. It is becoming apparent that epigenetic mechanisms, such as DNA methylation, histone modifications, and noncoding RNA expression, are spatio-temporally regulated and that these intracellular programs, in concert with extracellular signals, ensure appropriate gene activation. Here we summarize recent advances in understanding of the epigenetic regulation of human NSCs directly isolated from the brain or produced from pluripotent stem cells (embryonic and induced pluripotent stem cells, respectively).

5.1 Introduction

Multipotent mammalian neural stem cells (NSCs) have abilities to self-renew and to give rise to three major cell types in the nervous system, i.e., neurons and glial cells: astrocytes and oligodendrocytes (Temple 2001). However, this does not necessarily mean that NSCs have the potential to differentiate into all of these cell types from the beginning. During the early stage of neural development, NSCs undergo symmetric divisions to expand their own numbers. In the mid-gestational stage, NSCs switch their division mode to an asymmetric one to generate neurons first, and following this neurogenic phase, NSCs start producing astrocytes and oligodendrocytes in the late-gestational to early postnatal stages. To date, a variety of factors that induce the differentiation of human (h) NSCs have been identified, such as Wingless/int (Wnt) (to neurons) (Bengoa-Vergniory et al. 2014), ciliary neurotrophic factor and bone morphogenetic protein (BMP) (to astrocytes) (Shaltouki et al. 2013), and platelet-derived growth factor (PDGF) and triiodothyronine (T3) (to oligodendrocytes) (Douvaras and Fossati 2015; Stacpoole et al. 2013). Besides temporal changes in

M. Honda · K. Nakashima · S. Katada (✉)
Department of Stem Cell Biology and Medicine, Graduate School of Medical Sciences, Kyushu University, Fukuoka, Japan
e-mail: sakatada@scb.med.kyushu-u.ac.jp

L. Buzanska (ed.), *Human Neural Stem Cells*, Results and Problems in Cell Differentiation 66, https://doi.org/10.1007/978-3-319-93485-3_5

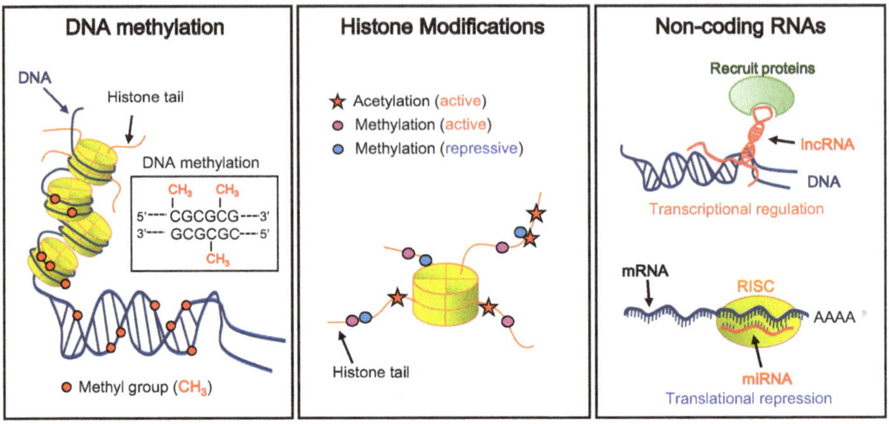

Fig. 5.1 Schematic representation of epigenetic regulations. Methylation of DNA (red circle) occurs predominantly at cytosine within cytosine-guanine dinucleotides (CpG) in mammals. Histone tail modifications occur in various combinations to fine-tune gene expression (either active or repressive). Noncoding RNAs, including lncRNA and miRNA, regulate gene expression at the transcriptional and the posttranscriptional level. See the text for details

cytokines' and growth factors' expression, recent studies have indicated that the response of NSCs to these extracellular cues also changes during the course of development. Moreover, mounting evidence supports the notion that epigenetic changes contribute to the change in the NSC response to extracellular signals, playing essential roles in NSCs' regulation.

What is "epigenetics"? The term "epigenetics" refers to changes in gene expression caused by chemical modification or chromatin remodeling that is not accompanied by DNA sequence alterations (Kubota et al. 2012). Epigenetic gene regulation is conducted mainly by three distinct categories of modifications and molecules: DNA methylation, histone modifications, and noncoding RNAs (Fig. 5.1). Not only do epigenetic changes take place during development, but they also occur during the processes of aging and disease progression, and moreover, various environmental factors, such as smoking, diet, and drugs, are known to influence the formation of the epigenetic status of the genome (epigenome). Because of the reversible nature of epigenetic modifications, studies aiming to treat diseases by restoring aberrant modifications to harmless ones using drugs have made great progress recently, and several "epigenetic drugs" have already been approved by the US Food and Drug Administration (FDA) for cancer treatment (Mottamal et al. 2015).

In this chapter, we first introduce the molecular bases of representative epigenetic modifications and noncoding RNAs (Fig. 5.1) and then discuss detailed epigenetic mechanisms that regulate NSCs' differentiation into each cell lineage in the order of neurons, astrocytes, and oligodendrocytes.

5.2 Molecular Mechanisms of Epigenetics

Epigenetic mechanisms include DNA methylation, histone modifications, and noncoding RNAs (Fig. 5.1). DNA methylation is covalent addition of a methyl group to the C-5 position of the cytosine ring (5mC) mediated by DNA methyltransferases (DNMTs). In general, DNA methylation of gene promoters is associated with transcriptional repression. DNA methylation is catalyzed by two varieties of DNMTs: maintenance methylation by DNMT1 and de novo methylation by DNMT3a and 3b. The opposite process, DNA demethylation, can be carried out either passively or actively. Passive demethylation takes place during the DNA replication process if the maintenance DNMT1 is absent, leaving the newly synthesized DNA strands unmethylated. On the other hand, active demethylation has recently been the subject of intense research, triggered by the discovery of ten-eleven translocation (TET) oxygenase family proteins that oxidize 5-methylcytosine (5mC), 5-hydroxymethylcytosine (5hmC), and 5-formylcytosine in a sequential manner to eventually produce 5-carboxycytosine (Bhutani et al. 2011). These oxidized cytosine derivatives are then excised by thymine-DNA glycosylate and replaced with unmethylated cytosine through the base excision repair pathway (Bhutani et al. 2011). Recently, it has been considered that 5hmC may serve as a distinct epigenetic mark rather than simply as a DNA demethylation intermediate, since 5hmC and 5mC are deposited in completely distinct regions across the genome (Chen et al. 2014; Globisch et al. 2010). It is anticipated that further study will unveil the biological importance of each modification of the cytosine residue.

Histone modifications are involved in regulating patterns of gene expression as well. The amino (N)-terminal tails of histone proteins are subject to a variety of modifications, including acetylation, methylation, phosphorylation, and ubiquitylation (Bannister and Kouzarides 2011). All of these chemical modifications influence the chromatin structure and gene expression. Histone acetylation is known to be associated with transcriptional activation by neutralizing the positive charge of the histone tails, resulting in a decrease of the affinity between histone and DNA, which may help to make more space for access by transcriptional machinery. Conversely, histone deacetylation is known to be associated with transcriptional repression. Both histone acetylation and deacetylation occur at lysine residues on the N-terminal tails of histones, and they are catalyzed by histone acetyltransferases (HATs) and histone deacetylases (HDACs), respectively. To date, the FDA has approved three HDAC inhibitors (HDACis) (vorinostat, romidepsin, and belinostat) for the treatment of cutaneous/peripheral T-cell lymphoma, and many more HDACis are in various stages of clinical trials (Mottamal et al. 2015).

Histone methylation occurs at lysine and arginine residues on the N-terminal tails of histone and regulates transcriptional activity positively or negatively depending on the locations and the number of methyl groups (lysine can have three different methylation statuses, mono-, di-, and trimethylation, whereas arginine can have mono- and di-methylation). To date, about a hundred histone lysine methyltransferases and arginine methyltransferases have been identified in human (Wigle and Copeland 2013).

Noncoding RNAs (ncRNAs), including long ncRNAs (lncRNAs) and microRNAs (miRNAs), are functional RNAs that are transcribed from DNA but not translated into proteins and that regulate gene expression at the transcriptional and posttranscriptional levels (Guil and Esteller 2012). In general, lncRNAs are longer than 200 nucleotides and form complexes with a variety of chromatin-modifying proteins or transcription factors and recruit them to specific gene loci, which results in transcriptional activation or repression depending on their context. On the other hand, miRNAs are phylogenetically conserved small noncoding RNAs, 18–25 nucleotides in length, and bind preferentially to the 3′ untranslated region (3′UTR) of target mRNAs and negatively regulate gene expression by inhibiting translation or promoting mRNA degradation. It is noteworthy that miRNAs might be biomarkers for various human diseases, including Alzheimer's and Parkinson's diseases, since changes of the expression level of several disease-specific miRNAs have been observed in these neurological disorders (Guo et al. 2014b; Mor and Shomron 2013).

5.3 Epigenetic Mechanism Regulating Generation of NSCs from Pluripotent Stem Cells and Neuronal Differentiation of NSCs

Since HDACis are convenient and powerful tools for changing epigenetic states, many differentiation-related studies have been performed using HDACis. It is well known that HDACis such as valproic acid (VPA) and trichostatin A (TSA) can promote neuronal differentiation of NSCs isolated from adult rat and mouse embryonic brains (Hsieh et al. 2004; Balasubramaniyan et al. 2006). Yang et al. reported that HDACis promote generation of hNSCs not only from hESCs but also from hiPSCs, while they did not observe the promotion of neuronal differentiation of hNSCs with HDACis (Yang et al. 2014). Thus, the effect of HDACis seems to be different among species. So far, 18 HDAC family genes have been identified in human. A knockdown study in H9 hESCs demonstrated that HDAC3 but not HDAC1 or HDAC2 negatively regulates NSC generation from hESCs (Yang et al. 2014). Since HDACs act by forming multiprotein complexes with their cognate cofactors, each HDAC could have distinct functions. For instance, HDAC1 and HDAC2 form a complex with corepressor RE1-silencing transcription factor (CoREST) complexes to repress target gene expression, while HDAC3 does so with nuclear receptor corepressor (NCoR)/silencing mediator of retinoic acid and thyroid hormone receptors (SMART) (Yang and Seto 2008). In this context, Yang et al. suggested that the specification of NSCs from hiPSCs is repressed by a complex composed of HDAC3 and its cognate corepressor SMART (Yang et al. 2014).

In comparison with histone acetylation studies, studies of histone methylation are still limited in human. However, in mouse, the enhancer of zeste homolog2 (Ezh2), a histone H3 lysine 27 (H3K27) methyltransferase involved in the polycomb group protein complex 2 (PRC2), has been shown to regulate the timing of NSC differentiation in the embryonic cortex (Pereira et al. 2010). That study showed that loss of function of Ezh2 results in a marked upregulation of gene expression, probably via removal of the repressive histone modifications in mouse NSCs, tipping the balance from self-renewal toward differentiation (Pereira et al. 2010). Mixed-lineage leukemia 1 (Mll1), a H3K4 methyltransferase, is required for adult neurogenesis in the mouse brain (Lim et al. 2009). NSCs have a bivalent chromatin domain characterized by high levels of both H3K27 trimethylation (me3) (repressive mark) and H3K4me3 (active mark) at key neurogenic genes such as *Dlx2*. MLL family members form complexes with H3K27-specific histone demethylases such as ubiquitously transcribed X chromosome tetratricopeptide repeat protein (Utx) and Jumonji domain containing 3 (Jmjd3). Therefore, MLL1 recruitment on the *Dlx2* promoter has been suggested to induce both H3K4 methylation and H3K27me3 demethylation, resulting in neuronal differentiation of NSCs (Lim et al. 2009).

Lysine-specific demethylase 1 (LSD1) also plays important roles in the regulation of mouse and human NSC functions. LSD1 has been reported to catalyze the demethylation of H3K4 me1/2 (active mark) and H3K9me1/2 (repressive mark), and thus it can repress or activate transcription (Shi et al. 2004). The molecular mechanism underlying this dual substrate specificity has remained largely unknown. Early studies showed that LSD1 is required for proliferation and maintenance of mouse NSCs (Sun et al. 2010; Zhang et al. 2014). A recent study suggested that LSD1 is essential for the neuronal differentiation of hNSCs isolated from the fetal neocortex (Hirano and Namihira 2016). Moreover, Laurent et al. reported that expression of an LSD1 isoform, LSD1-8a, is upregulated during the neuronal differentiation phase of the human neuroblastoma cell line SH-SY5Y (Laurent et al. 2015). They demonstrated that LSD1-8a does not have an intrinsic activity for K3K4me1/2 demethylation, but rather possesses only H3K9me1/2 demethylase activity. These findings clearly indicated that LSD1 has selective substrate specificity during the neuronal differentiation process, and regulates different sets of gene expression during the course of this process.

miRNAs are abundantly expressed in the nervous system, and about half of known miRNA species are detected in the human brain (Sun et al. 2013; Shao et al. 2010). miR-135b promotes NSC production from hiPSCs by targeting BMP type 2 receptor, transforming growth factor beta (TGFβ) receptor 1, and their downstream molecule SMAD5 (Bhinge et al. 2014). On the other hand, miR-125a/b promotes NSC production from hiPSCs by targeting SMAD4 (Boissart et al. 2012). Inhibition of the BMP/TGFβ signaling pathway induces the neuroectoderm from the ectoderm during development. Another miRNA, miR-9, is evolutionarily conserved from insects to human (Yuva-Aydemir et al. 2011) and plays important roles in NSC proliferation, migration, and differentiation, depending on the spatial and temporal context. In human, miR-9 was shown to promote proliferation of NSCs derived from hESCs and moreover to regulate migration of NSCs without affecting neuronal differentiation (Delaloy et al. 2010). STATHMIN is a microtubule-associated protein that regulates depolymerization of microtubules, and its mRNA was identified as an miR-9 target in that study. In contract, mouse miR-9 targets the orphan nuclear receptor TLX, which is necessary for the maintenance and proliferation of NSCs (Zhao et al. 2009). Overexpression of miR-9 in mouse NSCs inhibits proliferation and induces precocious neuronal differentiation.

Furthermore, gain- and loss-of-function analyses of miR-9 during development in mouse have shown that miR-9 suppresses NSC proliferation and promotes neuronal differentiation by the combined modulation of multiple target transcription factor genes, including *Foxg1*, *Nr2el*, *Gsh2*, and *Meis2* (Shibata et al. 2011). The differences between the findings of these human and mouse studies might be due to the difference of species or of the origin of the NSCs (prepared from ESCs versus brain).

miRNAs such as miR-124, miR-125, miR-181a/a* (miRNA with similar size as miRNA synthesized from the opposite strand), and miR-9/9* have been reported to promote neuronal differentiation of hNSCs. Recent studies have shown that miR-9/9* promotes neuronal differentiation of NSCs derived from hiPSCs via inhibition of Notch activity by targeting Notch receptor NOTCH2 and its downstream effector HES1 (Roese-Koerner et al. 2013, 2016). In contrast, miR-9/9* expression is induced by Notch signal activation. While Notch inhibits differentiation of mouse NSCs (Lutolf et al. 2002), it induces miR-9/9* through recruitment of the Notch intracellular domain (NICD)/RBPJ transcriptional complex to the miR-9/9* genomic locus. Taken altogether, these results indicate that the Notch signal pathway and miR-9/9* are reciprocally regulated, thereby controlling human NSCs' self-renewal and differentiation.

5.4 Epigenetic Mechanisms Regulating Astrocytic Differentiation of NSCs

During development, mammalian NSCs start to generate astrocytes after the neurogenic phase. The acquisition of gliogenic potential by NSCs is tightly linked with the timing of DNA demethylation on astrocyte-specific gene promoters such as *Gfap* and *S100β* (Takizawa et al. 2001). DNA demethylation of these promoters is triggered by dissociation of DNA methyltransferase (DNMT1) from them (Namihira et al. 2009). In NSC-specific *Dnmt1* conditional knockout mouse, global DNA hypomethylation and precocious astrocytic differentiation of NSCs were observed (Fan et al. 2005). These findings clearly demonstrate that DNA methylation regulates the timing of astrocytic differentiation in mouse in vivo. DNA methylation appears to be a critical cell-intrinsic determinant of astrocytic differentiation in human NSCs as well, since the *GFAP* promoter of early NSCs derived from hiPSCs is highly methylated (Ziller et al. 2015).

Histone modifications are also important for the acquisition of astrocytic differentiation capacity by NSCs. Indeed, at mid-gestation in mouse, the *Gfap* promoter is enriched in H3K9me2 (repressive mark); however, as gestation proceeds, the H3K9me2 level decreases, whereas transcriptional activation mark H3K4me2 is increased (Song and Ghosh 2004). We have also reported that ESCs deficient for three *Dnmts* (*Dnmt1*, *3a*, and *3b*) failed to induce *Gfap* expression in response to stimulation of a well-known astrocytic differentiation-inducing cytokine, leukemia inhibitory factor (LIF), despite DNA demethylation was absent on the *Gfap* promoter (Urayama et al. 2013). Micrococcal nuclease digestion experiments showed that the chromatin state of the *Gfap* promoter is loosened in E14.5 NSCs, which express *Gfap* in response to LIF, but not in ESCs or E11.5 NSCs, which cannot express *Gfap* even if stimulated by LIF (Urayama et al. 2013). Taken together, these results suggest that alteration of the chromatin accessibility around the *Gfap* promoter plays an important role in regulation of *Gfap* expression, in association with the acquisition of astrocytic differentiation potential by NSCs.

We discovered that oxygen tension affects the DNA methylation status of astrocytic genes in mid-gestational mouse NSCs (Mutoh et al. 2012). Indeed, E11.5 mNSCs cultured in hypoxia (2% O_2) showed promotion of *Gfap* promoter demethylation induced by the enhancement of Notch signal activation, leading to an increase in astrocytic differentiation (Mutoh et al. 2012). Another group reported that hypoxic culturing of hNSCs derived from ESCs led to decreases in *LIN28A*, *LIN28B*, and *HMGA2* expression and to an increase in *let-7* expression (Xie et al. 2014). All of these transcripts have been implicated in astrocytic differentiation of NSCs (Sanosaka et al. 2008; Patterson et al. 2014). LIN28A and LIN28B negatively regulate the level of mature *let-7* miRNA, whose critical target is *HMGA2* mRNA (Patterson et al. 2014). In fact, overexpression of *let-7* or knockdown of *HMGA2* promotes astrocyte differentiation of hNSCs (Patterson et al. 2014). These findings indicate that astrocytic differentiation of NSCs is regulated by not only DNA methylation but also other epigenetic mechanisms such as histone modifications and miRNAs.

5.5 Epigenetic Mechanisms Regulating Oligodendrocytic Differentiation of NSCs

In the central nervous system, oligodendrocytes are the myelin-forming cells, which assure the long-distance transmission of the action potential of neurons by producing a myelin sheath wrapped around adjacent neuronal axons. Aberrant oligodendrocytic differentiation and myelination are implicated in several neurodegenerative disorders such as multiple sclerosis, Huntington's disease, and Alzheimer's disease. Several groups have devised improved techniques for generating oligodendrocyte precursor cells and mature oligodendrocytes from hiPSCs and hESCs (Douvaras and Fossati 2015; Stacpoole et al. 2013). These techniques hold promise for elucidating the molecular mechanisms of oligodendrocytic differentiation in human. Compared with that of neuronal and astrocytic differentiation, epigenetic regulation of oligodendrocytic differentiation is still very unclear. However, a recent study with hESCs and hiPSCs demonstrated that expression levels of distinct sets of HMT, HAT, and HDAC family genes are changed during the course of oligodendrocytic differentiation (Douvaras et al. 2016). Furthermore, oligodendrocyte-specific knockout mouse studies reported by two different groups have revealed that *Hdac1* and *Hdac2* compound knockout mice as well as *Hdac3* single knockout mice show defects in the production and maturation of oligodendrocytes (Ye et al. 2009; Zhang et al. 2016). In addition, the importance of the repressive H3K27me3 mark seems to be conserved among species, since EZH2 (H3K27-specific KMT) critically regulates repression of neuronal genes during the transition from NSCs to oligodendrocyte precursor cells in both human and mouse (Douvaras et al. 2016; Sher et al. 2008).

5.6 Investigation of Neurological Disorders Using Patient-Derived iPSCs

Human iPSC technology has been facilitating our elucidation of molecular mechanisms of various types of diseases that lack appropriate animal models. To date, several studies have been performed with hiPSCs derived from patients suffering from neurological and neurodevelopmental diseases such as Rett syndrome (RTT). RTT is a neurodevelopmental disorder caused by mutations in the X-linked gene encoding methyl-CpG binding protein 2 (MeCP2) (Van den Veyver and Zoghbi 2001). It was previously suggested that the brains of RTT patients and RTT model mice exhibit abnormal epigenetic regulation, such as an increase in H3K9ac level and a decrease in H3K9me3 level (Thatcher and LaSalle 2006). A recent study showed that NSCs established from RTT patient-derived iPSCs generate fewer neurons and more astrocytes compared with those from healthy hiPSCs (Andoh-Noda et al. 2015). Because MeCP2 is generally known to function as a methylated DNA-binding transcriptional repressor, this aberrant differentiation of RTT-NSCs could be caused by abnormal epigenetic regulation, although we must await future investigations for the elucidation of its detailed molecular mechanisms. Although the mechanism has not yet been verified, we have reported that short-term VPA (HDACi) administration to MeCP2-deficient mice ameliorated RTT-like neurological symptoms (Guo et al. 2014a). This finding raises the possibility that artificial epigenetic regulation using epigenetic drugs such as HDACis can be used to treat congenital neurodevelopmental diseases.

5.7 Concluding Remarks and Perspective

Although many studies have proved the importance of epigenetic regulation in NSC fate specification (Fig. 5.2), most of them were performed in rodents, and human studies are still very limited. One of the reasons for this could be ethical issues regarding conducting human stem cell research (Sugarman 2008). However, the reprogramming of somatic cells to generate iPSCs or other types of cells can overcome this issue. iPSC-derived hNSCs have been used to unravel the molecular bases of human brain development. These studies have indicated that epigenetic mechanisms in fate specification are somewhat different among species. Therefore, a

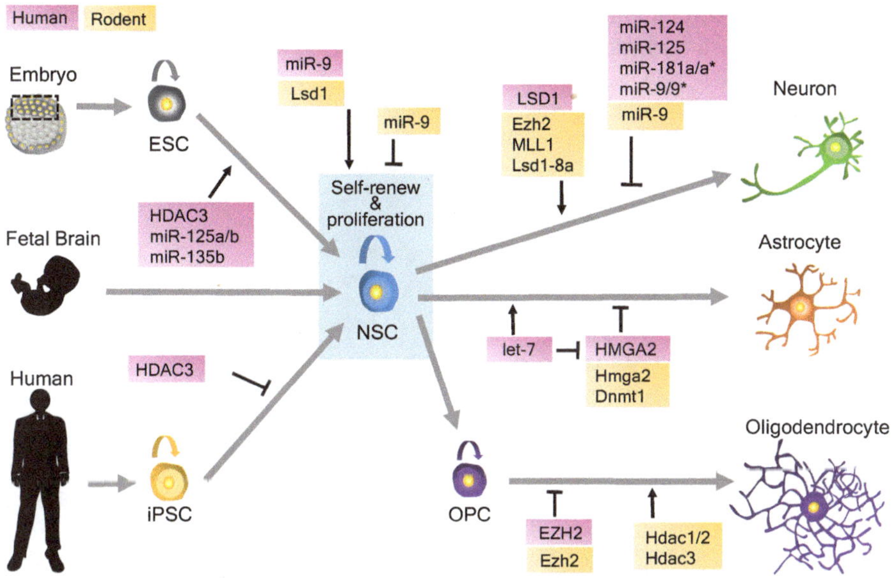

Fig. 5.2 Roles of epigenetic factors in human and rodent NSCs. Representative epigenetic factors involved in the regulation of self-renewal, proliferation, and differentiation of rodent (pink) and human (yellow) NSCs. →, promotion; ⊣, inhibition. See the text for details

more detailed understanding of epigenetic regulation of human NSCs will be an issue to be addressed in the future.

As described above, not only is epigenetics the key to understanding how fate determination of NSCs is precisely controlled, but such knowledge about epigenetic control should be applicable for developing strategies for clinical therapy. Although HDACis and DNMT inhibitors can effectively alter the epigenetic status in the genome, their effects are global rather than site-specific. To tackle this problem, other new approaches are being developed. For example, the clustered regularly interspaced short palindromic repeats (CRISPR)/CRISPR-associated proteins (Cas9) system is now known to be a powerful genome editing tool that can site-specifically edit the genome sequence when applied together with the expression of sequence-specific RNAs (guide RNAs) (Cong et al. 2013). We and others have recently shown that DNA methylation status can be site-specifically manipulated using a modified version of Cas9 molecules: "catalytically inactive Cas9 (dCas9)" fused to epigenetic modifying enzyme Tet1 or Dnmt3. dCas9-TET1 together with the expression of the respective gRNAs induced DNA demethylation on the *Gfap* and *brain-derived neurotrophic factor* (*Bdnf*) promoters in vitro and in vivo, which resulted in the activation of their transcription (Morita et al. 2016; Liu et al. 2016). In addition, histone acetylation and methylation have also been shown to be region-specifically modified by using dCas fused with p300 and Lsd1, respectively (Hilton et al. 2015; Kearns et al. 2015). Thus, utilizing these tools, we hope that we will be able to rectify aberrant epigenetic modifications and consequently correct problematic gene expression in various types of disorders in the not too distant future.

Acknowledgments We thank all of the members of the Laboratory of Molecular Neuroscience, Department of Stem Cell Biology and Medicine, Kyushu University, for valuable comments. This work was supported by the Takeda Science Foundation.

References

Andoh-Noda T, Akamatsu W, Miyake K et al (2015) Differentiation of multipotent neural stem cells derived from Rett syndrome patients is biased toward the astrocytic lineage. Mol Brain 8:31

Balasubramaniyan V, Boddeke E, Bakels R et al (2006) Effects of histone deacetylation inhibition on neuronal differentiation of embryonic mouse neural stem cells. Neuroscience 143:939–951

Bannister AJ, Kouzarides T (2011) Regulation of chromatin by histone modifications. Cell Res 21: 381–395

Bengoa-Vergniory N, Gorrono-Etxebarria I, Gonzalez-Salazar I et al (2014) A switch from canonical to noncanonical Wnt signaling mediates early differentiation of human neural stem cells. Stem Cells 32:3196–3208

Bhinge A, Poschmann J, Namboori SC et al (2014) MiR-135b is a direct PAX6 target and specifies human neuroectoderm by inhibiting TGF-beta/BMP signaling. EMBO J 33:1271–1283

Bhutani N, Burns DM, Blau HM (2011) DNA demethylation dynamics. Cell 146:866–872

Boissart C, Nissan X, Giraud-Triboult K et al (2012) miR-125 potentiates early neural specification of human embryonic stem cells. Development 139:1247–1257

Chen Y, Damayanti NP, Irudayaraj J et al (2014) Diversity of two forms of DNA methylation in the brain. Front Genet 5:46

Cong L, Ran FA, Cox D et al (2013) Multiplex genome engineering using CRISPR/Cas systems. Science 339:819–823

Delaloy C, Liu L, Lee JA et al (2010) MicroRNA-9 coordinates proliferation and migration of human embryonic stem cell-derived neural progenitors. Cell Stem Cell 6:323–335

Douvaras P, Fossati V (2015) Generation and isolation of oligodendrocyte progenitor cells from human pluripotent stem cells. Nat Protoc 10:1143–1154

Douvaras P, Rusielewicz T, Kim KH et al (2016) Epigenetic modulation of human induced pluripotent stem cell differentiation to oligodendrocytes. Int J Mol Sci 17

Fan G, Martinowich K, Chin MH et al (2005) DNA methylation controls the timing of astrogliogenesis through regulation of JAK-STAT signaling. Development 132:3345–3356

Globisch D, Munzel M, Muller M et al (2010) Tissue distribution of 5-hydroxymethylcytosine and search for active demethylation intermediates. PLoS One 5:e15367

Guil S, Esteller M (2012) Cis-acting noncoding RNAs: friends and foes. Nat Struct Mol Biol 19: 1068–1075

Guo W, Tsujimura K, Otsuka IM et al (2014a) VPA alleviates neurological deficits and restores gene expression in a mouse model of Rett syndrome. PLoS One 9:e100215

Guo Z, Maki M, Ding R et al (2014b) Genome-wide survey of tissue-specific microRNA and transcription factor regulatory networks in 12 tissues. Sci Rep 4:5150

Hilton IB, D'Ippolito AM, Vockley CM et al (2015) Epigenome editing by a CRISPR-Cas9-based acetyltransferase activates genes from promoters and enhancers. Nat Biotechnol 33:510–517

Hirano K, Namihira M (2016) LSD1 mediates neuronal differentiation of human fetal neural stem cells by controlling the expression of a novel target gene, HEYL. Stem Cells 34:1872–1882

Hsieh J, Nakashima K, Kuwabara T et al (2004) Histone deacetylase inhibition-mediated neuronal differentiation of multipotent adult neural progenitor cells. Proc Natl Acad Sci USA 101: 16659–16664

Kearns NA, Pham H, Tabak B et al (2015) Functional annotation of native enhancers with a Cas9-histone demethylase fusion. Nat Methods 12:401–403

Kubota T, Miyake K, Hirasawa T (2012) Epigenetic understanding of gene-environment interactions in psychiatric disorders: a new concept of clinical genetics. Clin Epigenetics 4(1):1

Laurent B, Ruitu L, Murn J et al (2015) A specific LSD1/KDM1A isoform regulates neuronal differentiation through H3K9 demethylation. Mol Cell 57:957–970

Lim DA, Huang YC, Swigut T et al (2009) Chromatin remodelling factor Mll1 is essential for neurogenesis from postnatal neural stem cells. Nature 458:529–533

Liu XS, Wu H, Ji X et al (2016) Editing DNA methylation in the mammalian genome. Cell 167: 233–247.e217

Lutolf S, Radtke F, Aguet M et al (2002) Notch1 is required for neuronal and glial differentiation in the cerebellum. Development 129:373–385

Mor E, Shomron N (2013) Species-specific microRNA regulation influences phenotypic variability: perspectives on species-specific microRNA regulation. Bioessays 35:881–888

Morita S, Noguchi H, Horii T et al (2016) Targeted DNA demethylation in vivo using dCas9 peptide repeat and scFv-TET1 catalytic domain fusions. Nat Biotechnol 34:1060

Mottamal M, Zheng S, Huang TL et al (2015) Histone deacetylase inhibitors in clinical studies as templates for new anticancer agents. Molecules 20:3898–3941

Mutoh T, Sanosaka T, Ito K et al (2012) Oxygen levels epigenetically regulate fate switching of neural precursor cells via hypoxia-inducible factor 1alpha-notch signal interaction in the developing brain. Stem Cells 30:561–569

Namihira M, Kohyama J, Semi K et al (2009) Committed neuronal precursors confer astrocytic potential on residual neural precursor cells. Dev Cell 16:245–255

Patterson M, Gaeta X, Loo K et al (2014) let-7 miRNAs can act through notch to regulate human gliogenesis. Stem Cell Rep 3:758–773

Pereira JD, Sansom SN, Smith J et al (2010) Ezh2, the histone methyltransferase of PRC2, regulates the balance between self-renewal and differentiation in the cerebral cortex. Proc Natl Acad Sci USA 107:15957–15962

Roese-Koerner B, Stappert L, Koch P et al (2013) Pluripotent stem cell-derived somatic stem cells as tool to study the role of microRNAs in early human neural development. Curr Mol Med 13: 707–722

Roese-Koerner B, Stappert L, Berger T et al (2016) Reciprocal regulation between bifunctional miR-9/9(*) and its transcriptional modulator notch in human neural stem cell self-renewal and differentiation. Stem Cell Rep 7:207–219

Sanosaka T, Namihira M, Asano H et al (2008) Identification of genes that restrict astrocyte differentiation of midgestational neural precursor cells. Neuroscience 155:780–788

Shaltouki A, Peng J, Liu Q et al (2013) Efficient generation of astrocytes from human pluripotent stem cells in defined conditions. Stem Cells 31:941–952

Shao NY, Hu HY, Yan Z et al (2010) Comprehensive survey of human brain microRNA by deep sequencing. BMC Genomics 11:409

Sher F, Rossler R, Brouwer N et al (2008) Differentiation of neural stem cells into oligodendrocytes: involvement of the polycomb group protein Ezh2. Stem Cells 26:2875–2883

Shi Y, Lan F, Matson C et al (2004) Histone demethylation mediated by the nuclear amine oxidase homolog LSD1. Cell 119:941–953

Shibata M, Nakao H, Kiyonari H et al (2011) MicroRNA-9 regulates neurogenesis in mouse telencephalon by targeting multiple transcription factors. J Neurosci 31:3407–3422

Song MR, Ghosh A (2004) FGF2-induced chromatin remodeling regulates CNTF-mediated gene expression and astrocyte differentiation. Nat Neurosci 7:229–235

Stacpoole SR, Spitzer S, Bilican B et al (2013) High yields of oligodendrocyte lineage cells from human embryonic stem cells at physiological oxygen tensions for evaluation of translational biology. Stem Cell Rep 1:437–450

Sugarman J (2008) Human stem cell ethics: beyond the embryo. Cell Stem Cell 2:529–533

Sun G, Alzayady K, Stewart R et al (2010) Histone demethylase LSD1 regulates neural stem cell proliferation. Mol Cell Biol 30:1997–2005

Sun AX, Crabtree GR, Yoo AS (2013) MicroRNAs: regulators of neuronal fate. Curr Opin Cell Biol 25:215–221

Takizawa T, Nakashima K, Namihira M et al (2001) DNA methylation is a critical cell-intrinsic determinant of astrocyte differentiation in the fetal brain. Dev Cell 1:749–758

Temple S (2001) The development of neural stem cells. Nature 414:112–117

Thatcher KN, LaSalle JM (2006) Dynamic changes in Histone H3 lysine 9 acetylation localization patterns during neuronal maturation require MeCP2. Epigenetics 1:24–31

Urayama S, Semi K, Sanosaka T et al (2013) Chromatin accessibility at a STAT3 target site is altered prior to astrocyte differentiation. Cell Struct Funct 38:55–66

Van den Veyver IB, Zoghbi HY (2001) Mutations in the gene encoding methyl-CpG-binding protein 2 cause Rett syndrome. Brain Dev 23(Suppl 1):S147–S151

Wigle TJ, Copeland RA (2013) Drugging the human methylome: an emerging modality for reversible control of aberrant gene transcription. Curr Opin Chem Biol 17:369–378

Xie Y, Zhang J, Lin Y et al (2014) Defining the role of oxygen tension in human neural progenitor fate. Stem Cell Rep 3:743–757

Yang XJ, Seto E (2008) The Rpd3/Hda1 family of lysine deacetylases: from bacteria and yeast to mice and men. Nat Rev Mol Cell Biol 9:206–218

Yang J, Tang Y, Liu H et al (2014) Suppression of histone deacetylation promotes the differentiation of human pluripotent stem cells towards neural progenitor cells. BMC Biol 12:95

Ye F, Chen Y, Hoang T et al (2009) HDAC1 and HDAC2 regulate oligodendrocyte differentiation by disrupting the beta-catenin-TCF interaction. Nat Neurosci 12:829–838

Yuva-Aydemir Y, Simkin A, Gascon E et al (2011) MicroRNA-9: functional evolution of a conserved small regulatory RNA. RNA Biol 8:557–564

Zhang F, Xu D, Yuan L et al (2014) Epigenetic regulation of Atrophin1 by lysine-specific demethylase 1 is required for cortical progenitor maintenance. Nat Commun 5:5815

Zhang L, He X, Liu L et al (2016) Hdac3 interaction with p300 histone acetyltransferase regulates the oligodendrocyte and astrocyte lineage fate switch. Dev Cell 37:582

Zhao C, Sun G, Li S et al (2009) A feedback regulatory loop involving microRNA-9 and nuclear receptor TLX in neural stem cell fate determination. Nat Struct Mol Biol 16:365–371

Ziller MJ, Edri R, Yaffe Y et al (2015) Dissecting neural differentiation regulatory networks through epigenetic footprinting. Nature 518:355–359

Chapter 6
Induced Pluripotent Stem Cells Reveal Common Neurodevelopmental Genome Deprograming in Schizophrenia

Sridhar T. Narla, Brandon Decker, Pinaki Sarder, Ewa K. Stachowiak, and Michal K. Stachowiak

Abstract Schizophrenia is a neurodevelopmental disorder characterized by complex aberrations in the structure, wiring, and chemistry of multiple neuronal systems. The abnormal developmental trajectory of the brain is established during gestation, long before clinical manifestation of the disease. Over 200 genes and even greater numbers of single nucleotide polymorphisms and copy number variations have been linked with schizophrenia. *How does altered function of such a variety of genes lead to schizophrenia?* We propose that the protein products of these altered genes converge on a common neurodevelopmental pathway responsible for the development of brain neural circuit and neurotransmitter systems. The results of a

---

We dedicate this chapter to Patrick W. Lee whose courage and comradery inspired our work.

S. T. Narla · B. Decker
Department of Pathology and Anatomical Sciences, Molecular and Structural Neurobiology and Gene Therapy Program, State University of New York, Buffalo, NY, USA

P. Sarder
Department of Pathology and Anatomical Sciences, Molecular and Structural Neurobiology and Gene Therapy Program, State University of New York, Buffalo, NY, USA

Department of Biomedical Engineering, State University of New York, Buffalo, NY, USA

E. K. Stachowiak
Department of Pathology and Anatomical Sciences, Molecular and Structural Neurobiology and Gene Therapy Program, State University of New York, Buffalo, NY, USA

Western New York Stem Cells Culture and Analysis Center, State University of New York, Buffalo, NY, USA

M. K. Stachowiak (✉)
Department of Pathology and Anatomical Sciences, Molecular and Structural Neurobiology and Gene Therapy Program, State University of New York, Buffalo, NY, USA

Western New York Stem Cells Culture and Analysis Center, State University of New York, Buffalo, NY, USA

Department of Biomedical Engineering, State University of New York, Buffalo, NY, USA
e-mail: mks4@buffalo.edu

L. Buzanska (ed.), *Human Neural Stem Cells*, Results and Problems in Cell Differentiation 66, https://doi.org/10.1007/978-3-319-93485-3_6

multichanneled investigation using induced pluripotent stem cell (iPSCs)- and embryonic stem cell (ESCs)-derived neuronal committed cells (NCCs) indicate an early (preneuronal) developmental-genomic etiology of schizophrenia and that the dysregulated developmental gene networks are common to genetically unrelated cases of schizophrenia. The results support a "watershed" mechanism in which mutations within diverse signaling pathways affect the common pan-ontogenic mechanism, integrative nuclear (n)FGFR1 signaling (INFS). Dysregulation of INFS in schizophrenia NCCs deconstructs coordinated gene networks and leads to formation of new networks by the dysregulated genes. This genome deprograming affects critical gene programs and pathways for neural development and functions. Studies show that the genomic deprograming reflect an altered nFGFR1—genome interactions and deregulation of miRNA genes by nFGFR1. In addition, changes in chromatin topology imposed by nFGFR1 may play a role in coordinate gene dysregulation in schizophrenia.

Abbreviations

3C	Chromatin conformation capture
CBP	CREB-binding protein
ChIPseq	Chromatin immunoprecipitation sequencing
CNVs	Copy number variations
ESCs	Embryonic stem cells
FGFR1(SP-/NLS)	Constitutively nuclear active variant of FGFR1
FGFR1(SP-/NLS)(TK-)	Dominant negative nuclear active variant of FGFR1
GO	Gene ontology
INFS	Integrative nuclear FGFR1 signaling
IPA	Ingenuity pathway analysis
iPSCs	Induced pluripotent stem cells
NCCs	Neuronal committed cells (integrative nuclear (n)FGFR1 signaling—INFS)
nFGFR1	Nuclear fibroblast growth factor receptor-1
RNAseq	Global RNA sequencing
SNPs	Single nucleotide polymorphisms
TAD	Chromatin topologically associated domains

6.1 Schizophrenia: A Disorder of Brain Development

Schizophrenia is one of the most debilitating mental illnesses worldwide (Hanzawa et al. 2013), with a lifetime prevalence of about 1.5–2% (Saha et al. 2005). There are currently no treatments that are completely effective or treat all the symptoms of schizophrenia (Blanchard et al. 2011; Rummel-Kluge et al. 2012). Schizophrenia is classified as a neurodevelopmental disorder, even though symptoms of the disease do not appear until puberty/young adulthood (Fatemi and Folsom 2009; Rehn and

Rees 2005). In males the peak onset of symptoms is between 10 and 25 years old, while in females it is between 25 and 35 years old (Rajji et al. 2009). A less frequent but particularly severe form of schizophrenia is accompanied by motor dysfunction that occurs during the prepubertal stage (Erlenmeyer-Kimling 2000). Another group, who do not show schizophrenia-like symptoms until after the age of 60, is defined as having a very late-onset schizophrenia-like psychosis (Howard et al. 2000; Keshavan 1999; Keshavan and Hogarty 1999). Together it is proposed that schizophrenia will occur by a two-hit model etiology in which early brain maldevelopment is followed by additional changes occurring adolescence (Keshavan 1999; Keshavan and Hogarty 1999). In this chapter, we will focus on changes thought to occur during the early brain development. Alterations in the schizophrenia brain are thought to occur during the first and early second trimester of development (Kneeland and Fatemi 2013) leading to improper clustering of neurons in layers II, III, and V of the cortex (Arnold et al. 1997), alteration in the number of nonpyramidal neurons in CA2, alteration in the shape of the hippocampus (Benes et al. 1998), hypoplastic development of dopamine neurons, and cerebellar atrophy (Akbarian et al. 1993; Bogerts et al. 1983; Connor et al. 2004; Schiller et al. 2006). These alterations in neuronal numbers and clustering are not due to neurodegeneration, as no neurodegenerative markers are observed in schizophrenia. In addition to neuronal alteration, changes in white matter structure have been observed (Davis et al. 2003), suggesting that even oligodendrocytes are effected. This widespread alteration of brain structure is thought to underlie the complexity of the clinical symptoms observed: positive symptoms (delusions and hallucinations), negative symptoms (affective flattening, amotivation, and anhedonia) (Blanchard et al. 2011; Foussias et al. 2011), and cognitive symptoms (disorganized speech and cognitive deficits) (DSM 4th edition). In addition, minor physical anomalies are associated with schizophrenia; these anomalies are consistent with abnormal development during the first trimester (Lloyd et al. 2008).

6.2 Schizophrenia: An Integrated Perspective on the Disease of Hundreds of Genes

While the symptoms of schizophrenia have been characterized well, the underlying causes have been difficult to pin down. Schizophrenia is a heritable familial disorder with a complex mode of inheritance and expression (Sullivan et al. 2003). Even in identical twins, the likelihood of both having schizophrenia is only up to 50%. This suggests that the disease could be a resultant of an interplay between genetic and environmental factors. Factors listed as acting during pregnancy which increases the frequency of the disease include infections (mother's immune attack hypothesis), episodes of hypoxia, and nicotinism. Possible environmental factors include being raised in a city, cannabis use during adolescence, certain infections, parental age, and poor nutrition during pregnancy.

While schizophrenia has been shown to be inheritable, its polygenetic nature and complexity make it difficult to dissect out the underlying genetic mechanisms. Several linkage studies have been carried out to better understand the schizophrenia genetics; however a lack of highly significant and consistently reproducible results has generally characterized those studies (Need et al. 2009). Next-generation sequencing technology has enabled researchers to look at the hundreds of thousands of single nucleotide polymorphisms (SNPs) simultaneously. However, similar to linkage studies, even though over 500 SNPs have been found to be significantly associated with schizophrenia, the overall results have been inconsistent (Welter et al. 2014). In addition to SNPs, copy number variations (CNVs) have also been associated with schizophrenia. A common theme is the enrichment of rare (<1% minor allele frequency) and large (>100 kb) CNVs (International Schizophrenia Consortium 2008; Malhotra et al. 2011; Walsh et al. 2008), and those that occur de novo (Kirov et al. 2012; Malhotra et al. 2011; Xu et al. 2008).

Each year brings reports of new candidate "schizophrenia genes" further complicating the picture of this polygenetic disease. Even though many genetic alterations have been associated with schizophrenia, no single alteration has been found to make up more then 1–2% of the schizophrenia population (International Schizophrenia Consortium 2008; Stefansson et al. 2008; Xu et al. 2008). Hence, the genetic causes of schizophrenia appear to be a multiplicity of rare risk alleles, and schizophrenia has been defined as a common, rare variant disease.

How do various mutations lead to a common disorder? One possible answer, proposed by Cannon and Keller, is the watershed hypothesis (Cannon and Keller 2006). According to this hypothesis, individual mutations dysregulate distinct biological pathways that in turn converge on a common ontogenic pathway(s) (Fig. 6.1). The common affected pathway should integrate signals from various pathways in which the individual schizophrenia gene mutations have been found and command the early stages of the brain development. The dysregulation of these common pathway would lead to brain malformations which increase the risk of the disease. However, the nature of such a central pathway and its organization has been yet unknown. This chapter will discuss evidence that the recently discovered integrative nuclear FGFR1 signaling (INFS) pathway could serve as candidate common pathway in schizophrenia.

6.3 IPSCs Model Cellular Developmental Abnormalities in Schizophrenia

One key proposition of the watershed hypothesis is that there is a common dysregulation of the developmental genome in schizophrenia. Identification of an early developmental gene dysregulation in adult brain tissue specimens may not be possible once the disease has progressed to a late-stage form. Additionally, tissue samples from schizophrenia patients in past studies have historically been limited to

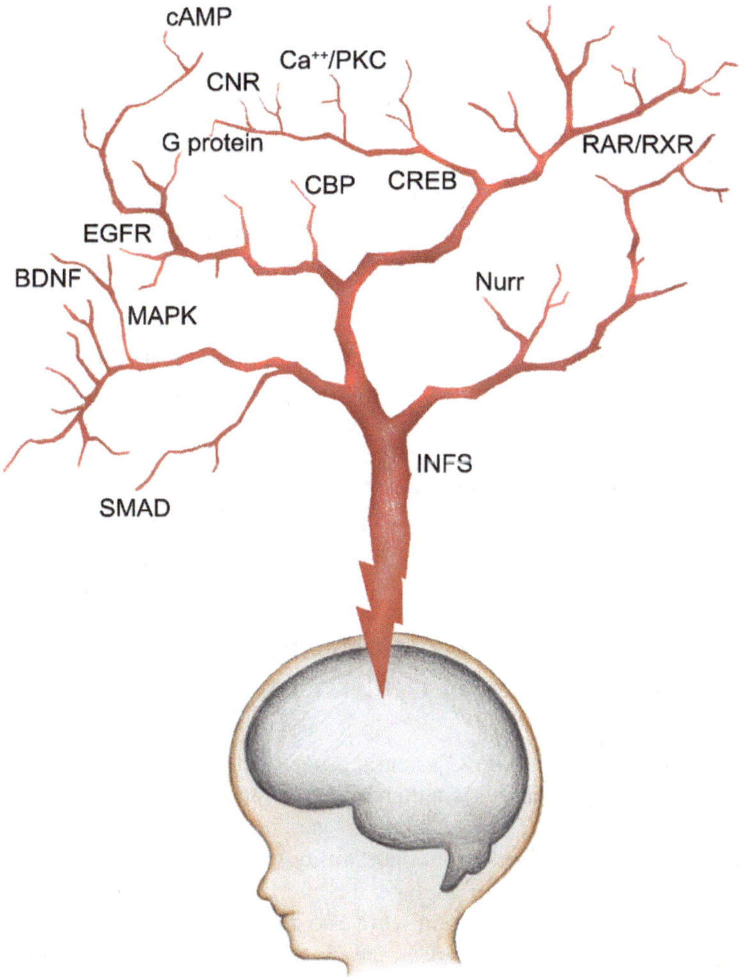

Fig. 6.1 Modified watershed hypothesis (Cannon and Keller 2006) of schizophrenia. In schizophrenia mutations are found in >200 genes of multiple signaling pathways which feed to a common pan-ontogenic mechanism the "Integrative Nuclear Fibroblast Growth Factor Receptor 1 Signaling" (INFS) pathway (for review see Stachowiak et al. 2011b, 2015; Stachowiak and Stachowiak 2016). In INFS, FGFR1 and its ligand, FGF-2, translocate into the nuclear interior, and through direct interaction with transcription gating factor CREB binding protein (CBP), nuclear (n) form of receptor, (n)FGFR1, directly controls the activation/inhibition of thousands of genes and epigenetic changes integral in ontogeny and brain development. INFS links downstream developmental gene programs to multiple upstream pathways such as cAMP, PKC, neurotrophins and MAPK, diverse growth factors, and nuclear retinoid and orphan Nur receptor-mediated pathways. Figure drawn by Sun Young Kang

postmortem individuals, whose samples were complicated by various factors such as substance abuse, drug treatment, postmortem interval, abnormal brain pH influenced by hypoxia, and nutritional deficiency (Deep-Soboslay et al. 2011). In recent years human-induced pluripotent stem cells (iPSCs) have emerged as new potential tools in testing the watershed hypothesis. In 2011, two laboratories reported successful development of iPSCs from schizophrenic patients (Brennand et al. 2011; Chiang et al. 2011). Brennand et al. had developed iPSCs from four patients diagnosed with schizophrenia or its schizoaffective variant and four from control non-schizophrenic subjects (Brennand et al. 2011). Schizophrenia hiPSC-derived NPCs have aberrant migration (Brennand et al. 2014b) and cellular polarity (Yoon et al. 2014), perturbed WNT signaling (Srikanth et al. 2015; Topol et al. 2015), increased oxidative stress (Brennand et al. 2014b; Paulsen et al. 2011; Robicsek et al. 2013), and altered responses to environmental stressors (Hashimoto-Torii et al. 2014), while schizophrenia hiPSC-derived neurons exhibit decreased neurite number (Brennand et al. 2011), reduced synaptic maturation (Brennand et al. 2011; Robicsek et al. 2013; Wen et al. 2014; Yu et al. 2014) and synaptic activity (Wen et al. 2014; Yu et al. 2014), and blunted activity-dependent response (Roussos et al. 2016).

6.4 Schizophrenia Patients' NCCs Share a Common Coordinate Pattern of Gene Dysregulation

The recent investigation into iPSC's neural progeny tested the watershed hypothesis of schizophrenia by examining whether patients with diverse genetic backgrounds and schizophrenia-linked copy number variants may show also common dysregulations of the genome. Such a possibility was suggested by broad transcription analysis using a microchip analysis method on mature neurons differentiated from different iPSC lines. A common set of 596 dysregulated genes in 4 patients was found (Brennand et al. 2011). Many of the changes in gene expression observed in mature neurons could reflect differences in the types of neurons that were generated from the patient and control IPSCs (Brennand et al. 2011, 2014a, 2015; Brennand and Gage 2011). Thus, to identify the genomic mechanism that leads to altered neuronal and brain development and therefore underlies the etiology of schizophrenia, we have focused on studying early neural development, i.e., the transition from iPSC-differentiated neural progenitor cells (NPCs) to neuron-committed cells (NCCs) induced by 2-day treatment with BDNF, GDNF, and cAMP.

Global RNA sequencing including small RNA has revealed a common set of 1349 dysregulated genes (FC $> \pm 1.5$ and q value > 0.05) in all 4 patients with diverse genetic backgrounds and different schizophrenia-linked copy number variants (Narla et al. 2017) (Fig. 6.2).

To determine whether this dysregulation of 1349 genes in schizophrenia NCCs represented a random or a correlated event, a pairwise correlation network analysis of all dysregulated genes was carried out (Fig. 6.3a1), for 909,226 potential

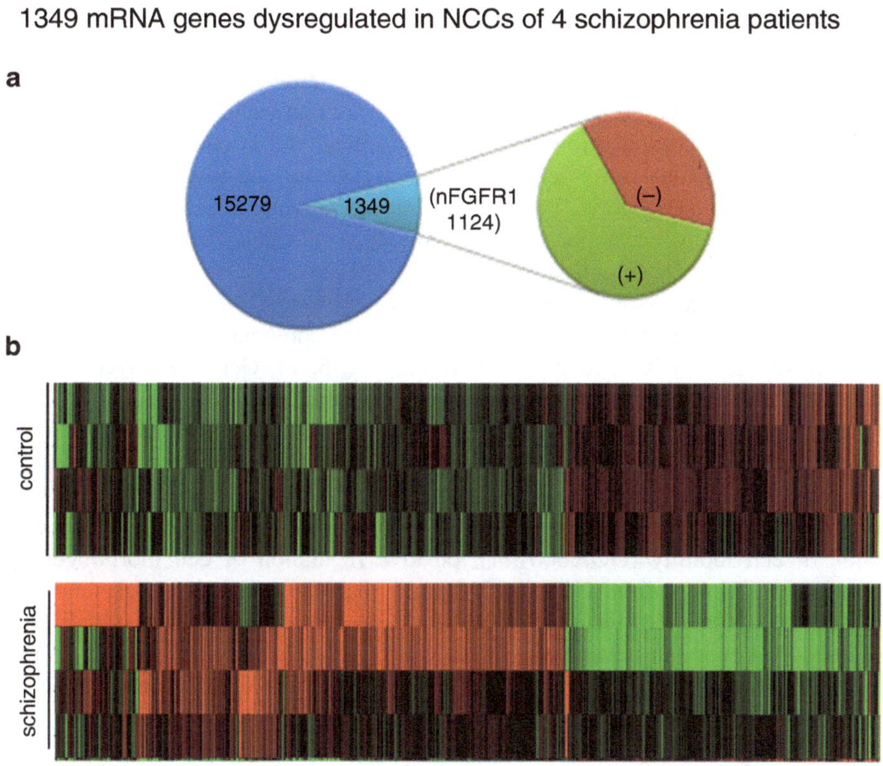

Fig. 6.2 RNAseq of control and schizophrenia neuronal committed cells (NCCs) derived from iPSCs of four schizophrenia patients and four control individuals (Narla et al. 2017). (**a**) Distribution of gene expression across eight samples: four control and four schizophrenia NCC lines. 15,279 expressed genes (mRNAs) were detected in all 8 samples, of which 1349 genes were dysregulated in all 4 schizophrenia NCC samples (FC $> \pm 1.5$ and q value >0.05). Among these the majority of genes, 63%, were upregulated. Nearly 84% (1124) of the dysregulated genes were targeted by nFGFR1 (ChIPseq analysis) (Narla et al. 2017). (**b**) Heatmap of 1349 genes that were dysregulated in all 4 schizophrenia NCC samples (FC $> \pm 1.5$ and q value > 0.05). Raw expression data were log transformed and then centered to the median of all eight samples. Red indicates higher value than median; green indicates lower value than median. This figure is based on the results from Narla et al. (2017)

relationships. Compared to control cells ($n = 4$), in which the distribution of correlations was flat (likelihood of genes having high or low correlation was similar), in patients' cells the numbers of positive correlations (genes changing in the same direction) and negative correlations (genes changing in opposite direction) were markedly increased.

The analysis of the highly interconnected nodes (genes which are highly correlated with a greater number of other genes) revealed that the networks formed by genes that were highly correlated in control cells were no longer found in the schizophrenia cells (Fig. 6.3b1) and that a new network of the highly correlated genes formed in schizophrenia cells (Fig. 6.3b2). Thus the control networks became disrupted in the patient NCCs, and a new network of connected genes formed in their place. The ontological gene categories represented by the control network, disrupted in schizophrenia, included genes involved with extracellular matrix, synapse formation, neuronal projection, and nervous system development. This network change suggests an enhanced push toward a neuronal phenotype in patient samples compared to controls. The control networks included both up- and downregulated genes. In the schizophrenia networks, however, the up- and downregulated genes distinctly segregated into separate networks. These findings indicated further the concerted gene dysregulation by singular factors.

To further characterize the observed gene dysregulation, we performed separate gene ontology (GO) analyses of all genes that were upregulated and all genes that were downregulated in schizophrenia NCCs (Fig. 6.1a). Genes involved in glial differentiation and axon ensheathment were present only in the downregulated category, while neuronal ontologies such as axonogenesis, neurotransmitter transport, and learning were overrepresented in the upregulated group (Narla et al. 2017). Importantly, genes involved in positive regulation of cell proliferation (GO:0008284), positive regulation of cell migration (GO:0030335), positive regulation of cell motility (GO:2000147), positive regulation of cell morphogenesis involved in differentiation (GO:0010770), and positive regulation of neuron differentiation (GO:0045666) were all in the upregulated category. These findings have established a genomic mechanism for the increased NPC proliferation, migration, and premature neuronal differentiation found recently in schizophrenia iPSC brain organoids (Stachowiak et al. 2017).

6.5 Dysregulation of miRNA and mRNA Interactive Networks

One category of factors that could elicit a concerted dysregulation of transcriptome in schizophrenia is miRNAs, which are known to influence overlapping gene sets in a coordinated fashion. miRNAs influence mRNA levels by promoting mRNA degradation, inhibiting mRNA translation, and acting at the transcription level (Bartel 2009; Younger and Corey 2011). The NCCs from 3 schizophrenia patients examined displayed a concerted dysregulation of 16 miRNAs, all of which were overexpressed, albeit to different extents. Within this group, mir-132 (Miller et al. 2012), mir-134 (Moreau et al. 2011; Santarelli et al. 2011), mir-218 (Perkins et al. 2007), and mir-17 (Shi et al. 2012) have previously been implicated in schizophrenia (Miller et al. 2012; Santarelli et al. 2011). TargetScan and MirTarBase analyses predicted that the overexpressed miRNAs may interact with >400 dysregulated mRNAs, in a largely overlapping manner as illustrated on Fig. 6.3c. In control NCCs these 16 miRNAs displayed a high degree of positive correlation consistent with the model in which different miRNAs controlled shared mRNA targets. In schizophrenia NCCs, all 16 miRNAs were upregulated, but to different degrees (1.5-

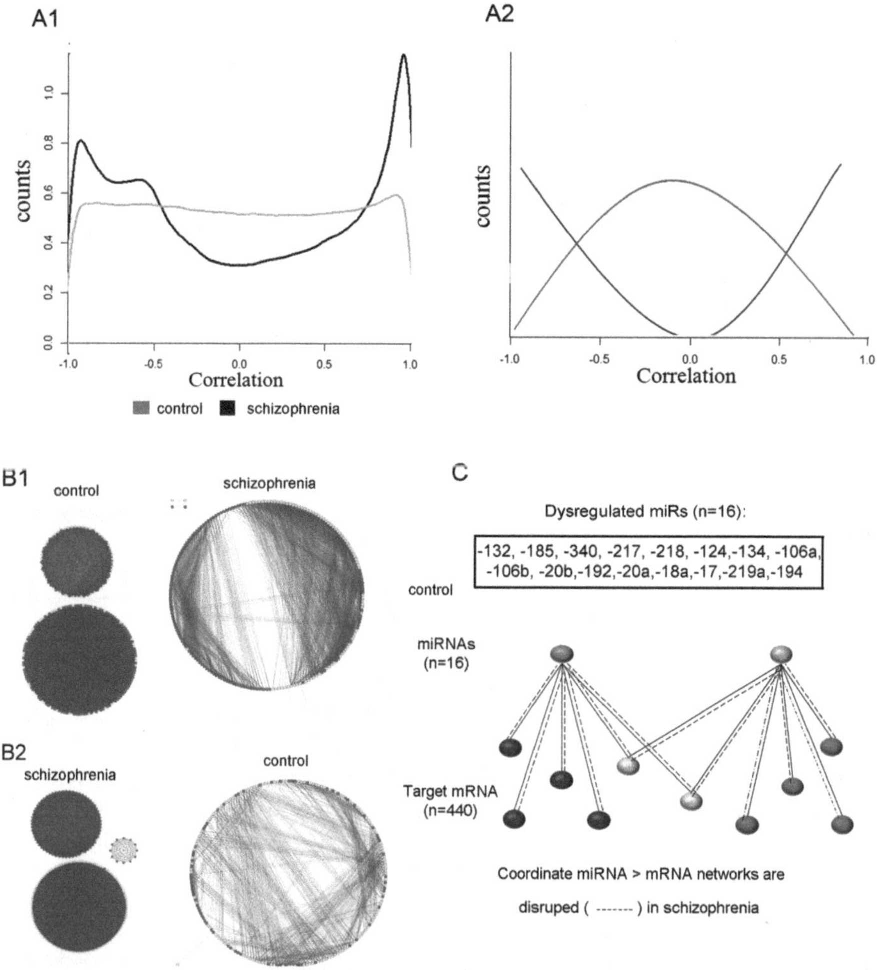

Fig. 6.3 Correlate gene networks in control NCCs are disrupted and replaced by new networks in schizophrenia [based on the results from Narla et al. (2017)]. (**a**) Analysis pairwise correlations among schizophrenia dysregulated 1349 mRNA genes. (**a1**) Correlation was performed using four control and four patient NCC samples. A flat distribution of correlation is observed in controls, while in patients an increase in the number of positively and negatively correlated genes was observed. (**a2**) Predicted forms of gene dysregulation: random-dyscoordinate (convex) and nonrandom-coordinate (concave). The dysregulation in schizophrenia (**a1**) follows the nonrandom-coordinate model. (**b**) Top 200 nodes (genes whose expression is highly positively correlated with that of multiple other genes) in control and in patient NCCs were identified (marked on the perimeters). (**b1**) Gray lines link pairs of genes whose correlation is >0.9. (**b1**) In the control set, two separate networks were observed, and each contained both upregulated and downregulated genes. These correlations were disrupted in schizophrenia. (**b2**) In the patient set, the upregulated and downregulated genes formed three separate networks which did not exist in control NCCs. (**c**) Dysregulation of miRNA in schizophrenia NCCs. NCCs from three control subjects and three patients were analyzed. In all 3 patients 16 miRNAs were dysregulated (all upregulated). Those dysregulated miRNAs target 440 mRNAs. The observed miRNA–mRNA correlations in control cells were eliminated in schizophrenia cells (indicated by interrupted lines) indicating a disassociation of the miRNA > mRNA networks (Narla et al. 2017). Results and (**b**) are from Narla et al. (2017)

to >70-fold), and the correlations among those 16 miRNAs found in control networks were lost. Analysis of the combined networks of miRNAs and mRNAs together demonstrated the loss of cooperation between the miRNAs and their target mRNAs in schizophrenia NCCs (Fig. 6.3c). This suggests that the normally tight miRNA to mRNA coordination is being overridden by a separate, possibly global mechanism, which is resulting in alterations in control of mRNA genes, and causing disruptions of correlated expression levels of normally interdependent miRNA.

One candidate pathway that could be involved is the pan-ontogenic integrative nuclear FGFR1 signaling (INFS) (Stachowiak et al. 2007, 2011b, 2015), which integrates signals from diverse pathways in which the schizophrenia-linked mutations have been found and which controls genes involved in neural development. A disruption of FGFR1 function in dopaminergic neurons of transgenic mice led to developmental brain malformation and behavioral changes that mimic the positive, negative, and cognitive deficits observed in humans (Stachowiak et al. 2013).

6.6 Pan-Ontogenic INFS in Brain Development

At the center of the INFS module are proteins that bear the historic name of fibroblast growth factors (FGF) and the high-affinity FGF receptors (FGFR). Neither FGFs nor FGFRs exist in single-cell organisms but are common to eumetazoans and are essential for the generation of tissues with specialized cells (Stachowiak et al. 2011b). Mutations of the single FGFR1 gene disrupt gastrulation and development of the central and peripheral nervous systems, mesodermal somites, muscles and bones, and the endoderm. These effects are accompanied by changes in the expression of genes (Ciruna and Rossant 2001; Ciruna et al. 1997; Dequeant and Pourquie 2008; Partanen et al. 1998) and microRNAs (Bobbs et al. 2012; Stuhlmiller and Garcia-Castro 2012) that control development. These findings firmly placed FGFR1 at the top of the developmental hierarchy; however, how could a single gene perform such a global ontogenic function was unknown.

FGFs emerged during early metazoan evolution equipped with nuclear localization signals (NLS), and their biological effects depend on nuclear accumulation (Popovici et al. 2006). In addition, NLS-lacking FGFs have evolved which act as extracellular secreted proteins. In the mammalian FGF family, NLS-containing FGFs act in the nucleus to promote cell speciation, whereas secreted FGFs act on the cell surface receptors as mitogens (Claus et al. 2003; Sherman et al. 1993; Stachowiak et al. 2007, 2011a). Individual FGF receptors (in mammals, FGFR1-4) likewise have adaptations directing them to different cellular compartments (Myers et al. 2003).

There are two separate pathways which have been characterized for FGFR1 processing. The newly synthesized FGFR1 can enter the constitutive membrane pathway (MP) in which receptor is processed and glycosylated in Golgi and accumulates in the plasma membrane. In the nuclear pathway, an atypical transmembrane domain in FGFR1 allows newly translated immobile receptor to be released from the pre-Golgi into the cytosol in a process that engages proteasome, FGF-2 ligand, and ribosomal S6 kinase activities. Nuclear transport of FGFR1 is mediated by importin-β (Stachowiak et al. 2007) and stimulated by a variety of developmental signals, including EGF, NGF, BDNF, BMP, retinoids, hormones and neurotransmitters, calcium, cyclic AMP, and PKC, and is inhibited by cell contact receptors. This is the reason that this pathway has been referred to as an integrative signaling (Stachowiak et al. 2007, 2011b).

The INFS mechanism is involved primarily in developmental transitions, most commonly the switches to differentiation and postmitotic development (Stachowiak et al. 2007, 2011b). In proliferating neural stem/progenitor cells (NS/PC) of the brain ventricles, FGFR1 is present in the cytoplasm, while in differentiating brain cortical cells or midbrain dopamine neurons, FGFR1 is located within the cell nucleus (Fang et al. 2005; Stachowiak et al. 2009a, b). As this development is completed, FGFR1 localization becomes again predominantly cytoplasmic. Nuclear accumulation of nFGFR1 occurs during differentiation of diverse stem cells and growth and differentiation of glial, neuronal, endothelial, and mesodermal cells as well as cancer cells.

In loss- and gain-of-function experiments, nFGFR1 was found to be essential for the control of the pluripotent state, necessary for neuronal programing by retinoic acid (RA), NGF, BDNF, or cAMP, and was sufficient to induce neuronal differentiation in the absence of additional stimulation (Lee et al. 2012). How can a single nuclear protein program the development of ESCs—a process that involves the coordinated regulation of thousands of genes that are located on different chromosomes and contain diverse regulatory elements?

nFGFR1, which lacks a DNA-binding domain, engages indirectly in gene regulation by binding to the domain of the CREB-binding protein (CBP). CBP is a common transcription co-regulator and histone-acetylating protein that interacts with multiple transcription factors (Fang et al. 2005; Hu et al. 2004; Kasper et al. 2006; Vo and Goodman 2001). The interaction with CBP allows FGFR1 to target a wide variety of genes. Next-generation sequencing has delineated global and direct gene programing by nFGFR1 and its partner CBP, which guide pluripotent embryonic stem cells (ESCs) toward development into multipotent neural progenitor cells (NPCs) and toward further differentiation (Stachowiak and Stachowiak 2016; Terranova et al. 2015). nFGFR1 cooperates with a multitude of transcription factors (TFs), including RXR, RAR, and orphan nuclear receptors, and targets thousands of genes (both mRNA and miRNA) across the entire genome in a nonrandom manner. Additionally nFGFR1's binding on the genome is increased during the transition into neuronal lineage underscoring its importance for neuronal development (Stachowiak and Stachowiak 2016; Terranova et al. 2015). nFGFR1 binds genes involved in pluripotency leading to their inactivation during the transition into neuronal stem cells (Terranova et al. 2015). In addition nFGFR1 binds to and activates Hox genes, which regulate spatial development of organs and tissues (Stachowiak and Stachowiak 2016; Terranova et al. 2015).

In addition to these genes, nFGFR1 has been found to both regulate the expression of a multitude of transcription factors and work with said transcription factors in order to control expression of various genes. Due to the global role nFGFR1 plays during development, INFS is a strong candidate for one of the downstream pathways theorized by Cannon and Keller's watershed hypothesis.

6.7 FGFR1 in Dysregulation of Schizophrenia Dysregulation of NCC Transcriptome

nFGFR1 is bound in a nonrandom fashion across all chromosomes in both control and schizophrenia NCC genomes. nFGFR1 binding across each chromosome was related to gene distribution. nFGFR1 binding was highly enriched in 5' UTR regions (>fivefold) and in promoters. Both control and schizophrenia NCCs and nFGFR1 almost exclusively associated with the promoters that were actively expressed. FGFR1 was found to be bound to promoters of >90% of 1378 genes dysregulated in schizophrenia but only to 55% of all genes (Fig. 6.4b). The majority of dysregulated genes had promoters targeted by nFGFR1, and the number targeted was higher in schizophrenia than in control NCCs. In addition, nFGFR1 was bound to more locations in schizophrenia compared to controls, and a large portion of the new binding sites were found in introns and distal intergenic regions. Potentially, such distal binding could be related to regulation of the 3D chromatin structure.

MACS2 analysis of the nFGFR1 binding score (the score reflects the abundance of nFGFR1 at a particular genomic locus in a cell population) showed that out of the 915 genes that bound nFGFR1 in both control and schizophrenia cells, the majority, 828 genes, showed stronger nFGFR1 binding in patient cells. Due to the broad genome binding alterations of nFGFR1 signaling in schizophrenia cell lines, nFGFR1 could be one possible factor in causing changes that are observed in schizophrenia.

Categories of dysregulated genes targeted by nFGFR1 were identified using gene ontology (GO), ingenuity pathway analysis (IPA), and reactome. As found for all 1349 dysregulated genes, the nFGFR1-targeted dysregulated gene (90%) NCCs overrepresented the pathways involved in axon guidance, neurotransmitter release, and glial cell differentiation (Table 6.1).

Many neuronal GO categories were overrepresented (Table 6.1) including genes involved in neural crest development, synaptic plasticity, learning, memory, and synapse organization. Several GO groups related to glial development were also overrepresented, including those involved in the processes of myelination, axon ensheathment, glial cell differentiation, and oligodendrocyte differentiation.

Transfection of the recombinant, constitutively nuclear variant of FGFR1 [FGFR1(SP-/NLS)], in which the cleavable SP is replaced with the NLS of FGF-2, and of dominant-negative variant FGFR1(SP-/NLS)(TK-), which lacks the tyrosine kinase (TK) domain, showed that nFGFR1 is sufficient and necessary for neuronal differentiation, both in the mouse brain (Bharali et al. 2005; Stachowiak et al. 2009a, b) and in cultured ESC or NPC cells treated with RA, NGF, BDNF, BMP, or cAMP (Fang et al. 2005; Horbinski et al. 2002; Lee et al. 2013; Stachowiak

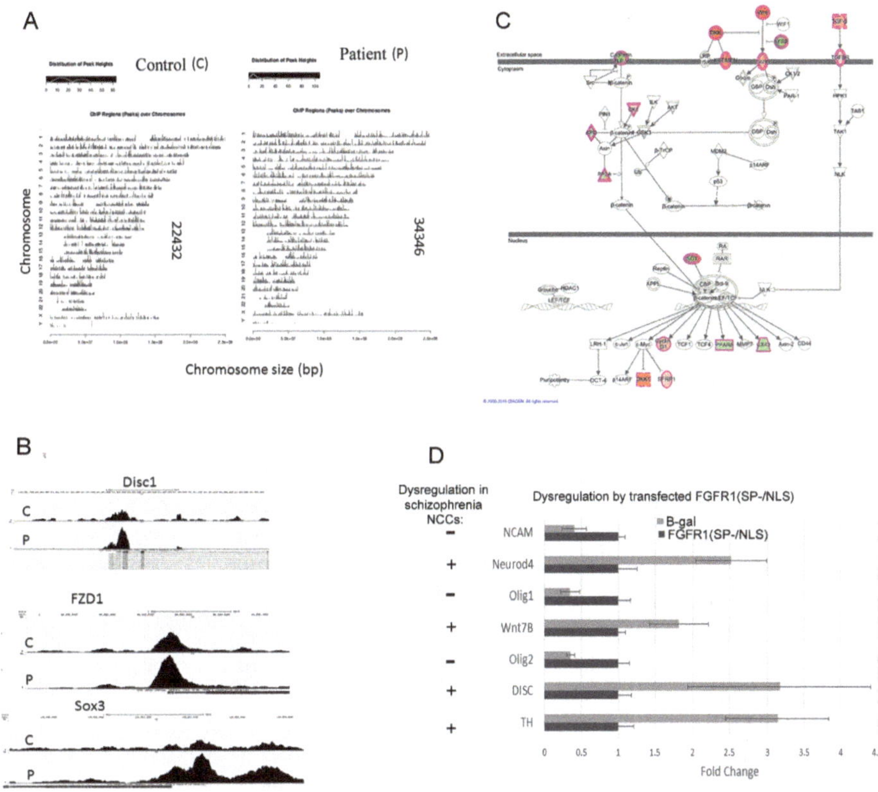

Fig. 6.4 (**a**) ChIPseq distribution of nFGFR1 peaks throughout the genomes of NCCs from control and schizophrenia iPSCs [based on the results from Narla et al. (2017)]. nFGFR1 binding sites (peaks) were enriched in the promoters but not in the intergenic regions. In patients, an increased nFGFR1 binding was observed in gene promoters, distal promoters ,and distal intergenic regions. (**b**) UCSC genome browser views of nFGFR1 binding for Discl, FZD1, and Sox3 genes. Tag distribution of nFGFR1—increased binding is observed in schizophrenia compared to control NCCs. (**c**) WNT signaling is dysregulated in schizophrenia patients. IPA pathway for Wnt signaling—pink outline represents genes that are dysregulated in schizophrenia. Green fill represents genes that are downregulated, and red fill represents genes that are upregulated. WNT, cadherins, Frizzled, and Sox3 are some of the genes dysregulated in this pathway.(**d**) Examples of nFGFR1-targeted genes (TH, Wnt7B, Neurod4, Olig2, Olig1, and NCAM) that were upregulated (+) or downregulated (−) in schizophrenia NCCs. Consistently, in control human NCCs, these genes were up- or downregulated by transfected constitutive active nuclear FGFR1(NLS/SP-). These findings are consistent with the model in which increased nFGFR1 gene targeting in schizophrenia leads to gene up- or downdysregulation. This modified figure is based on the results from Narla et al. (2017)

et al. 2003). In addition, studies have demonstrated that both full-length and truncated forms of FGFR1 accumulate in cancer cells and thereby promote metastasis (Chioni and Grose 2012; Coleman et al. 2014; Nguyen et al. 2013).

Reactome and IPA showed that the gene dysregulation in schizophrenia was centered on pathways controlling development of neuronal systems, neural genes,

Table 6.1 Selected gene otology (GO) terms for schizophrenia-dysregulated genes targeted by nFGFR1

GO term	Total genes	Genes dysregulated
Positive regulation of axon extension (GO:0045773)	36	10
Central nervous system neuron development (GO:0021954)	68	14
Positive regulation of axonogenesis (GO:0050772)	69	14
Regulation of cyclin-dependent protein kinase activity (GO:1904029)	99	18
Ensheathment of neurons (GO:0007272)	92	16
Axon ensheathment (GO:0008366)	92	16
Neurotransmitter secretion (GO:0007269)	101	17
Presynaptic process involved in synaptic transmission (GO:0099531)	105	17
Neurotransmitter transport (GO:0006836)	140	22
Glial cell differentiation (GO:0010001)	138	21
Cell cycle arrest (GO:0007050)	156	23
Eye morphogenesis (GO:0048592)	150	22
Regulation of synaptic plasticity (GO:0048167)	139	20
Regulation of synapse structure or activity (GO:0050803)	230	32
Extracellular matrix organization (GO:0030198)	374	52
Regulation of neurotransmitter levels (GO:0001505)	176	24
Negative regulation of nervous system development (GO:0051961)	265	36
Negative regulation of neurogenesis (GO:0050768)	243	33
Regulation of neuron differentiation (GO:0045664)	555	75
Telencephalon development (GO:0021537)	230	30
Regulation of neuron projection development (GO:0010975)	406	52
Positive regulation of neuron differentiation (GO:0045666)	306	39
Negative regulation of neuron differentiation (GO:0045665)	190	24
Positive regulation of neuron projection development (GO:0010976)	230	29
Regulation of neurogenesis (GO:0050767)	669	84
Axon guidance (GO:0007411)	567	71
Regulation of nervous system development (GO:0051960)	755	90
Axonogenesis (GO:0007409)	672	80
Neuron development (GO:0048666)	1024	115
Generation of neurons (GO:0048699)	1629	182
Regulation of MAPK cascade (GO:0043408)	783	74
Regulation of cell motility (GO:2000145)	715	67

extracellular organization, as well as other developmental genes. Examples of the affected pathways included pluripotency regulation, Notch signaling, Wnt/β-catenin signaling, PI3K/AKT signaling, eNOS signaling, VEGF signaling, and L1cam signaling, which play a role in axonal growth, axonal guidance pathway, glutamate receptor signaling, CREB signaling in neurons, various extracellular matrix pathways, and transcriptional regulation by TP-53 (Table 6.2). Pathways related to the release of dopamine, serotonin, norepinephrine, and glutamate neurotransmitters were also affected. Example of gene activity changes in Wnt pathway in schizophrenia is shown on Fig. 6.4c.

Together, these analyses revealed that the dysregulation of gene expression in NCCs derived from patients with schizophrenia was centered on neuronal genes as well as other developmental genes. The upregulated genes were found to be involved in (TP53-dependent) transcription of cell cycle genes, neuronal development, axon guidance, and cholesterol biosynthesis, whereas downregulated genes were involved in cell junction organization, cell–cell junctions, neurotransmitter receptor binding, and cell-cell communication including glutamate receptor signaling, CREB signaling in neurons, and dopamine degradation (Table 6.2).

Thus as predicted by the watershed hypothesis, one observes a common dysregulation of the fundamental developmental functions occurring already at a preneuronal stage.

The second key point of the watershed hypothesis is that dysregulation observed in schizophrenia would be found around specific pathways. Analysis of the 1349 dysregulated genes revealed many neuronal gene ontologies such as neural crest development, regulation of synaptic plasticity, learning, memory, and synapse organization overrepresentation. In addition to these neuronal functions, ontological groups related to glial function such as myelination, axon ensheathment, regulation of glial cell differentiation, and oligodendrocyte differentiation were also found overrepresented. In addition many of the dysregulated genes are involved in pathways centered around neurotransmitter release, developmental biology, axonal growth, and Notch signaling among others. These results suggest that dysregulation of genes observed in schizophrenia represents targeted dysregulation of pathways rather than a random set of genes.

Overexpression studies on select genes have shown that nFGFR1 can cause schizophrenia-like changes in genes such as TH, DISC, Olig2, Wnt7B, and others (Fig. 6.4d). Taken together these results suggest that nFGFR1 plays a strong role in schizophrenia dysregulation. Recent global studies verify that genes affected by overexpressing nFGFR1 in NCC derived from hESC include the same categories and pathways found to be dysregulated in schizophrenia cells (Fig. 6.4d).

6.8 Role of DNA Topology in Schizophrenia

How can expression of hundreds or thousands of genes, which during development are expressed in a coordinated manner, become simultaneously disrupted? How can complex transcriptional gene networks be replaced by new vastly different gene expression profiles? We have hypothesized that a disruption on the level of chromatin structure may be occurring in schizophrenia during the brain development, and as a result vast multigene programs are becoming dysregulated.

Table 6.2 Ingenuity pathway analysis (IPA) of schizophrenia-dysregulated genes targeted by nFGFR1

Ingenuity canonical pathways	Downregulated	Upregulated
Axonal guidance signaling	10/434 (2%)	41/434 (9%)
p53 signaling	3/98 (3%)	14/98 (14%)
ERK/MAPK signaling	6/187 (3%)	16/187 (9%)
Integrin signaling	8/207 (4%)	15/207 (7%)
Wnt/β-catenin signaling	6/169 (4%)	14/169 (8%)
Cyclins and cell cycle regulation	3/78 (4%)	9/78 (12%)
STAT3 pathway	4/73 (5%)	7/73 (10%)
PDGF signaling	3/77 (4%)	8/77 (10%)
Ephrin receptor signaling	4/174 (2%)	14/174 (8%)
PI3K/AKT signaling	4/123 (3%)	10/123 (8%)
NANOG in embryonic stem cell pluripotency	3/111 (3%)	10/111 (9%)
Actin cytoskeleton signaling	6/216 (3%)	14/216 (6%)
Gap junction signaling	5/155 (3%)	11/155 (7%)
Notch signaling	4/38 (11%)	3/38 (8%)
IGF-1 signaling	1/97 (1%)	10/97 (10%)
Synaptic long-term depression	7/142 (5%)	7/142 (5%)
Dopamine-DARPP32 feedback in cAMP signaling	8/161 (5%)	7/161 (4%)
Protein kinase A signaling	11/386 (3%)	17/386 (4%)
CREB signaling in neurons	9/171 (5%)	6/171 (4%)
Semaphorin signaling in neurons	0/53 (0%)	7/53 (13%)
Neurotrophin/TRK signaling	3/67 (4%)	5/67 (7%)
GDNF family ligand–receptor interactions	3/68 (4%)	5/68 (7%)
TR/RXR activation	1/85 (1%)	8/85 (9%)
ErbB2–ErbB3 signaling	1/57 (2%)	6/57 (11%)
Glutamate receptor signaling	4/57 (7%)	3/57 (5%)
Insulin receptor signaling	3/132 (2%)	9/132 (7%)
JAK/Stat signaling	1/72 (1%)	7/72 (10%)
Synaptic long-term potentiation	7/119 (6%)	4/119 (3%)
HGF signaling	3/105 (3%)	7/105 (7%)
Wnt/Ca+ pathway	2/56 (4%)	4/56 (7%)
EGF signaling	4/56 (7%)	2/56 (4%)
Prolactin signaling	3/73 (4%)	4/73 (5%)
Oct4 in mammalian embryonic stem cell pluripotency	1/46 (2%)	4/46 (9%)
G protein-coupled receptor signaling	7/256 (3%)	10/256 (4%)
NF-κB signaling	5/172 (3%)	7/172 (4%)
RhoA signaling	3/122 (2%)	6/122 (5%)
Neuregulin signaling	3/88 (3%)	4/88 (5%)
Glucocorticoid receptor signaling	2/275 (1%)	14/275 (5%)
Calcium signaling	4/178 (2%)	7/178 (4%)
BMP signaling pathway	0/76 (0%)	5/76 (7%)
cAMP-mediated signaling	4/219 (2%)	8/219 (4%)
Retinoic acid-mediated apoptosis signaling	2/61 (3%)	2/61 (3%)

(continued)

Table 6.2 (continued)

Ingenuity canonical pathways	Downregulated	Upregulated
Sonic hedgehog signaling	0/30 (0%)	2/30 (7%)
Telomere extension by telomerase	0/15 (0%)	1/15 (7%)
nNOS signaling in neurons	2/47 (4%)	0/47 (0%)

Selected IPA pathways and numbers of dysregulated genes relative to all genes in pathway are shown

Throughout cellular development, specific subsets of genes become active and can be found in de-condensed chromatin structures known as euchromatin, while transcriptionally inactive regions are tightly packed into complexes known as heterochromatin (Francastel et al. 2000). Temporal and positional DNA–protein interactions lead to the formation of chromatin topologically associated domains (TADs) within which coordinated regulation and expression of multiple loci take place. TADs contain looped together fragments of the same or different chromosomes, spanning distances that can be greater than 1 Mb.

Changes to chromatin structure occur as the cell transitions from one stage of development to another, such as is observed in ESCs differentiating into NCCs (Meshorer and Misteli 2006). Histone modifications, architectural proteins, and transcription factors together determine gene expression patterns and supervise delineation between active and repressed gene loci. Recent ChIP-seq studies in our laboratory have shown nFGFR1 to bind to genomic sites on every chromosome in both human and mouse ESCs. FGFR1-binding sites are significantly remodeled during retinoic acid (RA)-induced ESC stimulation toward neuronal cell differentiation. Gene ontology (GO) analysis revealed genes targeted by nFGFR1 are primarily involved in the maintenance and development of the stem cells in ESCs, while in RA-induced NCCs, nFGFR1 binds to promoters of genes engaged in the formation of mRNA transcripts and in the development of the nervous system. The nFGFR1 regulation of those genes and of neuronal differentiation was demonstrated by transfecting nuclear-active and dominant-negative nFGFR1 forms in ESC models (Stachowiak et al. 2011b, 2015).

We have hypothesized that nFGFR1, through its widespread binding across the genome, could be involved in dynamic organization of chromatin structure. We have considered two models of how nFGFR1 binding may elicit global gene regulation in brain development and dysregulation in schizophrenia and other developmental disorders. In the *cis* model, nFGFR1 acts by binding to transcription enhancer complexes at the promoter sites of individual regulated genes to influence their activities (Fig. 6.5a). In the *trans* model, nFGFR1 binding brings together distant DNA regions enabling their common regulation or dysregulation. We hypothesize that alternating DNA loops may be extruded by nFGFR1 allowing for the execution of the distinct gene programs.

We have begun testing our hypotheses by focusing on the activation of the HoxA genes, which govern the formation of different CNS regions and body parts. The HoxA gene cluster contains 12 HoxA genes A1, A2, A3, A4, A5, A6, A7, A9, A10,

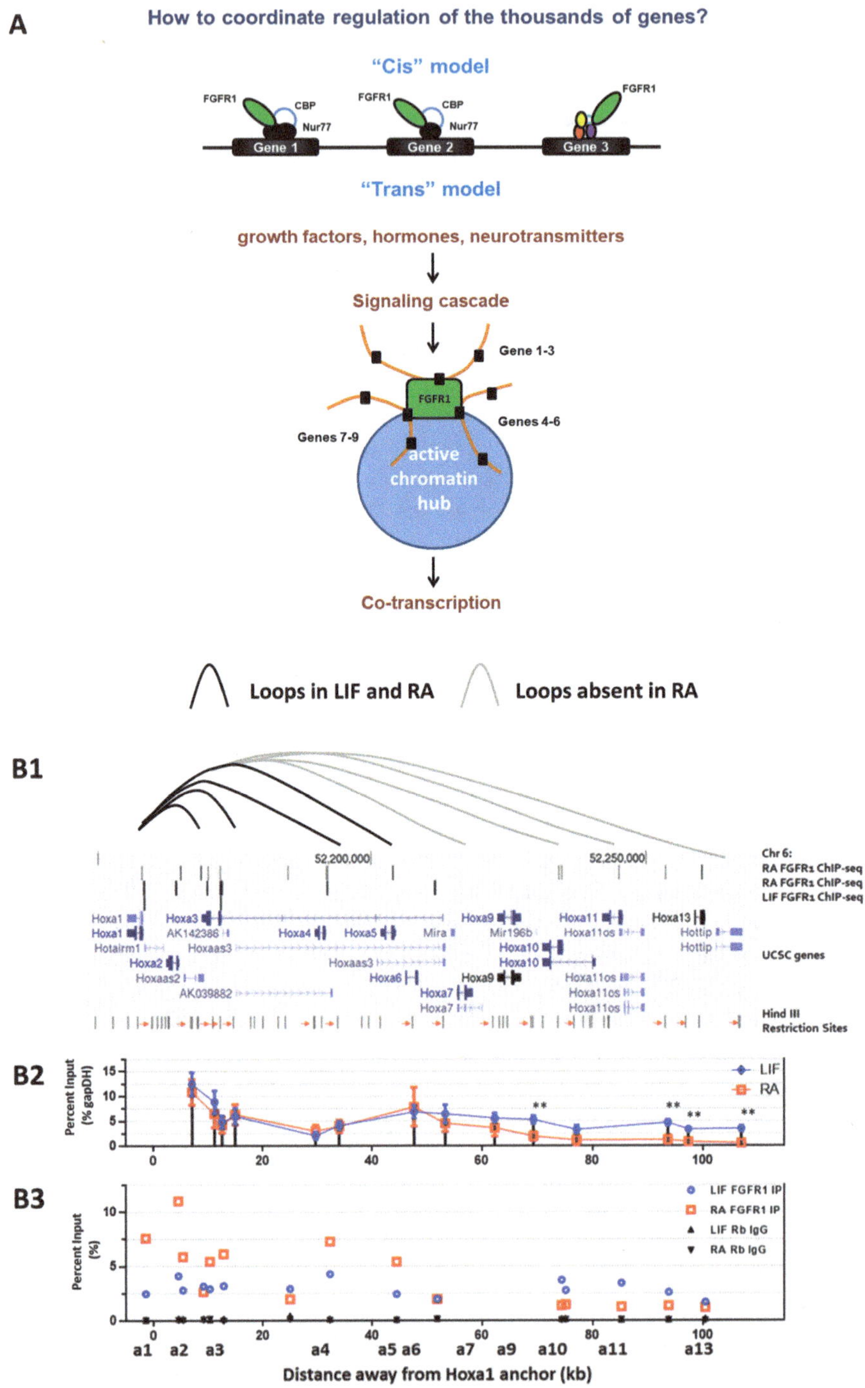

Fig. 6.5 (**a**) Models of global gene regulation by nFGFR1. In the cis model, the regulation of transcription by nFGFR1 occurs at the individual gene sites targeted by nFGFR1. In the trans model, nFGFR1 binds to sites which bring together distal chromatin and forms transcription-

A11, and A13, of which the 3′ genes (HoxA1–HoxA5) are involved in the progressive (head to tail) generation regions of the hindbrain regions and the remaining 5′ genes generate the spinal cord (reviewed in Stachowiak and Stachowiak 2016). nFGFR1 binds to several sites across the HoxA cluster, and during the RA-induced neuronal development, it activates predominantly the 3′ members (HoxA1–HoxA5) of the cluster (Terranova et al. 2015). To analyze the gene interactions within the HoxA cluster, we performed chromatin conformation capture (3C), a PCR-based technique, which estimates proximity between the selected gene loci (Dekker et al. 2002; Hagege et al. 2007). We have recently completed the 3C analysis in the HoxA cluster using HoxA1 as an anchor for measuring its interaction frequencies with downstream HoxA cluster members.

Within an inactive HoxA cluster of the pluripotent mESC, HoxA1 engages in the interactions with all downstream, HoxA2–HoxA13, genes, thus forming loops of different genomic lengths. During RA-induced neuronal differentiation, the HoxA1 locus maintains interactions only with the proximal 3′ HoxA2–HoxA5 genes. The interactions of HoxA1 with HoxA6, HoxA7 HoxA9, HoxA10, HoxA11, HoxA12, and HoxA13 are reduced, and thus the formation of the longer loops no longer occur. These structural changes correlate with nFGFR1 binding, which in RA-treated cells increase at the proximal (3′) HoxA genes but decrease at the distal (5′) HoxA genes. The exclusion of distal HoxA genes from the loops correlates also with their lack of or smaller activation by RA, compared to the proximal HoxA genes. Thus the observed changes in the loop formation isolate differences between the regulations of the hindbrain forming upstream HoxA genes from the spinal cord forming downstream HoxA genes.

These limited findings give backing to our proposed notion that nFGFR1 participates in the formation of the chromatin structures, which enable coordinated gene regulation during brain development and dysregulation in schizophrenia. Our continued experiments aim to identify the nFGFR1-associated chromatin interactions on a genome-wide scale. The results could provide insight into how chromatin topological programs form during development and how they may be disrupted in the schizophrenia leading to the brain malformations discussed in our accompanying chapter.

Fig. 6.5 (continued) associated domains, TADs, with coordinately regulated genes. (**b**) Structural regulation of the HoxA gene cluster—role of nFGFR1. (**b1**) UCSC genome browser tracks containing chromosome location, FGFR1 ChIP-seq binding data, gene location, and Hind III restriction enzyme site tracks. Loops forming between HoxA1 and other Hox genes in pluripotent (LIF) mESC and differentiated NCCs (RA) are indicated. (**b2**) 3C qPCR on nondifferentiated pluripotent mESC (LIF) and differentiated NCCs (RA) measuring the frequency of HoxA1 interacting with downstream HoxA cluster loci. (**b3**) ChIP-qPCR on LIF and RA conditions measuring nFGFR1 binding at ChIP-seq identified loci throughout the HoxA cluster. Control IgG are also indicated

6.9 Summary

In summary, schizophrenia is a developmental disorder characterized by complex aberrations in the structure, wiring, and chemistry of multiple neuronal systems. Over 200 genes, selected by their linkage, association, and expression, have been proposed to contribute to the etiology of the disease. However, there is no single gene whose expression is altered in a majority of schizophrenia patients (Rodriguez-Murillo et al. 2012; Sun et al. 2010). In the proposed transcriptional circuit, INFS integrates incoming developmental signals (St) transmitted by the diverse pathways in which the schizophrenia-linked genes reside. A disruption of any of the individual upstream signal leads to the dysregulation of nFGFR1 which in turn affects the diverse neuro-ontological regulations listed on Fig. 6.6. In addition FGFR1 binds to promoters of the unchanged schizophrenia-linked genes which may lead to their dysregulation as indicated in Fig. 6.6 by the nFGFR1 feedback loops. We propose that the alterations in nFGFR1 interactions with developmental gene networks, miRNA genes, and chromatin topology factors in schizophrenia may underlie the neurodevelopmental pathology of this disease.

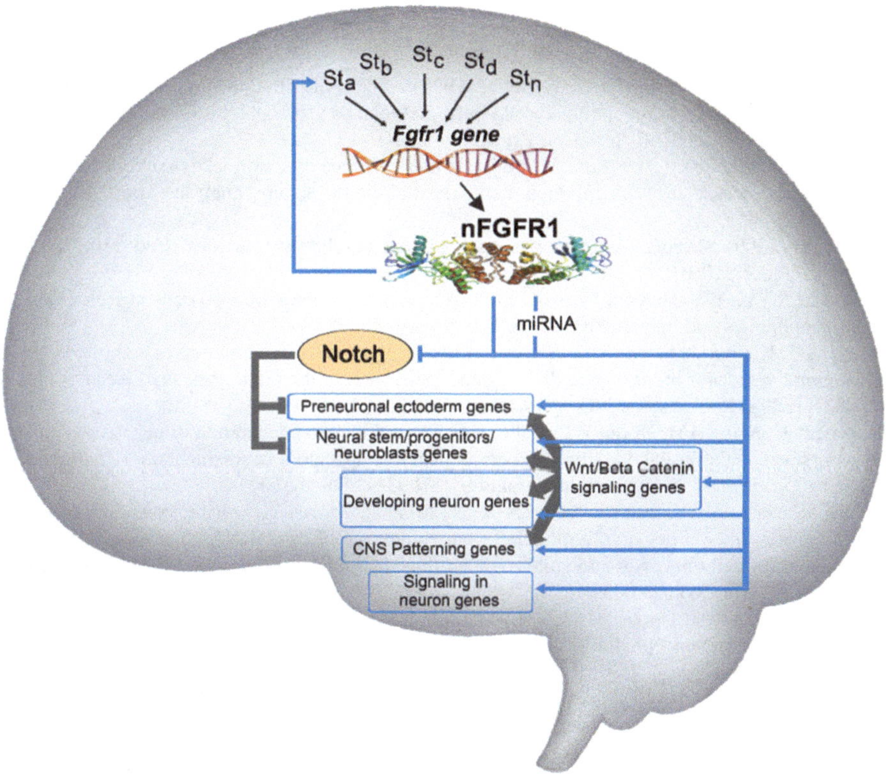

Fig. 6.6 Genetic experiments position the FGFR1 gene at the top of gene hierarchy that directs the development of multicellular animals. FGFR1 governs gastrulation, as well as development of the major body axes, neural plate, central and peripheral nervous systems, and mesoderm by affecting the genes and miRNAs that control the cell cycle, pluripotency, and differentiation (Stachowiak and Stachowiak 2016). This regulation is executed by nuclear protein, nFGFR1, which integrates diverse schizophrenia-linked genes and pathways (Sun et al. 2010). Signals generated by diverse developmental stimuli (St; neurotransmitters, hormones, growth factors, cell contact receptors, etc.,) in embryonic and brain stem cells are propagated by a newly synthesized nFGFR1 protein which translocates into the nucleus and "feeds forward" neurogenic signals to key mRNA and miRNA genes that program and execute different stages of neural development (based on the results of ChIPseq, ChIP, RNAseq, and RNA analyses). For example, nFGFR1 removes the "developmental road block" imposed by the anti-neural Notch1 gene. nFGFR1 targets and activates several master genes that initiate and instruct neural development. Those include proneural Ascl1, and multiple genes in the Wnt pathway. The nFGFR1 binding correlates the activation of genes that stimulate or transduce WNT signals with downregulation of the genes that inhibit Wnt receptors. nFGFR1 binding activates neuronal developmental genes Pax, Id3, Cdx1, IRX3, CREB/CBP signaling genes, and CNS patterning Hox genes. nFGFR1 targets activated axonal guidance genes, and genes involved in synaptic plasticity and development of dopamine and glutamate neurons [based on the Stachowiak and Stachowiak (2016)]. The ST represents diverse signaling pathways in which schizophrenia-linked genes have been found (see Fig. 6.1) and which are also regulated by nFGFR1. In schizophrenia, the mutations of these individual genes, including "weak" copy variations, are proposed to dysregulate this autoregulated genomic circuit and thus lead to broad molecular and developmental dysfunctions [figure is based on information in Narla et al. (2017), Stachowiak and Stachowiak (2016), and Terranova et al. (2015) and linked databases]. Figure drawn by Sun Young Kang

References

Akbarian S, Bunney WE Jr, Potkin SG et al (1993) Altered distribution of nicotinamide-adenine dinucleotide phosphate-diaphorase cells in frontal lobe of schizophrenics implies disturbances of cortical development. Arch Gen Psychiatry 50:169–177

Arnold SE, Ruscheinsky DD, Han LY (1997) Further evidence of abnormal cytoarchitecture of the entorhinal cortex in schizophrenia using spatial point pattern analyses. Biol Psychiatry 42:639–647

Bartel DP (2009) MicroRNAs: target recognition and regulatory functions. Cell 136:215–233. https://doi.org/10.1016/j.cell.2009.01.002

Benes FM, Kwok EW, Vincent SL et al (1998) A reduction of nonpyramidal cells in sector CA2 of schizophrenics and manic depressives. Biol Psychiatry 44:88–97

Bharali DJ, Klejbor I, Stachowiak EK et al (2005) Organically modified silica nanoparticles: a nonviral vector for in vivo gene delivery and expression in the brain. Proc Natl Acad Sci USA 102:11539–11544. https://doi.org/10.1073/pnas.0504926102

Blanchard JJ, Kring AM, Horan WP et al (2011) Toward the next generation of negative symptom assessments: the collaboration to advance negative symptom assessment in schizophrenia. Schizophr Bull 37:291–299. https://doi.org/10.1093/schbul/sbq104

Bobbs AS, Saarela AV, Yatskievych TA et al (2012) Fibroblast growth factor (FGF) signaling during gastrulation negatively modulates the abundance of microRNAs that regulate proteins required for cell migration and embryo patterning. J Biol Chem 287:38505–38514. https://doi.org/10.1074/jbc.M112.400598

Bogerts B, Hantsch J, Herzer M (1983) A morphometric study of the dopamine-containing cell groups in the mesencephalon of normals, Parkinson patients, and schizophrenics. Biol Psychiatry 18:951–969

Brennand KJ, Gage FH (2011) Concise review: the promise of human induced pluripotent stem cell-based studies of schizophrenia. Stem cells 29:1915–1922. https://doi.org/10.1002/stem.762

Brennand KJ, Simone A, Jou J et al (2011) Modelling schizophrenia using human induced pluripotent stem cells. Nature 473:221–225. https://doi.org/10.1038/nature09915

Brennand KJ, Landek-Salgado MA, Sawa A (2014a) Modeling heterogeneous patients with a clinical diagnosis of schizophrenia with induced pluripotent stem cells. Biol Psychiatry 75:936–944. https://doi.org/10.1016/j.biopsych.2013.10.025

Brennand KJ, Silvas J, Kim Y et al (2014b) Phenotypic differences in hiPSC NPCs derived from patients with schizophrenia. Mol Psychiatry 20:361–368

Brennand K, Savas JN, Kim Y et al (2015) Phenotypic differences in hiPSC NPCs derived from patients with schizophrenia. Mol Psychiatry 20:361–368. https://doi.org/10.1038/mp.2014.22

Cannon TD, Keller MC (2006) Endophenotypes in the genetic analyses of mental disorders. Annu Rev Clin Psychol 2:267–290. https://doi.org/10.1146/annurev.clinpsy.2.022305.095232

Chiang CH, Su Y, Wen Z et al (2011) Integration-free induced pluripotent stem cells derived from schizophrenia patients with a DISC1 mutation. Mol Psychiatry 16:358–360. https://doi.org/10.1038/mp.2011.13

Chioni AM, Grose R (2012) FGFR1 cleavage and nuclear translocation regulates breast cancer cell behavior. J Cell Biol 197:801–817. https://doi.org/10.1083/jcb.201108077

Ciruna B, Rossant J (2001) FGF signaling regulates mesoderm cell fate specification and morphogenetic movement at the primitive streak. Dev Cell 1:37–49. https://doi.org/10.1016/S1534-5807(01)00017-X

Ciruna BG, Schwartz L, Harpal K et al (1997) Chimeric analysis of fibroblast growth factor receptor-1 (Fgfr1) function: a role for FGFR1 in morphogenetic movement through the primitive streak. Development 124:2829–2841

Claus P, Doring F, Gringel S et al (2003) Differential intranuclear localization of fibroblast growth factor-2 isoforms and specific interaction with the survival of motor neuron protein. J Biol Chem 278:479–485. https://doi.org/10.1074/jbc.M206056200

Coleman SJ, Chioni AM, Ghallab M et al (2014) Nuclear translocation of FGFR1 and FGF2 in pancreatic stellate cells facilitates pancreatic cancer cell invasion. EMBO Mol Med 6:467–481. https://doi.org/10.1002/emmm.201302698

Connor SE, Ng V, McDonald C et al (2004) A study of hippocampal shape anomaly in schizophrenia and in families multiply affected by schizophrenia or bipolar disorder. Neuroradiology 46:523–534. https://doi.org/10.1007/s00234-004-1224-0

Davis KL, Stewart DG, Friedman JI et al (2003) White matter changes in schizophrenia: evidence for myelin-related dysfunction. Arch Gen Psychiatry 60:443–456. https://doi.org/10.1001/archpsyc.60.5.443

Deep-Soboslay A, Benes FM, Haroutunian V et al (2011) Psychiatric brain banking: three perspectives on current trends and future directions. Biol Psychiatry 69:104–112. https://doi.org/10.1016/j.biopsych.2010.05.025

Dekker J, Rippe K, Dekker M et al (2002) Capturing chromosome conformation. Science 295:1306–1311. https://doi.org/10.1126/science.1067799

Dequeant ML, Pourquie O (2008) Segmental patterning of the vertebrate embryonic axis. Nat Rev Genet 9:370–382. https://doi.org/10.1038/nrg2320

Erlenmeyer-Kimling L (2000) Neurobehavioral deficits in offspring of schizophrenic parents: liability indicators and predictors of illness. Am J Med Genet 97:65–71

Fang X, Stachowiak EK, Dunham-Ems SM et al (2005) Control of CREB-binding protein signaling by nuclear fibroblast growth factor receptor-1: a novel mechanism of gene regulation. J Biol Chem 280:28451–28462

Fatemi SH, Folsom TD (2009) The neurodevelopmental hypothesis of schizophrenia, revisited. Schizophr Bull 35:528–548. https://doi.org/10.1093/schbul/sbn187

Foussias G, Mann S, Zakzanis KK et al (2011) Prediction of longitudinal functional outcomes in schizophrenia: the impact of baseline motivational deficits. Schizophr Res 132:24–27. https://doi.org/10.1016/j.schres.2011.06.026

Francastel C, Schubeler D, Martin DI et al (2000) Nuclear compartmentalization and gene activity. Nat Rev Mol Cell Biol 1:137–143. https://doi.org/10.1038/35040083

Hagege H, Klous P, Braem C et al (2007) Quantitative analysis of chromosome conformation capture assays (3C-qPCR). Nat Protoc 2:1722–1733. https://doi.org/10.1038/nprot.2007.243

Hanzawa S, Bae JK, Bae YJ et al (2013) Psychological impact on caregivers traumatized by the violent behavior of a family member with schizophrenia. Asian J Psychiatry 6:46–51. https://doi.org/10.1016/j.ajp.2012.08.009

Hashimoto-Torii K, Torii M, Fujimoto M et al (2014) Roles of heat shock factor 1 in neuronal response to fetal environmental risks and its relevance to brain disorders. Neuron 82:560–572. https://doi.org/10.1016/j.neuron.2014.03.002

Horbinski C, Stachowiak EK, Chandrasekaran V et al (2002) Bone morphogenetic protein-7 stimulates initial dendritic growth in sympathetic neurons through an intracellular fibroblast growth factor signaling pathway. J Neurochem 80:54–63

Howard R, Rabins PV, Seeman MV et al (2000) Late-onset schizophrenia and very-late-onset schizophrenia-like psychosis: an international consensus. The International Late-Onset Schizophrenia Group. Am J Psychiatry 157:172–178. https://doi.org/10.1176/appi.ajp.157.2.172

Hu Y, Fang X, Dunham SM et al (2004) 90-kDa ribosomal S6 kinase is a direct target for the nuclear fibroblast growth factor receptor 1 (FGFR1): role in FGFR1 signaling. J Biol Chem 279:29325–29335. https://doi.org/10.1074/jbc.M311144200

International Schizophrenia Consortium (2008) Rare chromosomal deletions and duplications increase risk of schizophrenia. Nature 455:237–241. https://doi.org/10.1038/nature07239

Kasper LH, Fukuyama T, Biesen MA et al (2006) Conditional knockout mice reveal distinct functions for the global transcriptional coactivators CBP and p300 in T-cell development. Mol Cell Biol 26:789–809. https://doi.org/10.1128/MCB.26.3.789-809.2006

Keshavan MS (1999) Development, disease and degeneration in schizophrenia: a unitary pathophysiological model. J Psychiatr Res 33:513–521

Keshavan MS, Hogarty GE (1999) Brain maturational processes and delayed onset in schizophrenia. Dev Psychopathol 11:525–543

Kirov G, Pocklington AJ, Holmans P et al (2012) De novo CNV analysis implicates specific abnormalities of postsynaptic signalling complexes in the pathogenesis of schizophrenia. Mol Psychiatry 17:142–153. https://doi.org/10.1038/mp.2011.154

Kneeland RE, Fatemi SH (2013) Viral infection, inflammation and schizophrenia. Prog Neuro-Psychopharmacol Biol Psychiatry 42:35–48. https://doi.org/10.1016/j.pnpbp.2012.02.001

Lee YW, Terranova C, Birkaya B et al (2012) A novel nuclear FGF receptor-1 partnership with retinoid and Nur receptors during developmental gene programming of embryonic stem cells. J Cell Biochem 113:2920–2936. https://doi.org/10.1002/jcb.24170

Lee YW, Stachowiak EK, Birkaya B et al (2013) NGF-induced cell differentiation and gene activation is mediated by integrative nuclear FGFR1 signaling (INFS). PLoS One 8:e68931. https://doi.org/10.1371/journal.pone.0068931

Lloyd T, Dazzan P, Dean K et al (2008) Minor physical anomalies in patients with first-episode psychosis: their frequency and diagnostic specificity psychological medicine 38:71–77. https://doi.org/10.1017/S0033291707001158

Malhotra D, McCarthy S, Michaelson JJ et al (2011) High frequencies of de novo CNVs in bipolar disorder and schizophrenia. Neuron 72:951–963. https://doi.org/10.1016/j.neuron.2011.11.007

Meshorer E, Misteli T (2006) Chromatin in pluripotent embryonic stem cells and differentiation. Nat Rev Mol Cell Biol 7:540–546. https://doi.org/10.1038/nrm1938

Miller BH, Zeier Z, Xi L et al (2012) MicroRNA-132 dysregulation in schizophrenia has implications for both neurodevelopment and adult brain function. Proc Natl Acad Sci USA 109:3125–3130. https://doi.org/10.1073/pnas.1113793109

Moreau MP, Bruse SE, David-Rus R et al (2011) Altered microRNA expression profiles in postmortem brain samples from individuals with schizophrenia and bipolar disorder. Biol Psychiatry 69:188–193. https://doi.org/10.1016/j.biopsych.2010.09.039

Myers JM, Martins GG, Ostrowski J et al (2003) Nuclear trafficking of FGFR1: a role for the transmembrane domain. J Cell Biochem 88:1273–1291

Narla ST, Lee YW, Benson CA et al (2017) Common developmental genome deprogramming in schizophrenia – role of integrative nuclear FGFR1 signaling (INFS). Schizophr Res 185:17–32. https://doi.org/10.1016/j.schres.2016.12.012

Need AC, Attix DK, McEvoy JM et al (2009) A genome-wide study of common SNPs and CNVs in cognitive performance in the CANTAB. Hum Mol Genet 18:4650–4661. https://doi.org/10.1093/hmg/ddp413

Nguyen PT, Tsunematsu T, Yanagisawa S et al (2013) The FGFR1 inhibitor PD173074 induces mesenchymal-epithelial transition through the transcription factor AP-1. Br J Cancer 109:2248–2258. https://doi.org/10.1038/bjc.2013.550

Partanen J, Schwartz L, Rossant J (1998) Opposite phenotypes of hypomorphic and Y766 phosphorylation site mutations reveal a function for Fgfr1 in anteroposterior patterning of mouse embryos. Genes Dev 12:2332–2344

Paulsen BD, Maciel RD, Galina A et al (2011) Altered oxygen metabolism associated to neurogenesis of induced pluripotent stem cells derived from a schizophrenic patient. Cell Transplant. https://doi.org/10.3727/096368911X600957

Perkins DO, Jeffries CD, Jarskog LF et al (2007) microRNA expression in the prefrontal cortex of individuals with schizophrenia and schizoaffective disorder. Genome Biol 8:R27. https://doi.org/10.1186/gb-2007-8-2-r27

Popovici C, Fallet M, Marguet D et al (2006) Intracellular trafficking of LET-756, a fibroblast growth factor of C. elegans, is controlled by a balance of export and nuclear signals. Exp Cell Res 312:1484–1495. https://doi.org/10.1016/j.yexcr.2006.01.012

Rajji TK, Ismail Z, Mulsant BH (2009) Age at onset and cognition in schizophrenia: meta-analysis. Br J Psychiatry J Ment Sci 195:286–293. https://doi.org/10.1192/bjp.bp.108.060723

Rehn AE, Rees SM (2005) Investigating the neurodevelopmental hypothesis of schizophrenia. Clin Exp Pharmacol Physiol 32:687–696. https://doi.org/10.1111/j.1440-1681.2005.04257.x

Robicsek O, Karry R, Petit I et al (2013) Abnormal neuronal differentiation and mitochondrial dysfunction in hair follicle-derived induced pluripotent stem cells of schizophrenia patients. Mol Psychiatry. https://doi.org/10.1038/mp.2013.67

Rodriguez-Murillo L, Gogos JA, Karayiorgou M (2012) The genetic architecture of schizophrenia: new mutations and emerging paradigms. Annu Rev Med 63:63–80. https://doi.org/10.1146/annurev-med-072010-091100

Roussos P, Guennewig B, Kaczorowski DC et al (2016) Activity-dependent changes in gene expression in schizophrenia human-induced pluripotent stem cell neurons. JAMA Psychiatry 73:1180–1188. https://doi.org/10.1001/jamapsychiatry.2016.2575

Rummel-Kluge C, Komossa K, Schwarz S et al (2012) Second-generation antipsychotic drugs and extrapyramidal side effects: a systematic review and meta-analysis of head-to-head comparisons. Schizophr Bull 38:167–177. https://doi.org/10.1093/schbul/sbq042

Saha S, Chant D, Welham J et al (2005) A systematic review of the prevalence of schizophrenia. PLoS medicine 2:e141. https://doi.org/10.1371/journal.pmed.0020141

Santarelli DM, Beveridge NJ, Tooney PA et al (2011) Upregulation of dicer and microRNA expression in the dorsolateral prefrontal cortex Brodmann area 46 in schizophrenia. Biol Psychiatry 69:180–187. https://doi.org/10.1016/j.biopsych.2010.09.030

Schiller D, Zuckerman L, Weiner I (2006) Abnormally persistent latent inhibition induced by lesions to the nucleus accumbens core, basolateral amygdala and orbitofrontal cortex is reversed by clozapine but not by haloperidol. J Psychiatric Res 40:167–177. https://doi.org/10.1016/j.jpsychires.2005.03.002

Sherman L, Stocker KM, Morrison R et al (1993) Basic fibroblast growth factor (bFGF) acts intracellularly to cause the transdifferentiation of avian neural crest-derived Schwann cell precursors into melanocytes. Development 118:1313–1326

Shi W, Du J, Qi Y et al (2012) Aberrant expression of serum miRNAs in schizophrenia. J Psychiatr Res 46:198–204. https://doi.org/10.1016/j.jpsychires.2011.09.010

Srikanth P, Han K, Callahan DG et al (2015) Genomic DISC1 disruption in hiPSCs alters Wnt signaling and neural cell fate. Cell Rep 12:1414–1429. https://doi.org/10.1016/j.celrep.2015.07.061

Stachowiak MK, Stachowiak EK (2016) Evidence-based theory for integrated genome regulation of ontogeny—an unprecedented role of nuclear FGFR1 signaling. J Cell Physiol 231:1199–1218. https://doi.org/10.1002/jcp.25298

Stachowiak EK, Fang X, Myers J et al (2003) cAMP-induced differentiation of human neuronal progenitor cells is mediated by nuclear fibroblast growth factor receptor-1 (FGFR1). J Neurochem 84:1296–1312

Stachowiak MK, Maher PA, Stachowiak EK (2007) Integrative nuclear signaling in cell development: a role for FGF receptor-1. DNA Cell Biol 26:811–826. https://doi.org/10.1089/dna.2007.0664

Stachowiak EK, Roy I, Lee YW et al (2009a) Targeting novel integrative nuclear FGFR1 signaling by nanoparticle-mediated gene transfer stimulates neurogenesis in the adult brain. Integr Biol (Camb) 1:394–403. https://doi.org/10.1039/b902617g

Stachowiak EK, Roy I, Lee Y-W, Capacchietti M, Aletta JM, Prasad PN, Stachowiak MK (2009b) Targeting novel integrative nuclear FGFR1 signaling by nanoparticle-mediated gene transfer stimulates neurogenesis in adult brain. Integr Biol 1:394–403

Stachowiak EK, Roy I, Stachowiak MK (2011a) Triggering neuronogenesis by endogenous brain stem cells with DNA nanoplexes. In: Stachowiak ES Tzanakakis MK (ed) Stem cells: from mechanisms to technologies. World Scientific Publishing, Hackensack, NJ, pp 333–359

Stachowiak MK, Stachowiak EK, Aletta JM, Tzanakakis ES (2011b) A common integrative nuclear signaling module for stem cell development. In: Stachowiak MK, Tzanakakis ES (eds) Stem cells: from mechanisms to technologies. World Scientific, Hackensack, NJ, pp 87–132

Stachowiak MK, Kucinski A, Curl R et al (2013) Schizophrenia: a neurodevelopmental disorder--integrative genomic hypothesis and therapeutic implications from a transgenic mouse model. Schizophr Res 143:367–376. https://doi.org/10.1016/j.schres.2012.11.004

Stachowiak MK, Birkaya B, Aletta JM et al (2015) Nuclear FGF receptor-1 and CREB binding protein: an integrative signaling module. J Cell Physiol 230:989–1002. https://doi.org/10.1002/jcp.24879

Stachowiak EK, Benson CA, Narla ST et al (2017) Cerebral organoids reveal early cortical maldevelopment in schizophrenia—computational anatomy and genomics, role of FGFR1. Translational Psychiatry 7:6

Stefansson H, Rujescu D, Cichon S et al (2008) Large recurrent microdeletions associated with schizophrenia. Nature 455:232–236. https://doi.org/10.1038/nature07229

Stuhlmiller TJ, Garcia-Castro MI (2012) FGF/MAPK signaling is required in the gastrula epiblast for avian neural crest induction. Development 139:289–300. https://doi.org/10.1242/dev.070276

Sullivan PF, Kendler KS, Neale MC (2003) Schizophrenia as a complex trait: evidence from a meta-analysis of twin studies. Arch Gen Psychiatry 60:1187–1192. https://doi.org/10.1001/archpsyc.60.12.1187

Sun J, Jia P, Fanous AH et al (2010) Schizophrenia gene networks and pathways and their applications for novel candidate gene selection. PLoS One 5:e11351. https://doi.org/10.1371/journal.pone.0011351

Terranova C, Narla ST, Lee YW et al (2015) Global developmental gene programing involves a nuclear form of fibroblast growth factor receptor-1 (FGFR1). PLoS One 10:e0123380. https://doi.org/10.1371/journal.pone.0123380

Topol A, Zhu S, Tran N et al (2015) Altered WNT signaling in human induced pluripotent stem cell neural progenitor cells derived from four schizophrenia patients. Biol Psychiatry 78:e29–e34. https://doi.org/10.1016/j.biopsych.2014.12.028

Vo N, Goodman RH (2001) CREB-binding protein and p300 in transcriptional regulation. J Biol Chem 276:13505–13508. https://doi.org/10.1074/jbc.R000025200

Walsh T, McClellan JM, McCarthy SE et al (2008) Rare structural variants disrupt multiple genes in neurodevelopmental pathways in schizophrenia. Science 320:539–543. https://doi.org/10.1126/science.1155174

Welter D, MacArthur J, Morales J et al (2014) The NHGRI GWAS catalog, a curated resource of SNP-trait associations. Nucleic Acids Res 42:D1001–D1006. https://doi.org/10.1093/nar/gkt1229

Wen Z, Nguyen HN, Guo Z et al (2014) Synaptic dysregulation in a human iPS cell model of mental disorders. Nature. https://doi.org/10.1038/nature13716

Xu B, Roos JL, Levy S et al (2008) Strong association of de novo copy number mutations with sporadic schizophrenia. Nat Genet 40:880–885. https://doi.org/10.1038/ng.162

Yoon KJ, Nguyen HN, Ursini G et al (2014) Modeling a genetic risk for schizophrenia in iPSCs and mice reveals neural stem cell deficits associated with adherens junctions and polarity. Cell Stem Cell 15:79–91. https://doi.org/10.1016/j.stem.2014.05.003

Younger ST, Corey DR (2011) Transcriptional gene silencing in mammalian cells by miRNA mimics that target gene promoters. Nucleic Acids Res 39:5682–5691. https://doi.org/10.1093/nar/gkr155

Yu DX, Di Giorgio FP, Yao J et al (2014) Modeling hippocampal neurogenesis using human pluripotent stem cells. Stem Cell Rep 2:295–310. https://doi.org/10.1016/j.stemcr.2014.01.009

Chapter 7
Genome Editing in Human Neural Stem and Progenitor Cells

Raul Bardini Bressan and Steven M. Pollard

Abstract Experimental tools for precise manipulation of mammalian genomes enable reverse genetic approaches to explore biology and disease. Powerful genome editing technologies built upon designer nucleases, such as CRISPR/Cas9, have recently emerged. Parallel progress has been made in methodologies for the expansion and differentiation of human pluripotent and tissue stem cells. Together these innovations provide a remarkable new toolbox for human cellular genetics and are opening up vast opportunities for discoveries and applications across the breadth of life sciences research. In this chapter, we review the emergence of genome editing technologies and how these are being deployed in studies of human neurobiology, neurological disease, and neuro-oncology. We focus our discussion on CRISPR/Cas9 and its application in studies of human neural stem and progenitor cells.

7.1 Introduction

Genetic manipulations of mammalian embryonic stem cells or embryos have provided a wealth of gene knock-outs, conditional alleles, or reporter alleles. Such reverse genetic approaches—i.e., engineering-specific genetic changes and monitoring of the phenotypic consequences—have been the chief experimental tool to ascribe gene function. However, these techniques can be challenging as they require significant expertise and investment of resources for both generation of recombinant DNA constructs and downstream selection and screening for correctly engineered cells. Consequently, the precise engineering of primary human cells has not been widely used.

Recently, several parallel technological advances have vastly increased the ease and scope of reverse genetic approaches in human cells. First, the CRISPR/Cas9 platform now enables routine production of highly effective site-specific nucleases for genome editing. Second, induced pluripotent stem cell (iPS) technologies enable

R. B. Bressan · S. M. Pollard (✉)
MRC Centre for Regenerative Medicine and Edinburgh Cancer Research Centre, University of Edinburgh, Edinburgh, UK
e-mail: steven.pollard@ed.ac.uk

L. Buzanska (ed.), *Human Neural Stem Cells*, Results and Problems in Cell Differentiation 66, https://doi.org/10.1007/978-3-319-93485-3_7

routine production of human pluripotent cell cultures, from healthy or diseased donors. Third, continuous improvements in stem cell culture protocols allow more reliable expansion and differentiation of human cells. Finally, and often not given due prominence, is the relentlessly falling cost and increasing capacity of technologies for reading and writing DNA at scale. Faster, cheaper, and more elaborate targeting vectors, or libraries of plasmids, can be easily built.

These transformative technologies—CRISPR, human stem cell culture/differentiation, sequencing, and synthesis of DNA—are together driving a new era of functional genetic analysis in human pluripotent and somatic cells. This provides a solution to the long-standing issue of how to functionally annotate the coding and regulatory elements in the human genome. Likely, there will be significant new opportunities for discoveries in human cell biology without the confounding issues of genetic corruption associated with current "classic" cell lines, such as HeLa (Hyman and Simons 2011). These possibilities are generating considerable excitement across many branches of biology and biomedical science.

In this chapter, we discuss the ways in which the CRISPR/Cas9 genome editing technology can be deployed in primary human cells models to unravel genetic mechanisms involved in the development, function, and pathology of the human central nervous system (CNS).

7.2 Genome Editing BC (Before CRISPR): Gene Targeting and Programmable Nucleases

Eukaryotic genomes contain billions of DNA base pairs. Engineering changes at predetermined loci are therefore inherently challenging. Early transgenic technologies relied on random insertion of sequences delivered by viral vectors, with the obvious drawback of lack of control and risks of collateral damage to the genome. A major breakthrough in the field came with the development of gene targeting technology, which employs endogenous cellular homologous recombination (HR) mechanisms to replace a small portion of the genome with exogenously delivered donor DNA sequences (Smithies et al. 1985; Thomas and Capecchi 1987). This has supported the production of a wide repertoire of tools for mouse genetics and rapidly became the foundational technology; reviewed in Capecchi (2005).

Gene targeting is, however, extremely inefficient. Random integrations of exogenous constructs (also refered to as targeting vectors) occur more frequently than HR, and correct targeting events are only found in a handful of cells out of the millions transfected (Lin et al. 1985; Smithies et al. 1985; Thomas et al. 1986). This low efficiency means that selection strategies are needed, and significant numbers of clonal derivatives must be screened to isolate the correctly engineered cells. Consequently, conventional gene targeting experiments can only be achieved in cell cultures that are easily transfectable and clonally expandable. Moreover, the targeting vectors typically required large homology arms (5–10 kb), requiring handling of large bacterial artificial

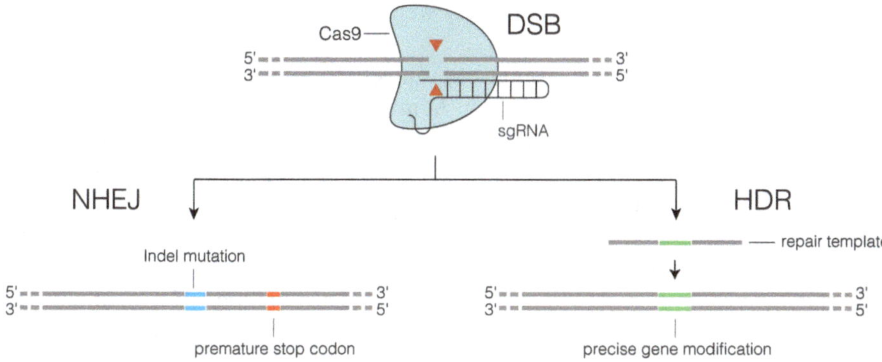

Fig. 7.1 Site-specific DSBs facilitate precise gene editing. DSBs generated by site-specific, programmable nucleases (depicted is Cas9) can be repaired in one of two ways by endogenous DNA repair machinery. In the error-prone NHEJ pathway, the ends of the DSB are processed and rejoined, normally resulting in random indel mutations at the site of junction. Indel mutations occurring within the coding region can change open reading frame and create premature stop codon, eventually leading to the inactivation of the gene. Alternatively, an exogenous repair template can be supplied to leverage the HDR pathway, allowing high fidelity and precise editing (which is referred to as gene targeting)

chromosome-based (BAC) targeting vectors. Although methods such as recombineering reduced some of the barriers, a significant level of technical expertise is still required (Court et al. 2002). In short, gene targeting as originally devised is technically challenging and time-consuming and remained largely restricted to mouse ES cell cultures. Improved efficiencies were needed to open up these approaches to human pluripotent and somatic cells.

This situation began to alter when it was demonstrated that gene targeting occurred with improved efficiency when accompanied with a double-stranded DNA break (DSB) at the target site (Rouet et al. 1994; Jasin 1996). Using a yeast rare cutting meganuclease (I-Sce I), Rouet et al. demonstrated that site-specific DSBs were able to promote endogenous HR-based DNA repair (Rouet et al. 1994), thereby increasing the efficiency of gene targeting by orders of magnitude. In addition to homology-directed repair (HDR), it was also found that these DSBs can be repaired by an alternative non-homologous end-joining (NHEJ) pathway. This is an error-prone pathway that results in the generation of random insertion or deletion mutations (indels) at the break site, and can therefore be used to disrupt gene function without the delivery of exogenous DNA (Fig. 7.1). These observations provided the foundations upon which recent genome editing technologies have been built.

Naturally occurring meganucleases, however, have a key limitation: the large recognition sequence is typically absent at the site of interest. How could DSBs be introduced at any desired site in the genome? This required the emergence of designer nucleases that could be produced by fusion of DNA binding domains (engineered to bind a specific sequence) to a restriction enzyme endonuclease domain (e.g., FokI). Two platforms emerged: zinc finger nucleases (ZFNs) and latterly TAL effector domains, to create TAL effector nucleases (TALENs) (Joung and Sander 2012). Such nucleases could be programmed to bind and cut at specific sites in the genome and worked relatively well in a variety of cell types and model organisms.

ZFNs and TALENs vastly improved the efficiencies of the gene targeting, allowing broad use of the technology in human pluripotent stem cells for the first time (reviewed in Hockemeyer et al. 2011; Joung and Sander 2012). This newfound reliability and flexibility in the types of precise genetic changes that can be engineered called for a new term in the lexicon of molecular genetics: *genome editing*. The process of generating precise insertions, deletions, or replacement modifications to the genome in cells or organisms in a tailor-made manner, analogous to a word processor, with cut, paste, and insert functions (Hsu et al. 2014).

However, widespread use of genome editing with ZFNs and TALENs was still hampered by technical difficulties in designing, assembling, and delivering these reagents to eukaryotic cells. It remained tricky to deploy by non-experts and was somewhat limited by poor specificity and risks of off-target effects (reviewed in Hsu et al. 2014). The true democratization of genome editing technologies has been realized with the advent of the CRISPR/Cas9 technology. This has provided a much simpler tool for the research community. It has rapidly been adopted by the field to become the prominent technology platform for genome editing.

7.3 The CRISPR/Cas9 System

CRISPR (clustered regularly interspaced short palindromic repeat) and their associated genes (*Cas*) are RNA-guided nucleases that cleave foreign genetic elements; reviewed in Bhaya et al. (2011). They can be viewed as a form of microbial adaptive immune system that memorizes previous infections by integrating short sequences of the invaders genome. Three major classes CRISPR/Cas systems (I–III) have been defined in bacteria and archaea. Each comprises a cluster of *Cas* genes that encode RNA-guided nucleases, noncoding RNAs, and distinctive array of repetitive elements. These repetitive elements are interspaced by short variable sequences, known as spacers, originally derived from an invading pathogen DNA (Makarova et al. 2011). Once transcribed, the CRISPR spacers direct Cas nucleases to cleave the re-invading pathogen DNA at complementary regions that contain a critical protospacer adjacent motif (PAM) (Makarova et al. 2011).

The Type II CRISPR system—also known as CRISPR/Cas9—is one of the best characterized and comprises a Cas9 nuclease, the crRNA array (encoding the CRISPR spacers), and an essential auxiliary trans-activating crRNA (tracrRNA) that facilitates the processing of the crRNA array into discrete units. In the case of *Streptococcus pyogenes* CRISPR/Cas9, each crRNA unit contains a 20-nt guide sequence that directs the Cas9 via Watson-Crick base pairing to a 20-bp DNA target that immediately precedes a 5′-NGG PAM. Once bound, the two separate nuclease domains catalyze a DSB at the target site (Jinek et al. 2012).

The biochemical characterization of the RNA–protein complex of the CRISPR/Cas9 system was reported in 2012 and immediately suggested a range of potential applications as a programmable nuclease in eukaryotic cells (Jinek et al. 2012). Soon after, in 2013, successful repurposing and application of the CRISPR/Cas9 for mammalian genome editing purposes was reported (Jinek et al. 2013; Cong et al. 2013; Mali et al. 2013; Cho et al. 2013). The successful implementation in eukaryotic cells involved heterologous expression of human codon-optimized Cas9 from *S. pyogenes* and the requisite RNA components, which were adapted by a chimeric fusion of the crRNA and a full-length tracrRNA to create a single-guide RNA (sgRNA) (Cong et al. 2013). For simplicity, we herein refer to this repurposed CRISPR/Cas9 system as CRISPR.

A key advantage of CRISPR when compared to ZFNs and TALENs is its simplicity. DNA-binding specificity is encoded exclusively by the sgRNA. There is no need for cumbersome protein engineering or elaborate plasmid assembly methods. Importantly, the design of the sgRNAs follows simple rules, and appropriate targets are found at high density throughout the genome. The resulting workflow to generate bespoke site-specific nucleases is shortened from weeks to days. CRISPR can therefore be readily implemented by any laboratory familiar with the basic tools of cell culture and molecular biology. Furthermore, CRISPR has the added advantage of enabling straightforward library screening as many thousands of gRNAs can be combined and simultaneously be delivered as pools (Koike-Yusa et al. 2013; Wang et al. 2014). In addition, engineering of large-scale chromosomal deletions and rearrangements can be introduced when sgRNAs are used as pairs to induce DSBs at sites that flank the region of interest (Choi and Meyerson 2014; Torres et al. 2014; Kraft et al. 2015). A myriad of applications are now emerging: lineage tracing reporters, RNA editing, epigenetic reprogramming, and tethering or tracking DNA topologies. This is a truly transformative technology with diverse applications.

7.4 CRISPR-Based Genome Editing in Human Pluripotent Stem Cells and Neural Stem Cells

In this section, we describe some of the most common genetic manipulations made possible by CRISPR and early examples of their value to explore the neural lineages. These include gene knock-outs, generation of conditional and reporter alleles, engineering of point mutations, large-scale chromosomal engineering, and high-throughput genetic screenings (Fig. 7.2). In most studies, CRISPR gene editing has been performed in human pluripotent cells (hESCs or iPSCs), followed by in vitro differentiation into the neural cell type of interest. Also, CRISPR techniques can be directly applied to neural stem cells (NSCs) expanded from human fetal tissues (Bressan et al. 2017), which sidesteps the need for ES/iPSC differentiation protocols. The genome editing approaches discussed below can also be performed in cancer stem cells; an example is patient-derived glioblastoma cells—the most common and lethal form of brain cancer (Bulstrode et al. 2017; Mohammad et al. 2017; Dewari et al. 2018).

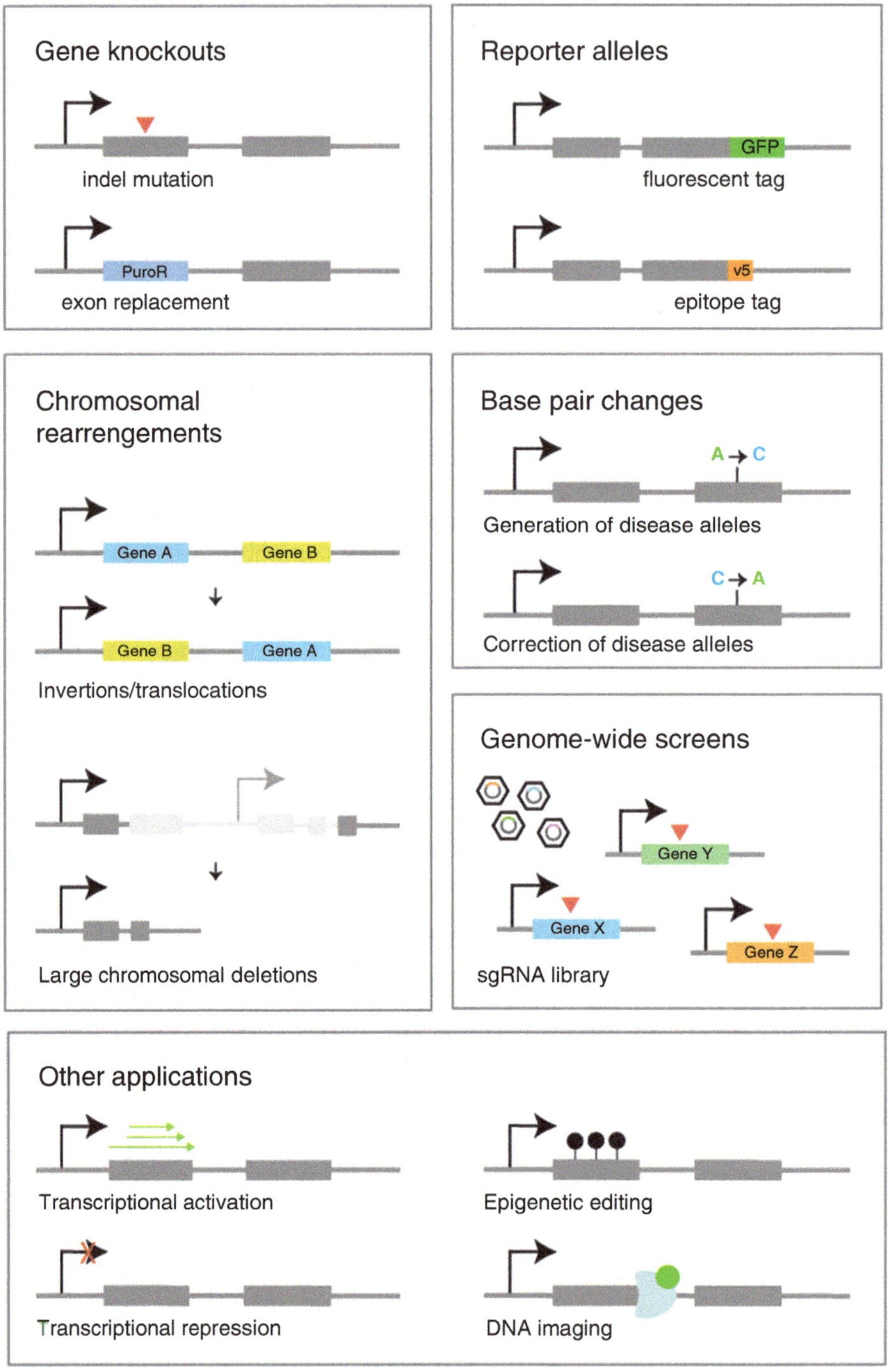

Fig. 7.2 The CRISPR/Cas9 genome editing toolkit. Examples of reverse genetic approaches to study human NS cells using genome editing include (i) generation of single and multiple gene knock-outs, (ii) chromosome inversions, deletions, and translocations, (iii) insertion of fluorescent

7.4.1 Transgene Insertion at Defined "Safe-Harbor" Loci

"Safe-harbor" loci are sites where inserted transgenes are known to have stable expression with less risk of silencing. This also avoids issues of random integration disturbing expression of endogenous genes. Using CRISPR, we and others have demonstrated efficient transgene insertion into widely used safe-habor locus *AAVS1* in human ES/iPS (Takayama et al. 2017), fetal-derived NSC (Bressan et al. 2017), and primary glioblastoma stem cell lines (Bulstrode et al. 2017). Gene targeting efficiencies are high, and stable expression of exogenous DNA elements can be achieved in the stem cells and their resulting progeny. This has been of great value for introducing expression cassettes for gain-of-function studies, or constitutive reporter systems enabling stable and consistent expression.

7.4.2 Gene Knock-outs

Targeted inactivation of endogenous genes via deletion or mutation of coding sequences is the gold standard experimental approach to defining gene function. Unlike RNAi technologies, loss of the gene product is unequivocal and permanent, while risks of nonspecific or off-target effects are minimized. The most straightforward strategy for gene ablation using CRISPR is to co-deliver the Cas9 protein with an sgRNA in order to create a DSB within a critical coding exon. Once repaired by NHEJ, indels are created that can disrupt the open reading frame and result in a nonfunctional protein product. For instance, in human NSC systems, this approach has been used to uncover the roles of *CHD8* loss-of-function mutation in the pathogenesis of autism spectrum disorders (Wang et al. 2015) as well as to test the effect of the putative attachment factor AXL on the microcephaly-causing Zika virus infection (Wells et al. 2016).

To achieve high efficiency of biallelic knock-out (i.e., disruption of both copies of the gene of interest), a simple strategy is to enrich a pool of transfected cells through the use of selectable markers. Various forms of selectable vectors (plasmid or viral) are currently available and allow enrichment of successful transfectants by fluorescence-activated cell sorting or drug selection (e.g., Cas9-2A-GFP) (Ding et al. 2013). Likewise, delivery of recombinant Cas9 protein and in vitro transcribed

Fig. 7.2 (continued) reporters and epitope tags into endogenous genes, and (iv) introduction of mutations or polymorphisms or repairing of disease-relevant mutations. CRISPR/Cas9 system also offers opportunities for forward genetic approaches, including mutagenic screens through viral delivery of Cas9 and pooled sgRNA libraries. Besides genome editing, CRISPR/Cas9 technologies can be used for modulation of gene expression (activation or repression, by binding to target promoters or enhancers of endogenous genes), epigenetic regulation using dCas9 fused with a histone or DNA modification domains, as well as genomic imaging using dCas9 fused with fluorescent tags

sgRNAs has also been shown to provide higher knock-out efficiencies in certain cell types (Kim et al. 2014). Use of these ribonucleoproteins has the added advantage of rapid "on" and "off" rates and removal of the risk of random plasmid integration.

The complementary approach to knock-out of endogenous genes is to use HR-based gene targeting. This offers more sophisticated control over the type of allele to be generated, enabling replacement or insertion of any desired sequence by an allelic variant and/or selectable marker. In contrast to conventional experiments using BAC-based vectors, the homology arm sequences in the targeting vectors used alongside CRISPR can now be made smaller (1 kb or less). Moreover, decreasing costs and increasing sizes/quality of commercial synthetic DNA removes most of the technical difficulties involved in the construction of bespoke targeting vectors. Our lab has demonstrated that CRISPR can be deployed for efficient generation of knock-outs via gene targeting in both mouse and human genetically normal and glioblastoma patient-derived NS cell cultures (Bressan et al. 2017; Bulstrode et al. 2017). In this case, CRISPR sgRNAs were delivered together with a targeting vector containing a drug-selection marker in order to replace target exons of interest. Because the targeted DSBs are more frequently repaired by NHEJ than by HDR mechanisms, this enabled replacement of one allele with the selectable marker and disruption of the remaining allele by indel formation. Biallelic mutant cells emerged as discrete resistant colonies that could be selectively propagated and genotyped. This shortens the workflow for the generation of knock-out NS cell lines, although does require the construction of targeting vectors.

7.4.3 Conditional Alleles

Conditional alleles enable inactivation of genes in a temporal or cell type-specific manner. Given most genes have multiple functions in distinct tissues or stages of development, this is a vital tool. One approach to achieve controllable editing using the CRISPR system is to conditionally express either CRISPR sgRNAs or the Cas9 nuclease. Dow et al. achieved high frequencies of biallelic disruptions in multiple target loci both in mouse ES cell cultures and in vivo using tetracycline-inducible Cas9 and stably expressed gRNAs (Dow et al. 2015). Also, Gonzales et al. demonstrated efficient biallelic gene knock-outs in engineered tetracycline-regulated Cas9 human ES/iPS cell lines transiently transfected with single or multiple in vitro-transcribed sgRNAs (González et al. 2014).

In order to investigate loss-of-function mutations involved in neurological disorders, Rubio et al. developed an editing pipeline that allows conditional gene knock-out simultaneously with in vitro generation of human neurons, by either differentiation of pluripotent stem cells or direct reprogramming of skin fibroblasts (Rubio et al. 2016). To achieve this, the authors employed human ES/iPS cell lines targeted with a doxycycline-inducible Cas9–sgRNA construct and coupled with an accelerated neuronal differentiation protocol. The strategy was used to model neuronal dysfunction in tuberous sclerosis and early-onset epilepsy through inactivation of *TSC2* and *KCNQ2* genes, respectively.

Despite the power and control afforded by the use of inducible Cas9 expression, there are some caveats. A major limitation is that it generates a heterogeneous pool of knock-outs and non-knockout cells with unpredictable genotypes due to the random nature of the resulting indels. A more elegant approach is the use of recombinase-based systems such as Cre-loxP and Flp-FRP, which have been widely used in mouse genetics (Kuhn et al. 1995). These provide an efficient method for homogenous acute genetic ablation. Indeed, using CRISPR gene targeting, integration of flanking loxP or FRT sites for Cre-lox recombination system has been reported in mouse zygotes (Yang et al. 2013) as well as human ES and iPS cells (Chen et al. 2015). In the latter study, the authors devised a multistep strategy that involves targeting FRT sites on each side of the exon of interest using a floxed drug-resistance cassette, followed by simultaneous insertion of an activity-controllable recombinase into the safe-harbor locus *AAVS1* (Chen et al. 2015). The system proved useful to address the function of genes with diverse roles and temporal expression patterns, such as *OTX2*, during different stages of neural development. More recently, an elegant one-step method for engineering conditional and reversible knock-outs has also been described (Andersson-Rolf et al. 2017). This combines an invertible intronic selection cassette with high-efficiency Cas9-assisted gene editing and is widely applicable in established cell lines and human ES/iPS cells.

7.4.4 Introduction of Point Mutations

The precise introduction of small genetic modifications and point mutations via HDR is an experimental approach used widely to define the functional importance of a specific protein domain or critical amino acid residue in proteins of interest. Also, many human genetic studies have uncovered somatic mutations or risk alleles linked to neurological diseases, and these can now be engineered into genetically normal control cells. This newfound ability to create isogenic matched pairs of mutants and controls enables insights into disease mechanisms. This has become the standard strategy for determining the role of disease-associated mutations from patient-derived lines, as further discussed below.

Because *knock-ins* usually occur at relatively low frequencies, conventional strategies for inserting these small genetic modifications often necessitate antibiotic selectable markers to enrich for correctly targeted cells (An et al. 2014; Savic et al. 2015). Although efficient, these can be laborious as they require the construction of elaborate targeting donor vectors and a second round of manipulations to remove the selectable marker after the identification of correctly targeted clones.

An alternative approach is to make use of synthetic single-strand DNA (ssDNA) templates of up to 200 nucleotides, which can be used in conjunction with CRISPR for efficient selection-free gene editing (Wang et al. 2013; Yang et al. 2013). This involves the delivery of CRISPR/Cas9 components together with an ssDNA template harboring the desired nucleotide change flanked by a short sequence homologous to the genomic locus. These are easy to produce and commercially available. Successfully edited clones can then be screened by PCR or restriction fragment length polymorphisms if the engineered event alters a restriction site. Detailed protocols describing optimal delivery methods and design of ssDNA donor template have been published (Richardson et al. 2016).

7.4.5 Reporter Alleles and Protein Tagging

Gene targeting can also be employed to insert large DNA sequences such as fluorescent markers or biochemical tags at the ends of coding sequences in order to generate in-frame fusion proteins. The resulting knock-in lines are particularly useful to monitor levels and localization of a specific gene product and have been widely devised for real-time observation of gene-expression dynamics, cell-lineage tracing, and isolation of a specific cell population of interest from differentiating cultures or embryos. Moreover, endogenous protein fusions—particularly using small epitope tags such as FLAG, HA, V5, and MYC—can leverage biochemical studies and are particularly desirable when high-quality antibodies are not available (Dewari et al. 2018).

Recently, endogenous protein fusions have been achieved in murine neural progenitor cells in vivo, thereby providing a platform to map localization dynamics of many endogenous proteins in various cell types, regions, and ages in the mammalian brain (Mikuni et al. 2016; Uemura et al. 2016). In human ES/iPSC cultures, several strategies for generation of reporter alleles have also been reported. These generally involve the co-transfection of Cas9-sgRNA expression plasmids and a donor targeting vector carrying the homology arms, the reporter gene, and an antibiotic selection marker that allows enrichment of targeted events (Merkle et al. 2015). The approach has been applied, for instance, for the generation of NEUROG2-mCherry and GAD67-tdTomato reporter lines, allowing prospective isolation of neuronal progenitors and GABAergic neurons, respectively, from differentiating human iPSC cultures (Li et al. 2015; Liu et al. 2016). Similarly, the construction of FOXA2-eGFP and NKX2.2-eGFP reporter lines has been deployed to monitor the emergence of ventral neural progenitors and oligodendrocyte precursors, respectively, following neural differentiation of human ES cultures (Chi et al. 2016; Rodrigues et al. 2017).

To avoid the need for subsequent excision of the section marker, selection-free approaches have also been devised for tagging expressed and silent genes (Ratz et al. 2015; Zhu et al. 2015; Schmid-Burgk et al. 2016). An application of this strategy has been provided by Li et al. who demonstrated the creation of endogenous FMR1-Luciferase reporter iPSC lines derived from Fragile X syndrome (FXS) patients (Li et al. 2016b). FXS is caused by mutational-based hypermethylation and silencing of FMR1 in neural progenitor cells, and, therefore, one potential therapeutic strategy is to reactivate the silenced FMR1 gene through the use of small molecules. As well as their utility for testing potential genetic reactivation strategies and mechanistic studies, these reporter lines can be applied in high-throughput cell-based chemical screening (Li et al. 2016b).

While most approaches to date have employed HDR-based knock-in strategies using homology arms-containing donor vectors, recent reports also suggest that NHEJ-mediated knock-in could represent an alternative platform for endogenous gene tagging in hPSCs (Lackner et al. 2015; He et al. 2016). Instead of homology arms, these approaches use donor vectors in which the reporter gene of interest is flanked by sgRNA recognition sites that allow excision of the reporter sequence from the plasmid. Consequently, upon co-expression of Cas9 and sgRNAs, the sequence encoding the tag of interest is released in the cell and spontaneously integrates at the genomic cutting site by a homology-independent mechanism (Lackner et al. 2015; He et al. 2016).

7.4.6 Chromosomal Rearrangements

Many human genetic disorders are caused by structural variants such as deletions, translocations, and inversions of genomic regions involving several genes. These rearrangements result from DNA breakage at two different locations followed by NHEJ-based processing of the broken ends within the same (leading to inversions and deletions) or different chromosomes (leading to translocations). Establishing experimental models of these more complex genomic alterations in human PSCs and neural stem cells can be used to explore cellular mechanisms of disease (e.g., DiGeorge and Down's syndromes, schizophrenia, microcephaly, and brain tumors).

Two independent sgRNAs targeting the desired break points on either the same or different chromosome result in Cas9 cleavage at the two genomic loci. This results in loss of the intervening sequence, or can instead induce rearrangements between the targeted regions when on separate chromosomes. This approach has been used in human cell lines and primary cells to model inversions, translocations, and large chromosomal deletions observed in different cancer types (Choi and Meyerson 2014; Torres et al. 2014; Byrne et al. 2015), as well as to correct disease-associated chromosomal inversions in patient-derived human iPS cells (Park et al. 2015b). Recently, a general strategy for chromosomal engineering has also been devised in murine NSCs, which allows interrogation of oncogenic potential of genomic rearrangements identified in human brain cancers (Cook et al. 2017).

7.4.7 Genome-Wide Screenings

Genome-wide loss-of-function screening is a powerful hypothesis-free approach to identify novel genes and pathways involved in biological processes. Although RNA interference (RNAi)-based screens in mammalian cells have been valuable, this approach involves knock-down rather than permanent knock-out of the gene product. The transient and incomplete loss of protein alongside well-acknowledged off-target effects can limit its use. CRISPR gRNA libraries, therefore, represent a powerful alternative to overcome the limitations of RNAi screenings (Morgens et al. 2016) and are being used in an increasing number of studies. Similar screening strategies also enable interrogation of noncoding regulatory elements in the genome (Korkmaz et al. 2016).

CRISPR screens are typically performed using pooled sgRNA libraries containing from 10^3 to 10^5 sgRNAs, with multiple sgRNAs targeting the same site (Koike-Yusa et al. 2013; Wang et al. 2014). Using multiplex parallel oligonucleotide synthesis technologies, CRISPR libraries are produced as a complex pool of oligonucleotides containing the sgRNA sequences and directly cloned into a plasmid enabling lentiviral production. The Cas9 nuclease can be stably constitutively expressed in the target cell line, or alternatively encoded within the same lentiviral vector expressing the sgRNAs. Following lentiviral transduction, cells are selected for stable transgene integration and subjected to phenotypic screening.

To minimize off-target effects and increase efficiency of gene knock-out, CRISPR sgRNA library design algorithms have been optimized through numerous bioinformatics platforms (Heigwer et al. 2014; Doench et al. 2014; Chari et al. 2015). These optimized libraries provide a remarkably versatile discovery tool for human PSC- and NSC-based models that can be applied to both positive and negative screens in order to identify survival-enhancing and lethal mutations, respectively. A good exemplar of this was reported by Toledo et al. (2016), who performed genome-wide CRISPR-Cas9 knock-out screens in patient-derived glioblastoma stem-like cells (GSCs) and non-neoplastic human NSCs in order to identify genes required for their in vitro growth (Toledo et al. 2015). Combined with further in vitro and in vivo validation, such an approach was able to uncover several genes that are essential to GSCs but nonessential in NSCs (Toledo et al. 2016). Other opportunities provided by CRISPR libraries include the use of custom sgRNA libraries to assess the function of individual protein domains in a specific biological process (Shi et al. 2015), or to identify functional enhancers responsible for the regulation of a specific gene, as recently demonstrated in human cancer lines (Korkmaz et al. 2016).

7.5 Modeling CNS Development and Exploring Mechanisms of Stem Cell Biology

For human neurobiology, the emergence of genome editing using CRISPR alongside improved methods to derive, culture, and differentiate human NSCs and their progeny has opened up new experimental avenues. Molecular and cellular mechanisms and etiology of neurological disorders can now be explored in the appropriate human cellular context (Heidenreich and Zhang 2016). This is now enabling cellular genetic analysis of human development and disease modeling, thereby offering great potential to advance both basic and translational neuroscience research. This will complement ongoing discoveries made in classical model organisms (e.g., fly, zebrafish, and worm).

A range of distinct neural stem and progenitor cells can be obtained through ES/iPS cell differentiation, primary culture of human fetal tissues, or experimentally induced via direct reprogramming/transdifferentiation (Dolmetsch and Geschwind 2011; Kelava and Lancaster 2016a). These can also be engineered with disease alleles, or directly generated from patients to explore mechanisms of pathology and use in phenotypic drug discovery (O'Duibhir et al. 2016).

In parallel to investigating fundamental stem cell biology, the recent developments in three-dimensional cerebral organoid culture systems (Lancaster et al. 2013) now open up a wealth of opportunity to also model the cellular complexity of human brain development, particularly in aspects that involve organ structure formation, acquisition of regional cell identity, and cell–cell interactions (Kelava and Lancaster 2016b). An example of the utility of CRISPR/Cas9 gene editing combined with human cerebral organoid cultures has been recently provided by Matsui et al., who applied CRISPR knock-out to assess the role of the tumor suppressor retinoblastoma protein (RB) during brain development (Matsui et al. 2017). The authors uncover the roles of RB in the control of survival and cell cycle progression in different cell compartments as well as in neuronal migration, which would be otherwise difficult to investigate using dissociated adherent cultures.

7.6 Modeling Human CNS Disease Using Genome Editing

Neurological disorders have been traditionally studied using animal models or human immortalized human cell lines (Gottlieb 2002). While these systems have been useful for certain research topics, there is inevitable risk of artifactual findings due to their transformed nature. Novel approaches based on non-transformed diploid human ESCs and iPSC technology have changed the way cellular and molecular processes are modeled under normal and pathological conditions in vitro. As a result, ESC/iPSC-based systems have largely replaced the previous experimental systems for ex vivo modeling of human genetic diseases.

iPSCs can be derived from patient donor tissues, carrying known genetic variants in order to study their effects under the relevant cell-type context. For neuroscience, this was particularly valuable as it has made it possible to explore effects monogenic and polygenic disorders in the relevant context of patient-derived neural cells—a system previously not easily available for experimental research. Indeed, iPSC-based disease models have been generated for several neurological disorders, including Parkinson's, Alzheimer, Rett syndrome, and Huntington diseases, and have been proven to faithfully recapitulate cellular and molecular aspects of the human diseases (Sandoe and Eggan 2013).

Despite the successes, one recurring caveat has been the phenotypic variability between individual iPS cell lines; iPS cells from different sources have considerable line-to-line variability in their properties. This variation is unpredictable and arises due to the many differences in the genetic background (e.g., millions of SNPs within each genome). There is also variation depending on the type of reprogramming protocol and strategy used, or a risk of only partial reprogramming. Thus, in efforts to model a genetic disease, small phenotypic differences driven by the disease allele may not be detectable above the line-to-line variation and lead to false conclusions (Soldner et al. 2011). One very effective way to establish whether the phenotypes observed in patient-specific lines are caused by a specific genetic alteration is to use gene editing technology to remove or correct disease-associated mutations, i.e., genetic rescue. The resulting matched pairs of cell lines are isogenic, except for the disease-relevant mutation. This eliminates variation arising from stem cell line derivation and genetic background (Soldner et al. 2011) and enables firmer conclusions to be drawn.

The versatility of iPSC technology combined with the ease of CRISPR genome editing is thus extremely powerful. Indeed, genome editing in human iPSCs has rapidly become the standard experimental approach in stem cell research and human disease modeling (Hockemeyer and Jaenisch 2016) and is leading to many new opportunities to examine the genetic link between risk variants and cellular pathways involved in complex neurological disorders.

An early example of the power of this strategy is from a seminal study using ZFNs to engineer into human iPSCs candidate susceptibility variants identified from genome-wide association studies of Parkinson's disease patients (Soldner et al. 2011). More recently, these authors went on to show that CRISPR genome editing can also be used to identify critical regulatory elements that affect synuclein (Soldner et al. 2016). Genome editing can therefore help validate risk alleles identified in human genetic analysis. Importantly, the experimental paradigm established is not only relevant for Parkinson's disease, but is widely applicable for mechanistic studies of the molecular consequences of risk alleles associated with other diseases.

Another example is the exploration of molecular and cellular mechanisms underlying frontotemporal dementia (FTD)—a familial neurodegenerative disorder often caused by mutations in the gene encoding the microtubule-associated protein TAU (*MAPT*). By studying neurons differentiated from FTD patient iPS cell lines and their CRISPR gene edited isogenic controls, Silva et al. (2016) showed that MAPT mutations are associated with enhanced stress-inducible markers and vulnerability to proteotoxic, excitotoxic, and mitochondrial stressors (Silva et al. 2016). Another recent study also found that FTD iPSC-derived neurons display dysregulated excitability and accumulation and extracellular release of misfolded TAU followed by neuronal death. These phenotypes could be rescued upon CRISPR-mediated gene correction (Imamura et al. 2016). Similarly, genome editing in FTD patient-derived iPS cells has also helped to uncover the role of MAPT mutations in astrocytes (Hallmann et al. 2017), thereby providing insights into intrinsic and extrinsic mechanisms of neuronal degeneration in FTD pathogenesis.

Other successful examples of CRISPR-based gene editing coupled with neural lineage differentiation of patient iPS cell lines include reversion of loss-of-function mutations in *MECP2* in Rett syndrome (Bu et al. 2017), as well as correction of single-amino changes within *SOD1* gene in amyotrophic lateral sclerosis (Bhinge et al. 2017) and citron kinase in primary microcephaly (Li et al. 2016a). Removal of the exonic CAG- or CGG-repeat expansions—causative mutations of Huntington disease and Fragile X syndrome, respectively—has also been reported in human patient iPSC lines (Xu et al. 2017; Park et al. 2015a).

Similarly, the strategy has been further deployed to mimic disruptive mutations in the *CHD8* gene—one of the top genetic risk factors in autism spectrum disorders (ASDs) (Wang et al. 2015). By using CRISPR to knock-out one copy of *CHD8* in iPSCs derived from healthy control, authors showed that *CHD8* regulates the expression of ASD-associated genes and controls multiple cell processes critical for neural functions, thereby shedding light on the molecular links between *CHD8* functions and ASD.

CRISPR gene editing is not limited to studies of the coding sequences in the genome. Regulatory elements can also be modified. Genome-wide association studies (GWASs) have begun to uncover the plethora of coding or noncoding genetic variations—using microarrays that contain millions of single-nucleotide polymorphisms (SNPs)—that are associated with many human complex diseases or traits. Efforts are now being made to tease out the individual variants' contributions to disease onset as well as to model the additive effects that underlie complex polygenic disorders. Thus, investigators now have capabilities to use CRISPR to functionally annotate these numerous candidate risk alleles and to test for combinatorial effects by multiplexing.

7.7 Beyond Genome Editing: Epigenetic Manipulations

The aforementioned successful applications of the CRISPR/Cas9 systems demonstrate that functional genetics approaches—so successful in mouse genetic analysis—can now be deployed for the study of human neural stem cells and their derivatives. However, by manipulating its components, the CRISPR system can be repurposed to exert a variety of novel functions as transcriptional regulators. For instance, catalytically inactive forms of Cas9 (dCas9) can be fused with transcriptional activation (e.g., VP64) or repressive domains (e.g., KRAB), to create synthetic transcription factors, capable of activating or repressing target genes (Perez-Pinera et al. 2013). This indeed has proved useful for direct activation of endogenous genes and reprogramming of fibroblasts into neuronal cells (Black et al. 2016). Epigenome editing can be achieved by fusing dCas9 with a histone (e.g., LSD1) or DNA modification domains (DNMTs, TETs). Other uses include imaging of chromosome topology using dCas9-GFP fusions, or pulldown of the chromatin and associated bound factors using epitope-tagged dCas9. Comprehensive discussion of these alternative uses of CRISPR is beyond the scope of this chapter, but is reviewed elsewhere (Hsu et al. 2014; Black et al. 2017).

7.8 Concluding Remarks

In this chapter, we have discussed how precise genetic modifications introduced by genome editing in human cells are opening up new approaches to studying human brain development and disease modeling. A major bottleneck in the functional annotation of the human genome has been removed. Many of the core tools and strategies of transgenic research and molecular genetics that have been so successful in animal models can therefore now be exploited in human stem cells. There will likely be significant new advances in our understanding of how the human nervous system is built and maintained and how it is compromised by age, injury, and disease.

Acknowledgements We thank Kirsty Ferguson and Faye Robertson for helpful comments on the manuscript. R.B.B. is supported by a fellowship from the Science Without Borders Program (CAPES, Brazil). S.M.P. is a Cancer research UK Senior Research Fellow (A17368).

References

An MC, O'Brien RN, Zhang N et al (2014) Polyglutamine disease modeling: epitope based screen for homologous recombination using CRISPR/Cas9 system. PLoS Curr. https://doi.org/10.1371/currents.hd.0242d2e7ad72225efa72f6964589369a

Andersson-Rolf A, Mustata RC, Merenda A et al (2017) One-step generation of conditional and reversible gene knockouts. Nat Methods 14:287–289. https://doi.org/10.1038/nmeth.4156

Bhaya D, Davison M, Barrangou R (2011) CRISPR-Cas systems in bacteria and archaea: versatile small RNAs for adaptive defense and regulation. Annu Rev Genet 45:273–297. https://doi.org/10.1146/annurev-genet-110410-132430

Bhinge A, Namboori SC, Zhang X et al (2017) Genetic correction of SOD1 mutant iPSCs reveals ERK and JNK activated AP1 as a driver of neurodegeneration in amyotrophic lateral sclerosis. Stem Cell Rep 8:856–869. https://doi.org/10.1016/j.stemcr.2017.02.019

Black JB, Adler AF, Wang H-G et al (2016) Targeted epigenetic remodeling of endogenous loci by CRISPR/Cas9-based transcriptional activators directly converts fibroblasts to neuronal cells. Cell Stem Cell 19:406–414. https://doi.org/10.1016/j.stem.2016.07.001

Black JB, Perez-Pinera P, Gersbach CA (2017) Mammalian synthetic biology: engineering biological systems. Annu Rev Biomed Eng 19:249–277. https://doi.org/10.1146/annurev-bioeng-071516-044649

Bressan RB, Dewari PS, Kalantzaki M et al (2017) Efficient CRISPR/Cas9-assisted gene targeting enables rapid and precise genetic manipulation of mammalian neural stem cells. Development. https://doi.org/10.1242/dev.140855

Bu Q, Wang A, Hamzah H et al (2017) CREB signaling is involved in Rett syndrome pathogenesis. J Neurosci 37:3671–3685. https://doi.org/10.1523/JNEUROSCI.3735-16.2017

Bulstrode H, Johnstone E, Marqués-Torrejón MÁ et al (2017) Elevated FOXG1 and SOX2 in glioblastoma enforces neural stem cell identity through transcriptional control of cell cycle and epigenetic regulators. Genes Dev 31:757–773. https://doi.org/10.1101/gad.293027.116

Byrne SM, Byrne SM, Ortiz L et al (2015) Multi-kilobase homozygous targeted gene replacement in human induced pluripotent stem cells. Nucleic acids Res 43:e21–e21. https://doi.org/10.1093/nar/gku1246

Capecchi MR (2005) Gene targeting in mice: functional analysis of the mammalian genome for the twenty-first century. Nat Rev Genet 6:507–512. https://doi.org/10.1038/nrg1619
Chari R, Mali P, Moosburner M, Church GM (2015) Unraveling CRISPR-Cas9 genome engineering parameters via a library-on-library approach. Nat Methods 12:823–826. https://doi.org/10.1038/nmeth.3473
Chen Y, Cao J, Xiong M et al (2015) Engineering human stem cell lines with inducible gene knockout using CRISPR/Cas9. Cell Stem Cell 17:233–244. https://doi.org/10.1016/j.stem.2015.06.001
Chi L, Fan B, Zhang K et al (2016) Targeted differentiation of regional ventral neuroprogenitors and related neuronal subtypes from human pluripotent stem cells. Stem Cell Rep 7:941–954. https://doi.org/10.1016/j.stemcr.2016.09.003
Cho SW, Kim S, Kim JM, Kim J-S (2013) Targeted genome engineering in human cells with the Cas9 RNA-guided endonuclease. Nat Biotechnol 31:230–232. https://doi.org/10.1038/nbt.2507
Choi PS, Meyerson M (2014) Targeted genomic rearrangements using CRISPR/Cas technology. Nat Commun 5:3728. https://doi.org/10.1038/ncomms4728
Cong L, Ran FA, Cox D et al (2013) Multiplex genome engineering using CRISPR/Cas systems. Science 339:819–823. https://doi.org/10.1126/science.1231143
Cook PJ, Thomas R, Kannan R et al (2017) Somatic chromosomal engineering identifies BCAN NTRK1 as a potent glioma driver and therapeutic target. Nat Commun 8. https://doi.org/10.1038/ncomms15987
Court DL, Sawitzke JA, Thomason LC (2002) Genetic engineering using homologous recombination. Annu Rev Genet 36:361–388. https://doi.org/10.1146/annurev.genet.36.061102.093104
Dewari PS, Southgate B, Mccarten K et al (2018) An efficient and scalable pipeline for epitope tagging in mammalian stem cells using Cas9 ribonucleoprotein. elife 7:e35069. https://doi.org/10.7554/eLife.35069
Ding Q, Regan SN, Xia Y et al (2013) Enhanced efficiency of human pluripotent stem cell genome editing through replacing TALENs with CRISPRs. Stem Cell 12:393–394. https://doi.org/10.1016/j.stem.2013.03.006
Doench JG, Hartenian E, Graham DB et al (2014) Rational design of highly active sgRNAs for CRISPR-Cas9-mediated gene inactivation. Nat Biotechnol 32:1262–1267. https://doi.org/10.1038/nbt.3026
Dolmetsch R, Geschwind DH (2011) The human brain in a dish: the promise of iPSC-derived neurons. Cell 145:831–834. https://doi.org/10.1016/j.cell.2011.05.034
Dow LE, Fisher J, O'Rourke KP et al (2015) Inducible in vivo genome editing with CRISPR-Cas9. Nat Biotechnol 33:390–394. https://doi.org/10.1038/nbt.3155
González F, Zhu Z, Shi Z-D et al (2014) An iCRISPR platform for rapid, multiplexable, and inducible genome editing in human pluripotent stem cells. Cell Stem Cell 15:215–226. https://doi.org/10.1016/j.stem.2014.05.018
Gottlieb DI (2002) Large-scale sources of neural stem cells. Annu Rev Neurosci 25:381–407
Hallmann A-L, Araúzo-Bravo MJ, Mavrommatis L et al (2017) Astrocyte pathology in a human neural stem cell model of frontotemporal dementia caused by mutant TAU protein. Sci Rep 7:42991. https://doi.org/10.1038/srep42991
He X, Tan C, Wang F et al (2016) Knock-in of large reporter genes in human cells via CRISPR/Cas9-induced homology-dependent and independent DNA repair. Nucleic Acids Res 44:e85–e85. https://doi.org/10.1093/nar/gkw064
Heidenreich M, Zhang F (2016) Applications of CRISPR-Cas systems in neuroscience. Nat Rev Neurosci 17:36–44. https://doi.org/10.1038/nrn.2015.2
Heigwer F, Kerr G, Boutros M (2014) E-CRISP: fast CRISPR target site identification. Nat Methods 11:122–123. https://doi.org/10.1038/nmeth.2812
Hockemeyer D, Jaenisch R (2016) Induced pluripotent stem cells meet genome editing. Stem Cell 18:573–586. https://doi.org/10.1016/j.stem.2016.04.013
Hockemeyer D, Wang H, Kiani S et al (2011) Genetic engineering of human pluripotent cells using TALE nucleases. Nat Biotechnol 29:731–734. https://doi.org/10.1038/nbt.1927

Hsu PD, Lander ES, Zhang F (2014) Development and applications of CRISPR-Cas9 for genome engineering. Cell 157:1262–1278. https://doi.org/10.1016/j.cell.2014.05.010

Hyman AH, Simons K (2011) The new cell biology: beyond HeLa cells. Nature 480:34–34. https://doi.org/10.1038/480034a

Imamura K, Sahara N, Kanaan NM et al (2016) Calcium dysregulation contributes to neurodegeneration in FTLD patient iPSC-derived neurons. Sci Rep 6:130. https://doi.org/10.1038/srep34904

Jasin M (1996) Genetic manipulation of genomes with rare-cutting endonucleases. Trends Genet 12:224–228. https://doi.org/10.1016/0168-9525(96)10019-6

Jinek M, Chylinski K, Fonfara I et al (2012) A programmable dual-RNA-guided DNA endonuclease in adaptive bacterial immunity. Science 337:816–821. https://doi.org/10.1126/science.1225829

Jinek M, East A, Cheng A et al (2013) RNA-programmed genome editing in human cells. eLife 2: e00471. https://doi.org/10.7554/eLife.00471

Joung JK, Sander JD (2012) TALENs: a widely applicable technology for targeted genome editing. Nat Rev Mol Cell Biol. https://doi.org/10.1038/nrm3486

Kelava I, Lancaster MA (2016a) Stem cell models of human brain development. Stem Cell 18: 736–748. https://doi.org/10.1016/j.stem.2016.05.022

Kelava I, Lancaster MA (2016b) Dishing out mini-brains: current progress and future prospects in brain organoid research. Dev Biol 420:199–209. https://doi.org/10.1016/j.ydbio.2016.06.037

Kim S, Kim S, Kim D et al (2014) Highly efficient RNA-guided genome editing in human cells via delivery of purified Cas9 ribonucleoproteins. Genome Res 24:1012–1019. https://doi.org/10.1101/gr.171322.113

Koike-Yusa H, Li Y, Tan E-P et al (2013) Genome-wide recessive genetic screening in mammalian cells with a lentiviral CRISPR-guide RNA library. Nat Biotechnol 32:267–273. https://doi.org/10.1038/nbt.2800

Korkmaz G, Lopes R, Ugalde AP et al (2016) Functional genetic screens for enhancer elements in the human genome using CRISPR-Cas9. Nat Biotechnol 34:192–198. https://doi.org/10.1038/nbt.3450

Kraft K, Geuer S, Will AJ et al (2015) Deletions, inversions, duplications: engineering of structural variants using CRISPR/Cas in mice. Cell Rep 10:833–839. https://doi.org/10.1016/j.celrep.2015.01.016

Kuhn R, Schwenk F, Aguet M, Rajewsky K (1995) Inducible gene targeting in mice. Science 269: 1427–1429

Lackner DH, eacute AC, Guzzardo PM et al (2015) A generic strategy for CRISPR-Cas9-mediated gene tagging. Nat Commun 6:1–7. https://doi.org/10.1038/ncomms10237

Lancaster MA, Renner M, Martin CA (2013) Cerebral organoids model human brain development and microcephaly. Nature 501(7467):373–379. https://doi.org/10.1038/nature12517

Li S, Xue H, Wu J et al (2015) Human induced pluripotent stem cell NEUROG2 dual knockin reporter lines generated by the CRISPR/Cas9 system. Stem Cells Dev 24:2925–2942. https://doi.org/10.1089/scd.2015.0131

Li H, Bielas SL, Zaki MS et al (2016a) Biallelic mutations in citron kinase link mitotic cytokinesis to human primary microcephaly. Am J Hum Genet 99:501–510. https://doi.org/10.1016/j.ajhg.2016.07.004

Li M, Zhao H, Ananiev GE et al (2016b) Establishment of reporter lines for detecting fragile X mental retardation (FMR1) gene reactivation in human neural cells. Stem Cells 35:158–169. https://doi.org/10.1002/stem.2463

Lin FL, Sperle K, Sternberg N (1985) Recombination in mouse L cells between DNA introduced into cells and homologous chromosomal sequences. Proc Natl Acad Sci USA 82:1391–1395

Liu J, Gao C, Chen W et al (2016) CRISPR/Cas9 facilitates investigation of neural circuit disease using human iPSCs: mechanism of epilepsy caused by an SCN1A loss-of-function mutation. Transl Psychiatry 6:e703. https://doi.org/10.1038/tp.2015.203

Makarova KS, Haft DH, Barrangou R et al (2011) Evolution and classification of the CRISPR-Cas systems. Nat Rev Microbiol 9:467–477. https://doi.org/10.1038/nrmicro2577

Mali P, Yang L, Esvelt KM et al (2013) RNA-guided human genome engineering via Cas9. Science 339:823–826. https://doi.org/10.1126/science.1232033

Matsui T, Nieto-Estévez V, Kyrychenko S et al (2017) Retinoblastoma controls growth, survival and neuronal migration in human cerebral organoids. Development 144:1025–1034. https://doi.org/10.1242/dev.143636

Merkle FT, Neuhausser WM, Santos D et al (2015) Efficient CRISPR-Cas9-mediated generation of knockin human pluripotent stem cells lacking undesired mutations at the targeted locus. Cell Rep 11:875–883. https://doi.org/10.1016/j.celrep.2015.04.007

Mikuni T, Nishiyama J, Sun Y et al (2016) High-throughput, high-resolution mapping of protein localization in mammalian brain by in vivo genome editing. Cell 165:1803–1817. https://doi.org/10.1016/j.cell.2016.04.044

Mohammad F, Weissmann S, Leblanc B et al (2017) EZH2 is a potential therapeutic target for H3K27M-mutant pediatric gliomas. Nat Med 23:483–492. https://doi.org/10.1038/nm.4293

Morgens DW, Deans RM, Li A, Bassik MC (2016) Systematic comparison of CRISPR/Cas9 and RNAi screens for essential genes. Nat Biotechnol 34:634–636. https://doi.org/10.1038/nbt.3567

O'Duibhir E, Carragher NO, Pollard SM (2016) Accelerating glioblastoma drug discovery: convergence of patient-derived models, genome editing and phenotypic screening. Mol Cell Neurosci. https://doi.org/10.1016/j.mcn.2016.11.001

Park C-Y, Halevy T, Lee DR et al (2015a) Reversion of FMR1 methylation and silencing by editing the triplet repeats in fragile X iPSC-derived neurons. Cell Rep 13:234–241. https://doi.org/10.1016/j.celrep.2015.08.084

Park C-Y, Kim DH, Son JS et al (2015b) Functional correction of large factor VIII gene chromosomal inversions in hemophilia A patient-derived iPSCs using CRISPR-Cas9. Cell Stem Cell 17:213–220. https://doi.org/10.1016/j.stem.2015.07.001

Perez-Pinera P, Kocak DD, Vockley CM et al (2013) RNA-guided gene activation by CRISPR-Cas9–based transcription factors. Nat Methods 10:973–976. https://doi.org/10.1038/nmeth.2600

Ratz M, Testa I, Hell SW, Jakobs S (2015) CRISPR/Cas9-mediated endogenous protein tagging for RESOLFT super-resolution microscopy of living human cells. Sci Rep 5:340. https://doi.org/10.1038/srep09592

Richardson CD, Ray GJ, DeWitt MA et al (2016) Enhancing homology-directed genome editing by catalytically active and inactive CRISPR-Cas9 using asymmetric donor DNA. Nat Biotechnol:1–7. https://doi.org/10.1038/nbt.3481

Rodrigues GMC, Gaj T, Adil MM et al (2017) Defined and scalable differentiation of human oligodendrocyte precursors from pluripotent stem cells in a 3D culture system. Stem Cell Rep:1–14. https://doi.org/10.1016/j.stemcr.2017.04.027

Rouet P, Smih F, Jasin M (1994) Expression of a site-specific endonuclease stimulates homologous recombination in mammalian cells. Proc Natl Acad Sci USA 91:6064–6068

Rubio A, Luoni M, Giannelli SG et al (2016) Rapid and efficient CRISPR/Cas9 gene inactivation in human neurons during human pluripotent stem cell differentiation and direct reprogramming. Sci Rep 6:109. https://doi.org/10.1038/srep37540

Sandoe J, Eggan K (2013) Opportunities and challenges of pluripotent stem cell neurodegenerative disease models. Nat Neurosci 16:780–789. https://doi.org/10.1038/nn.3425

Savic D, Smith SB, Mendenhall EM (2015) CETCh-seq: CRISPR epitope tagging ChIP-seq of DNA-binding proteins. Genome Res 25:1581–1589. https://doi.org/10.1101/gr.193540.115

Schmid-Burgk JL, Höning K, Ebert TS, Hornung V (2016) CRISPaint allows modular base-specific gene tagging using a ligase-4-dependent mechanism. Nat Commun 7:12338. https://doi.org/10.1038/ncomms12338

Shi J, Wang E, Milazzo JP et al (2015) Discovery of cancer drug targets by CRISPR-Cas9 screening of protein domains. Nat Biotechnol 33:661–667. https://doi.org/10.1038/nbt.3235

Silva MC, Cheng C, Mair W et al (2016) Human iPSC-derived neuronal model of tau-A152T frontotemporal dementia reveals tau-mediated mechanisms of neuronal vulnerability. Stem Cell Rep 7:325–340. https://doi.org/10.1016/j.stemcr.2016.08.001
Smithies O, Gregg RG, Boggs SS et al (1985) Insertion of DNA sequences into the human chromosomal beta-globin locus by homologous recombination. Nature 317:230–234
Soldner F, Laganière J, Cheng AW et al (2011) Generation of isogenic pluripotent stem cells differing exclusively at two early onset Parkinson point mutations. Cell 146:318–331. https://doi.org/10.1016/j.cell.2011.06.019
Soldner F, Stelzer Y, Shivalila CS et al (2016) Parkinson-associated risk variant in distal enhancer of α-synuclein modulates target gene expression. Nature 533:95–99. https://doi.org/10.1038/nature17939
Takayama K, Igai K, Hagihara Y et al (2017) Highly efficient biallelic genome editing of human ES/iPS cells using a CRISPR/Cas9 or TALEN system. Nucleic Acids Res 45:5198–5207. https://doi.org/10.1093/nar/gkx130
Thomas KR, Capecchi MR (1987) Site-directed mutagenesis by gene targeting in mouse embryo-derived stem cells. Cell 51:503–512
Thomas KR, Folger KR, Capecchi MR (1986) High frequency targeting of genes to specific sites in the mammalian genome. Cell 44:419–428. https://doi.org/10.1016/0092-8674(86)90463-0
Toledo CM, Ding Y, Hoellerbauer P et al (2015) Genome-wide CRISPR-Cas9 screens reveal loss of redundancy between PKMYT1 and WEE1 in glioblastoma stem-like cells. Cell Rep 13: 2425–2439. https://doi.org/10.1016/j.celrep.2015.11.021
Toledo et al. (2016) doi: 10.1016/j.celrep.2015.11.021
Torres R, Martin MC, Garcia A et al (2014) Engineering human tumour-associated chromosomal translocations with the RNA-guided CRISPR-Cas9 system. Nat Commun 5:3964. https://doi.org/10.1038/ncomms4964
Uemura T, Mori T, Kurihara T et al (2016) Fluorescent protein tagging of endogenous protein in brain neurons using CRISPR/Cas9-mediated knock-in and in utero electroporation techniques. Sci Rep 6:35861. https://doi.org/10.1038/srep35861
Wang H, Yang H, Shivalila CS et al (2013) One-step generation of mice carrying mutations in multiple genes by CRISPR/Cas-mediated genome engineering. Cell 153:910–918. https://doi.org/10.1016/j.cell.2013.04.025
Wang T, Wei JJ, Sabatini DM, Lander ES (2014) Genetic screens in human cells using the CRISPR-Cas9 system. Science 343:80–84. https://doi.org/10.1126/science.1246981
Wang P, Lin M, Pedrosa E et al (2015) CRISPR/Cas9-mediated heterozygous knockout of the autism gene CHD8 and characterization of its transcriptional networks in neurodevelopment. Mol Autism 6:55. https://doi.org/10.1186/s13229-015-0048-6
Wells MF, Salick MR, Wiskow O et al (2016) Genetic ablation of AXL does not protect human neural progenitor cells and cerebral organoids from Zika Virus infection. Cell Stem Cell 19: 703–708. https://doi.org/10.1016/j.stem.2016.11.011
Xu X, Tay Y, Sim B et al (2017) Reversal of phenotypic abnormalities by CRISPR/Cas9-mediated gene correction in Huntington disease patient-derived induced Pluripotent stem cells. Stem Cell Rep 8:619–633. https://doi.org/10.1016/j.stemcr.2017.01.022
Yang H, Wang H, Shivalila CS et al (2013) One-step generation of mice carrying reporter and conditional alleles by CRISPR/Cas-mediated genome engineering. Cell 154:1370–1379. https://doi.org/10.1016/j.cell.2013.08.022
Zhu Z, Verma N, Gonzalez F et al (2015) A CRISPR/Cas-mediated selection-free knockin strategy in human embryonic stem cells. Stem Cell Rep 4:1103–1111. https://doi.org/10.1016/j.stemcr.2015.04.016

Chapter 8
Brain Organoids: Expanding Our Understanding of Human Development and Disease

L. B. Chuye, A. Dimitri, A. Desai, C. Handelmann, Y. Bae, P. Johari, J. M. Jornet, I. Klejbor, M. K. Stachowiak, and E. K. Stachowiak

Abstract Stem cell-derived brain organoids replicate important stages of the prenatal human brain development and combined with the induced pluripotent stem cell (iPSC) technology offer an unprecedented model for investigating human neurological diseases including autism and microcephaly. We describe the history and birth

---

L. B. Chuye, A. Dimitri, A. Desai, and C. Handelmann are co-first authors.

L. B. Chuye · A. Desai · C. Handelmann · Y. Bae
Department of Pathology and Anatomical Sciences, State University of New York, Buffalo, NY, USA

A. Dimitri
Department of Pathology and Anatomical Sciences, State University of New York, Buffalo, NY, USA

Department of Biological Sciences,, State University of New York, Freedonia, NY, USA

P. Johari · J. M. Jornet
Department of Electrical Engineering, State University of New York, Buffalo, NY, USA

I. Klejbor
Medical University of Gdansk, Gdańsk, Poland

M. K. Stachowiak
Department of Pathology and Anatomical Sciences, State University of New York, Buffalo, NY, USA

Western New York Stem Cells Culture and Analysis Center, State University of New York, Buffalo, NY, USA

Department of Biomedical Engineering, State University of New York, Buffalo, NY, USA

E. K. Stachowiak (✉)
Department of Pathology and Anatomical Sciences, State University of New York, Buffalo, NY, USA

Western New York Stem Cells Culture and Analysis Center, State University of New York, Buffalo, NY, USA
e-mail: eks1@buffalo.edu

L. Buzanska (ed.), *Human Neural Stem Cells*, Results and Problems in Cell Differentiation 66, https://doi.org/10.1007/978-3-319-93485-3_8

of organoids and their application, focusing on cerebral organoids derived from embryonic stem cells and iPSCs. We discuss new insights into organoid-based model of schizophrenia and shed light on challenges and future applications of organoid-based disease model system. This review also suggests hitherto unrevealed potential applications of organoids in combining with new technologies such as nanophotonics/optogenomics for controlling brain development and atomic force microscopy for studying mechanical forces that shape the developing brain.

Abbreviations

AFM	Atomic force microscopy
CNVs	Copy number variations
ESCs	Embryonic stem cells
INFS	Integrative nuclear FGFR1 signaling
iPSCs	Induced pluripotent stem cells
NCCs	Neural committed cells
nFGFR1	Nuclear fibroblast growth factor receptor-1
NPCs	Neural progenitor cells
SNPs	Single nucleotide polymorphisms

8.1 Expanding Our Understanding of Human Neurological Disease

From the early descriptions of epilepsy found in the book of Charaka Samhita (1500–500 BCE) to Broca's aphasias discovered by Pierre Paul Broca during the 1800s (Lazar and Mohr 2011), to the more recently described phantom limb syndrome, various neurological diseases have always captivated human imagination. Yet, for all the predicament they cause to human kind, and despite all the remarkable progress in technology, advances in this area are hard to come by. Truthfully, the human brain with its anatomical complexity and connectivity is a more abstruse system to study; arguably though, part of the problem in understanding its intricate ways dwells in the fact that scientists lack appropriate models to study it (Quadrato et al. 2016).

Studies in common rodent models, phylogenetic differences aside, have provided us a head start and have prompted us to elucidate the basic layout of the mammalian brain. As some researchers have indicated, critical differences between species ought to be taken into consideration when modeling a human disease into an animal system (Muotri 2016; Seok et al. 2013; DeFelipe et al. 2002). For instance, Quiñones-Hinojosa and Chaichana (2007) have demonstrated substantial differences between the cortical organization of human and rodent brains, particularly in the subventricular zone (SVZ) and layer II of the cortex. Located laterally in the wall of the ventricles, the SVZ is one of the main neurogenic areas in the mammalian brain. The SVZ of the rodents though has a particular rostral migratory stream that runs parallel to the basal ventricular surface, for which there is no known analog in humans. In the same vein, layer II in humans is a hypocellular area, but this layer in rodents is tightly packed with astrocytes. DeFelipe et al. (2002) have reported other significant differences between clades. They discovered double bouquet interneurons present in humans and macaques, which has not been detected in rodents (DeFelipe et al. 2002). This study also demonstrated differences in the neuronal migratory patterns between species (DeFelipe et al. 2002). As Marin et al. (2001) described, up to 65% of the GABAergic (Dlx1/2+ and Mash1+) neurons in primates originate from the ventricular and subventricular dorsal telencephalon ,while the other 35% (Dlx1/2+ and Mash1−) of neurons originate in the ganglionic eminence of the ventral telencephalon (Marin et al. 2001). Yet, most of the GABAergic neurons in rodents originate in the ganglionic eminence of the ventral telencephalon (Anderson et al. 1999; Tan et al. 1998). In addition, the GABAergic neurons make up 15% of all cortical neurons in rodents, but they are more abundant in the primate cortex, where they constitute approximately 25% of all neurons (DeFelipe et al. 2002). This observation suggested a more prevalent inhibitory signaling in the primate cortex. Researchers have also found that the percentage of asymmetrical synapses, on which the postsynaptic density is prominent, was higher in human cortex compared to rodents, particularly in layers IV, V, and VI (Peters and Palay 1996; Peters et al. 1991).

The effects of pathologically altered proteins, which in humans cause diseases, may vary between species. In an exemplary study, Thomas et al. managed to recreate mutations of DNA repair exonuclease 1 (TREX1) in a mouse model, which in humans leads to the Aicardi–Goutiéres Syndrome (Thomas et al. 2017). In humans, the TREX1 mutations cause microcephaly, intracerebral calcifications, and global developmental delay (Abdel-Salam et al. 2017). Surprisingly, mice with the same mutations failed to develop significant neurological manifestations.

8.2 From Cancer and Epilepsy to Neural Stem Cells

Cognizant of the limitations posed by phylogenetic differences, many researchers have argued in favor of transitioning toward using more apparent human systems (Quadrato et al. 2016; Muotri 2016). In the last couple of decades, emergence of new technologies has enabled scientists to isolate and manipulate human stem cells to grow 3D organ-like structures, including cerebral-like organs. However, the connection between the stem cells and the brain had begun long before these technologies were even conceived.

Altman et al. and Nottebohm et al. reported the occurrence of adult neurogenesis in rats, cats, and birds' brains as early as 1962 and 1985 (Altman and Das 1965; Nottebohm 1985). Subsequently, other groups had tried to identify, to no avail, stem cells in the adult brains of higher primates (Rakic 1985; Eckenhoff and Rakic 1988). The publication of two key papers in the field of neuroscience would eventually change and advance this field. By 1994, Kirschenbaum and Goldman were able to culture human brain stem cells in vitro (Kirschenbaum et al. 1994). The brain samples for their research were obtained from 11 patients from the age of 15 to 52 who underwent anterior temporal lobectomy. Out of the 11 sets of brain specimens, several contained ventricular tissues. By culturing these disassociated tissues, the researchers were able to identify fiber-projecting neuron-like cells. Some of these cells were pulse-prelabeled with 3′H-thymidine and displayed depolarization-induced calcium fluxes typical to mature neurons. These observations indicated that neurogenesis was taking place. These significant findings were not broadly acknowledged by the scientific community until 1998, when Eriksson et al. discovered in vivo neurogenesis occurring in the adult human brain (Eriksson et al. 1998). In this study, Eriksson used a nonradioactive bromodeoxyuridine (BrdU), a synthetic analog of thymidine, to monitor neuronal proliferation in terminal cancer patients. Five patients between the ages of 52 and 72 accepted to participate in this study. The brain autopsy samples were collected from the hippocampal area, and Eriksson ensured no sign of metastasis in any of the brains sampled. The tissue samples were fixed, sectioned, and immunostained with anti-BrdU antibody. The results showed that all the samples obtained from patients treated with BrdU were undergoing mitosis. This is how, thanks to those generous epileptic and cancer patients who donated their brains to science, scientists were able to demonstrate that neurogenesis in the brain of adult human indeed occurred. This is also how the long-standing Santiago Ramon y Cajal's dogma of "Everything may die, nothing may be regenerated" came to an end (Stahnisch and Nitsch 2002).

8.3 From ESCs to iPSCs

The discovery of the neural stem cell in adult human brain led to a renaissance in the fields of neuroscience and disease. There was no controversy about the use of the brain stem cells for research. As long as the patients and their families were informed about the risks and had their consent, scientists were able to obtain and use those cells in their laboratories (Eriksson et al. 1998; Kirschenbaum et al. 1994). However, the problem was with the scarcity of the brain-derived neural stem cells, and, as latter learned, they were not a suitable starting material for developing cerebral organoids.

This problem was overcome by using the embryonic stem cells (ESCs) and learning how to differentiate these pluripotent cells into the multipotent neural stem cells. The human ESC (hESC) lines provided ample material for developing organoid technologies, but they were not universally admitted across the world scientific community. Presenting the arguments in favor or against the use of hESCs for research makes for a lengthy discussion and far outreaches the objectives of this review. However, a breakthrough occurred in 2006 when Takahashi and Yamanaka were able to reprogram and induce pluripotency in differentiated cell lineages (Takahashi and Yamanaka 2006). By infecting human fibroblasts with retroviruses expressing few transcription factors such as Oct3/4, Sox2, Klf4, and c-Myc, Yamanaka group was able to prime the fibroblasts to dedifferentiate and express markers of the pluripotent stem cells. The reprogrammed cells were referred to as the induced pluripotent stem cells (iPSCs). A similar mechanism was used by herpesviruses to hijack the host molecular machinery in order to synthetize their own viral proteins and to proliferate (McBride 2017; Sadeghipour and Mathias 2017). Takahashi and his colleagues were also able to reprogram the iPSCs into becoming neural stem cells and cardiomyocytes using protocols developed earlier by Kawasaki et al. (2000) and Laflamme et al. (2007).

Natural follow-up questions emerged after Takahashi's discovery: Are there any differences between the ESCs and the iPSCs? Are there any advantages in working with one versus the other? Like ESCs, iPSCs can express stem cell markers and can derive three germ layers, but despite the similarities, small difference does exist. By comparing the reported transcriptional profiles of different human iPSCs and ESCs available on the Gene Expression Omnibus (GEO) repository (http://www.ncbi.nlm.nih.gov/geo/), Ghosh et al. found few differences, which in part could reflect different analytical methodologies (Lee et al. 2013). The results suggested that younger the tissue of origin, potentially shorter the distance between the iPSCs and ESCs.

8.4 The Birth of the Organoids

In this section, we will briefly review some of the techniques used by different groups to grow human tissues, focusing in particular on the recent developments of cerebral organoids. At the end of the section, we will present brief annotations on the importance of adopting these new technologies as a surrogate model for the study of human embryology, human diseases, for drug development, and the implications this new technology might have for the future of medicine.

Tissue culture techniques were devised more than a century ago to study animal cells. The early approach was to observe aggregated tissue samples. The methods of disaggregation and replating of tissue cells were developed later on (Carrel 1912; Carrel and Burrows 1911; Freshney and ebrary Inc. 2010). In 1952 as Renato Dulbecco was trying to grow viruses in a petri dish on a bed of fibroblasts, he found that adding trypsin would detach cells from the dish. Then, he reseeded these free fibroblasts in the new dish, which would repopulate by forming a monolayer of single cells (Dulbecco 1952). If the cells are allowed to reaggregate, in agar, in a collagen matrix, or in other media, the resulting 3D formation is referred to as a histotypic culture, in which the aggregating cells belong to the same type and lineage. In the organotypic culture, on the other hand, the aggregating cells belong to different types/lineages that form an organ. Organotypic cultures are useful for studying the relations among different cells (Abdel-Salam et al. 2017; Muotri et al. 2005; Narla et al. 2017; Muotri 2016; Ruzzo and Geschwind 2016; Renner et al. 2017; Sutcliffe and Lancaster 2017; Giandomenico and Lancaster 2017; Eiraku and Sasai 2012; Sasai 2013) and prepare the ground for developing organ-like structures, the organoids. For instance, Dr. Hans Clevers has used organotypic cultures for growing intestinal, liver, stomach, lungs, and prostate organoids (Clevers 2013; Bartfeld and Clevers 2017; Clevers 2009; Barker et al. 2008).

Parallel to the discovery of neurogenesis in the adult human brain, a breakthrough new method for tissue culture occurred during the 1990s. In 1990, Dr. Mina Bissell at the Lawrence Berkeley National Laboratory in California was attempting to grow murine mammary gland tissue in a petri dish. She had transplanted mammary acinus from an early pregnant mouse to the culture dish, but these tissues disorganized and stopped producing milk. Dr. Bissell then modified the microenvironment of the tissue culture by adding extracellular matrix (ECM) extracts and allowing cells to reorganize themselves into a three-dimensional (3D) structure analogous to the acinus. This modification resulted in the production of milk and led to the hypothesis that the microenvironment was activating genomic programs for cell interactions and aggregations to become functional tissues and organs (Foley 2017). This was a seminal finding, ultimately leading to the development and growing of the human organoids.

Early organotypic cultures used to study neural development were referred to as neurospheres (Lancaster and Knoblich 2014; Eiraku and Sasai 2012). The protocol for culturing neurospheres consists of laying NSCs in a nonadhesive substrate and adding mitogenic growth factors such as epidermal growth factor (EGF) and extracellular FGF-2 (Fig. 8.1). Later on, after removing growth factors, the multipotent NSCs begin to differentiate into neuronal, astrocytic, and oligodendrocytic lineages. Although neurospheres self-assemble into structures that resemble simple brain structures, most of the cell lines remain undifferentiated. The neurospheres tend to have clonal specificity toward a particular cell lineage (Gil-Perotin et al. 2013). As Lancaster noted, neurosphere cytostructure lacked the organization when compared with more advanced organoids (Lancaster and Knoblich 2014). Moreover, other researchers have also been growing organotypic culture from tumor cells, to which they refer as tumorspheres (Kaushik et al. 2017; Yang et al. 2017).

An important step in growing cerebral organoids was 2D structures referred to as neural rosettes. They develop from progenitor cells through an anchorage-dependent growth, but its architecture represents early stages of the neural tube organization (Broccoli et al. 2014). The neural progenitors in the rosettes present apicobasal morphology and undergo interkinetic nuclear migration. However, significant limitation to this 2D model was the lack of cell type differentiation that as some researchers have suggested might be related to anchorage dependence characteristic of the neural progenitors (Lancaster and Knoblich 2014). Another type of organotypic culture was the serum-free floating culture of embryoid body-like aggregates with quick reaggregation (SFEBq) method. This method requires the PSCs to be placed in low attachment substrate; growth factors are used to stimulate cell differentiation and regional specification (Fig. 8.1). The SFEBq method has been used to grow and differentiate specific brain tissues such as substantia nigra, septum, striatum, cerebellum, and dentate gyrus, cerebral cortex, pituitary gland, and retina (Brewer 1995). A critical limitation of this technique is the low survival rate of the tissues in culture (Wataya et al. 2008). Improvements in this technique such as

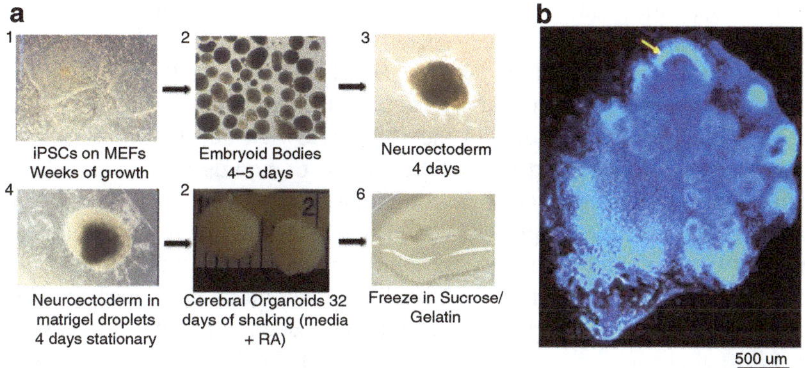

Fig. 8.1 Generation of brain organoid from human iPSCs (Stachowiak et al. 2017). (**a**) Mouse embryonic fibroblasts (MEFs) are used as feeder cells for iPSC culture (panel 1). The iPSCs are grown in a tissue-culture plate to reach confluence. Subsequently, three wells of iPSC are combined into a single well, and, after an additional 24 h, embryoid bodies (Peters et al.) are developed (panel 2). The EBs are maintained in N2/B27 media with dual SMAD inhibitors (which rapidly increases neural differentiation) exchanged daily for 4–5 days. The EBs are transferred into a 24-well low attachment plate, 1 EB per well, and incubated in neural induction medium forming neuroepithelial tissue (neuroectoderm) (panel 3). The neural induction medium is changed daily for 4 days. Beads of matrigel are made by placing a droplet of matrigel onto dented parafilm, the neuroepithelial tissue is placed onto the matrigel droplet, and then the unit is incubated at 37 °C to polymerize. The formed neuroectoderms in matrigel beads (panel 4) are removed from the parafilm and incubated in differentiation medium, without vitamin A, for 4 days without shaking. After 4 days, the growing neuroepithelial beads are referred to as day 1 cerebral organoids (panel 5), are moved to an orbital shaker, and are shaken in differentiation media with all-trans retinoic acid (RA). The differentiation media with RA is exchanged every 3–4 days. The organoids shown here have been grown for 32 days. Subsequently, the brain organoids are fixed in 4% paraformaldehyde and embedded in 10% gelatin/7.5% sucrose (panel 6) and slowly frozen in liquid N_2 to mitigate the cell damage caused through the freezing process. The organoids are stored at −80 °C. (**b**) Section through cerebral organoid—tile scanning of DAPI. A yellow arrow points to cortical rosettes. The organoids formed polarized structures with a distinguishable border that separated a forebrain-like region containing multiple rosettes from a hindbrain-like structure, which typically lacked rosettes (bottom pole), as reported previously (Lancaster and Knoblich 2014). The rosettes contained a ventricle-like lumen surrounded by distinct layers of cells (see Fig. 8.2)

implementation of Rho-associated protein kinase (ROCK) inhibitor to reduce apoptosis and B27-supplemented neurobasal medium optimization have been adapted from other protocols (Brewer et al. 1993).

A critical breakthrough in the generation of cerebral organoids were brain-like aggregates of the ESC-derived neural cells described for the first time by the Yoshiki Sasai laboratory in Japan in 2008 (Eiraku et al. 2008). Later, Juergen A. Knoblich and Madeline Lancaster working at the Institute for Molecular Biotechnology in Vienna (Austria; Lancaster and Knoblich 2014; Lancaster et al. 2013) have published protocols of generating the brain-like organotypic cultures. Their intention was to grow a monolayer of ESCs, but much to their disappointment the explants would not attach to the substrate. ESCs in culture were organized into spherical structures. Madeline Lancaster would eventually call these spherical structures "cerebral organoids" and others had referred to them as "mini-brains." Nature unveiled in this way an important secret about pluripotency, a secret that was previously passed on to Mina Bissell (Howlett and Bissell 1990) more than 20 years ago, but that for some reason had been forgotten. As we mentioned before, Dr. Bissell discovered organoids in the 1990s (Howlett and Bissell 1990). Yet, it took more than 20 years and the rediscovery of organoids by Yoshiki Sasai and Madeline Lancaster for the scientific community to catch up and to appreciate the magnitude of this breakthrough. The conclusion driven by the research of these two scientists was simple: for organoids, the context matters (Eiraku et al. 2011). The 3D expansion so brilliantly noted by Dr. Bissell and later on by Dr. Lancaster was only possible after extracellular matrix or matrigel was added to the mix (Lancaster et al. 2013; Lowenthal and Gerecht 2016). By using RT-PCR, Lancaster's group was able to identify the formation of different brain regional entities in the organoids such as forebrain (BF1+ and Six3+) and hindbrain (Krox20+, Isl1+). By using immunohistochemistry, they were able to identify dorso-cortical forebrain (Emx1+), ventral forebrain, hippocampus, and choroid plexus, among others (Lowenthal and Gerecht 2016; Lancaster et al. 2013).

The method for growing organoids, as described by Lancaster et al. (Lancaster and Knoblich 2014), consists of two phases: (1) induction of neural identity and differentiation and (2) 3D self-assembly of the early brain structure. Induction of neural identity and differentiation can be achieved by placing the PSCs into ESC medium with ROCK inhibitor. This process leads to the formation of embryoid bodies (Peters et al.), the outer surface of which contains ectoderm, which in turn will give origin to the neural tissue. In the center of the EBs, a mesoderm tissue forms but does not develop. Neural ectoderm is subsequently replated in a differentiation medium containing B27 supplement, 2-mercaptoethanol, and insulin. Later on, phase 1 retinoic acid (RA) is added to the mixture. In the second phase, the neuroectoderms are grown in a simulated low gravity condition for the 3D self-organization to take place.

The developing cerebral organoids displayed apicobasal cortical polarity (12) and for the first time offered a great example of corticogenesis occurring and being observed *in vitro* (Lancaster and Knoblich 2014) and in Fig. 8.1. This new technology opened the early period of human brain development to a direct scientific exploration.

Alongside organoid studies using ESCs, the development of iPSCs brought the organoid technology to the forefront of biomedical research for its potential in observing human disorders. Combining the organoid technology with iPSCs begins to shed light on early human brain development and its perturbation in diverse disorders. It offers a novel ex vivo diagnostic tool and potentially new preventive and corrective treatments that might eradicate the disease. Also, in the transplantation therapy, the risk of immune rejection will be greatly reduced if patients receive tissues that are derived from their own cells (autologous iPSCs).

8.5 Cerebral Organoids from ESCs and iPSCs to Recapitulate Cellular Processes of Cortical Development

Stachowiak laboratory routinely develops organoids using a modified version of the protocol established by Lancaster group. This modified protocol launched in our laboratory is outlined in Fig. 8.1 (Stachowiak et al. 2017). Once the shaking cultures are established, in the following weeks, the organoids gradually increase in size and the number of developed rosettes increases as well. The 21–32 days organoids form polarized structures with a distinguishable border that separates a forebrain-like region containing multiple circular rosettes from a hindbrain-like structure, which typically lacks the rosettes (Fig. 8.1; also Lancaster and Knoblich 2014).

During corticogenesis, neural progenitor cells originate in the ventricular zone (VZ) and proliferate in the subventricular zone (SVZ). Once proliferation ends, the immature, postmitotic neuronal cells migrate outward using established radial glia as a scaffold, with each new round of migrating cells moving past the pervious cell layer formed in an "inside-out" pattern. Distinct layers within the cortex are the intermediate zone (IZ), the cortical plate (CP), and the marginal zone (MZ) (Kriegstein and Noctor 2004): Similar to *in vivo* corticogenesis, organoid cortical rosettes develop three major zones which can be distinguished by nuclear staining with 4′,6-diamidino-2-phenylindole (DAPI) and by immunocytochemistry. Figure 8.2b shows stained sections of the organoids developed in our lab using modified protocol (Stachowiak et al. 2017). In DAPI-stained organoid sections (Fig. 8.2a), the VZ contains a ventricle-like lumen surrounded by compact layers of vertically aligned elongated cells. The area outside the VZ, the IZ, contains uniform, predominantly round cells, and the outermost CZ contains horizontally aligned cortical layers. Staining reveals cellular organization consistent with the inside-out pattern of human neocortex development (Kriegstein and Noctor 2004). In addition, proliferative marker Ki67 antigen-expressing NPCs and GFAP-expressing radial glia (Fig. 8.2b, c) are mostly present in the VZ. This is similar to the developing brain in the ventricular and subventricular zones, where the generation of new cells by the brain stem and progenitor cells takes place. Few Ki67-positive cells are also found in the IZ, and proliferating cells are not detected in the CZ. The doublecortin-positive neuroblasts are present in the IZ and the CZ, and βIII-tubulin is expressed by neurogenic radial glia in the VZ and by neuron committed cells (NCCs) in the IZ and young neurons in the CZ (Stachowiak et al. 2017).

8.6 Modeling Human Neurological Disorders

Growing cerebral organoids to the stage where they further replicate tissue organization and neuronal connections and communications provides new means of determining the development and underlying causes of the human neurological

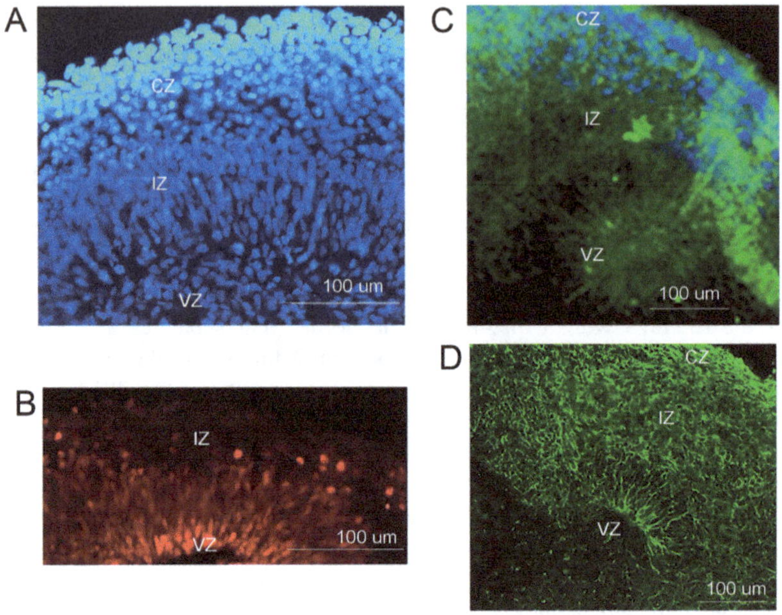

Fig. 8.2 Enlarged view of cortical rosette of the brain organoids (Stachowiak et al. 2017). (**a**) Figure shows three major zones distinguished by DAPI staining. The ventricular zone (VZ) contains a ventricle-like lumen surrounded by compact layers of vertically aligned elongated cells. The area outside the VZ, the intermediate zone (IZ), contains uniform, predominantly round cells. The outermost cortical zone (CZ) contains horizontally aligned cortical layers. (**b**) Ki67 immunostaining shows proliferating neural progenitor cells in the VZ. (**c**) Immunostaining with anti-GFAP antibody reveals GFAP expressing radial glia concentrated in the VZ. (**d**) βIII-tubulin was expressed by neurogenic radial glia in the VZ and by young neurons in the IZ and CZ

disorders. The ability to recapitulate fetal brain development is a critical step to reveal irregular developmental patterns that cause an early developmental disease (i.e., microcephaly and autism), a disease with a delayed developmental onset (i.e., schizophrenia), or leaves the individual with a predisposition to developing a disease later on. Generating organoids from iPSC of the Alzheimer's or Parkinson's patients and healthy subjects and observing changes in tissue structure or protein developmental expression could potentially reveal how certain individuals may be predisposed to the disease later in life and whether this predisposition comes from fetal development (Raja et al. 2016). The organoid model can also be combined with drug therapies to see if certain drugs can reverse or counteract changes in protein expression and enhance proper function (Choi et al. 2016).

The first disease model in human cerebral organoids was genetically based microcephaly. Organoid modeling of microcephaly has been successfully done and studied by the same group of researchers who developed iPSC cerebral organoid protocols (Lancaster et al. 2013). Their studies illustrated successful development of microcephaly from iPSC cultures with a key developmental feature, i.e., premature neuronal differentiation (Lancaster et al. 2013) that could be a cause of the disease.

One of the organoid research areas expected to lead to clinical treatments are studies of the effects of infectious agents and the immune factors on brain development. Brain organoids were used by Patricia P. Garcez et al. to identify the effects of Zika virus (ZIKV) on the ongoing development of the prepubertal brains (Garcez et al. 2016). ZIKV infection being present during gestation, in the placenta and embryo, had only been correlated with the occurrence of microcephaly in the infant. The controlled infection of neural cells and brain organoid permitted a direct examination of the ZIKV-infected developing brain. Human iPSC-derived NSCs were infected with ZIKV and then developed into neurospheres and brain organoids alongside the uninfected cells. The model allowed observing structural changes in developing neurospheres while quantifying the presence of ZIKV in the membranes, mitochondria, and vesicles of the infected cells. Infected neurospheres showed increased apoptosis and reduced neurosphere growth, with cell detachment and smooth membrane structures occurring in the neurosphere infected by ZIKV. When already formed brain organoids were exposed to ZIKV, the investigators found 40% reduction in organoid growth. These results were consistent with the pathological and clinical manifestation of the ZIKV-associated human microcephaly. These findings will likely aid in the development of new treatments of the Zika infections and of other related diseases. Similarly, our team has initiated new cerebral organoid studies to investigate how the immune factors produced during pregnancy may affect early human brain development. Here also, this type of research may lead to new preventive treatments.

8.7 Polygenic Neurodevelopmental Diseases

In schizophrenia and autism spectrum disorder (Stefansson et al.), hundreds of genomic traits have been identified as possible drivers for the disease phenotype (described further below) causing the study of each individual mutation a daunting task (Sanders et al. 2012; He et al. 2013). Hence, some researchers have adopted a different approach for studying these conditions. Instead of trying to understand how each one of these individual mutations can contribute to the general disorganization of the brain function, they are now identifying main pathways affected by these mutations as a whole, in an approach they call the watershed-hypothesis approach (Narla et al. 2017; Cannon and Keller 2006). A similar approach is also being used to study the ASD. Here too, scientists are trying to identify the convergent pathways affected by these mutations (He et al. 2013).

Based on certain clinical and genetic similarities between schizophrenia and ASD, it has been broadly believed that these two conditions might be related (Canitano and Pallagrosi 2017). It has been theorized that at the base of the social and cognitive disturbances, common to both conditions, there is an imbalance between excitatory/inhibitory neuronal activities. This imbalance can be a consequence of disrupted cortical architecture, mainly between the pyramidal glutamatergic neurons and inhibitory GABAergic parvalbumin interneurons. Such a hypothesis is supported by findings reported by multiple groups. For instance, Zielinski et al. reported an increase in the thickness of the frontal lobe cortex in early childhood and a subsequent arrest in development during late childhood in autistic patients (Donovan and Basson 2017; Zielinski et al. 2014). These findings are supported by a recent study, which used the iPSC-derived brain organoids from patients with autism. The study suggests an increased production of inhibitory neurons caused by increased FOXG1 gene expression (Mariani et al. 2015).

Whether schizophrenia and ASD are related or not, ought to be explored. Yet, if what is causing the misbalance between excitatory and inhibitory neuronal activity is indeed a disarray of the cortical structure, using a mouse model will limit the insights, given the important differences in the cortical microanatomy between species (Quiñones-Hinojosa and Chaichana 2007; DeFelipe et al. 2002; Marin et al. 2001; Anderson et al. 1999; Tan et al. 1998)

8.8 Schizophrenia iPSC Cerebral Organoids

A groundbreaking knowledge that comes from the studies employing iPSC cerebral organoids concerns schizophrenia (Stachowiak et al. 2017). Schizophrenia is the most severe mental illness affecting 1.5% of the world population that has been plaguing the mankind throughout its history. Written references to the schizophrenia-like illness can be traced to the old Pharaonic Egypt, where the thought disturbances typical to schizophrenia, depression, and dementia were regarded as symptoms of an ill mind, which at that time had been synonymous with the heart.

In the Middle Ages and in the Renaissance, mental illness was described as religious-like phenomenon and in later times as a phenomenon of civilization and culture. At some point, the mentally ill were considered to have freely chosen the path of mistake and against reason and morality. The perspective was ethical, not medical, and the treatment included constraints and rewards. This continued until mental illness started to be perceived as a product of natural causalities and an object of medical inquiry. Schizophrenia was classified as a distinct mental disorder, a "dementia praecox" by Kraepelin in 1887 (Jablensky et al. 1993), and the term schizophrenia, a fragmented mind, was coined by Eugen Bleuler, in 1911 (Nuechterlein and Dawson 1984).

On one hand, schizophrenia has been explained as an effect of an external "milieu" and, on the other, as an effect of the physical state of the brain. Even as recent as 1975 defective upbringing by parents was thought to be ultimately responsible for the disease, by affecting functioning of the adolescent brain (Arieti 1975).

8.9 When Does Schizophrenia Begin?

Typically, the onset of symptoms occurs during adolescence, suggesting that the pathology develops in parallel with symptoms, and is precipitated by stress and other environmental influences. Schizophrenia was looked upon as a functional disorder caused by a dysregulation of neuronal communications, a disorder of the communication software. Eliminating bad external influences and retuning neuronal circuits by psychotherapy and/or drugs were considered as viable means to cure schizophrenia. Meanwhile, an alternative idea has emerged that schizophrenia begins in utero and involves a malconstruction of brain circuits (a hardware disorder), which reveals itself later in life as these circuits become fully functional and utilized. It began with the isolated reports of the structural changes in the brain, especially cortex, observed postmortem in adult schizophrenia patients. An improper clustering of neurons in the cortical layers II, III, and V (Arnold et al. 1997) could arise during the first and early second trimester when cortical structure is laid down (Kneeland and Fatemi 2013). These alterations in neuronal numbers or clustering were not due to neurodegeneration, as no neurodegenerative markers are observed in Schizophrenia.

Schizophrenia appears to be specifically a human condition, a disorder of the association cortices, with especially prominent deficiencies in the dorsolateral prefrontal cortex (PFC). True dorsolateral PFC is found only in higher primates, and especially in humans, it is characterized by highly elaborate pyramidal cells with extensive recurrent connections. Schizophrenia is now recognized as an inheritable familial disorder, however one that appears to be the result of interplay between genetic and environmental factors.

While schizophrenia has been shown to be inheritable, the exact genetics behind it is less understood. From over 600 single nucleotide polymorphisms (SNPs), 200 genes and multitudes of copy number variations (CNV) have been found to be significantly associated with schizophrenia (Need et al. 2009; Welter et al. 2014; Malhotra et al. 2011; Kirov et al. 2012; Xu et al. 2008); however, no single alteration makes up more then 1–2% of the schizophrenia population (Xu et al. 2008; International Schizophrenia Consortium 2008; Stefansson et al. 2008). Hence, the genetic causes of schizophrenia appear to be a multiplicity of rare risk alleles and schizophrenia has been defined as a common, rare-variant disease. Also, some environmental factors acting during pregnancy appear to correlate positively with disease incidence (viral infections, nicotinism, etc.)

As mentioned earlier, to explain how various mutations can lead to a common disorder, Cannon and Keller proposed a watershed model (Cannon and Keller 2006) in which individual mutations dysregulate distinct biological pathways, which converge into a common ontogenic pathway(s). The dysregulation of such common pathways was proposed to lead to brain malformations, which increase the risk of the disease. The nature of such pathways has been unknown. Recent genomics-bioinformatics investigation has shown that pan-ontogenic integrative nuclear FGFR1 signaling (INFS) may serve as such pathways (Narla et al. 2017).

8.10 Brain Organoid Study of Schizophrenia (Stachowiak et al. 2017)

It is only now becoming evident, largely due to the organoid-based investigation that schizophrenia indeed entails early developmental malformation of the brain cortex, which is shared by unrelated patients with different genetic backgrounds. More importantly, recent organoid studies carried by our team gave unprecedented retrospective view of schizophrenia. Studies revealed for the first time that the trajectory of the illness was determined at the early stages of human brain development when, its basic structures ventricles, and cortex and the tissue in between are laid down (Stachowiak et al. 2017).

After establishing the protocol for the generation of hESC cerebral organoids, we applied this procedure to human iPSCs lines reprogrammed from schizophrenia and control individuals (Stachowiak et al. 2017), in which common dysregulated transcriptomes have been recently identified (Narla et al. 2017). Below, we describe few of the findings reported in our recent publication. In general, the iPSC cerebral organoids followed the developmental pattern observed in ESC organoids. However, a detailed cellular analysis revealed several significant differences between control and schizophrenia organoids (Stachowiak et al. 2017) (Fig. 8.3). The control iPSC organoids, similar to hESC organoids, contained few layers of NPCs expressing Ki67, a marker protein of the proliferating cells which were restricted largely to the VZ. In contrast, in schizophrenia organoids, the Ki67+ cells were strikingly relocated from the VZ into the IZ, as well as into the CZ (Fig. 8.3a). Computational analyses revealed an increased proliferation and migration of the schizophrenia NPCs in multiple patients' organoids (Stachowiak et al. 2017).

The transcription factor T-Box Brain 1 (TBR1) is expressed by developing neuroblasts, which migrate to and provide the first pioneer neurons of the developing cerebral cortex (Kolk et al. 2006). TBR1 is necessary for neuronal differentiation of NPCs and is a potential master regulator in autism spectrum disorders (Chuang et al. 2015). At 5 weeks of control iPSC organoid development, cells expressing nuclear TBR1 were distributed throughout the entire CZ and IZ (example on Fig. 8.3b). In contrast, in schizophrenia organoids, TBR1+ cells were absent from the upper cortical region, while cells expressing high levels of TBR1 were found concentrated predominantly in deep organoid layers. Consistent with these findings, staining with Pan-Neu Ab which reacts with key somatic, nuclear, dendritic, and axonal proteins of the pan-neuronal architecture revealed differentiated Pan-Neu+ neurons concentrated in the CZ of the control iPSC organoids, forming a distinct cortical layer at 5 weeks (Fig. 8.3). These mature neurons formed a dense network of long processes parallel and perpendicular to the cortical surface. At 5 weeks, the overall density of the Pan-Neu fibers in schizophrenia cortex appeared reduced. This decrease was verified by quantitative measurements (Stachowiak et al. 2017). Instead, the schizophrenia organoids displayed differentiated Pan-Neu+ neurons deep within the IZ and VZ regions (Fig. 8.3d). These mature subcortical neurons were found at 2 weeks in the schizophrenia organoids, at the time when no such neurons were observed in the

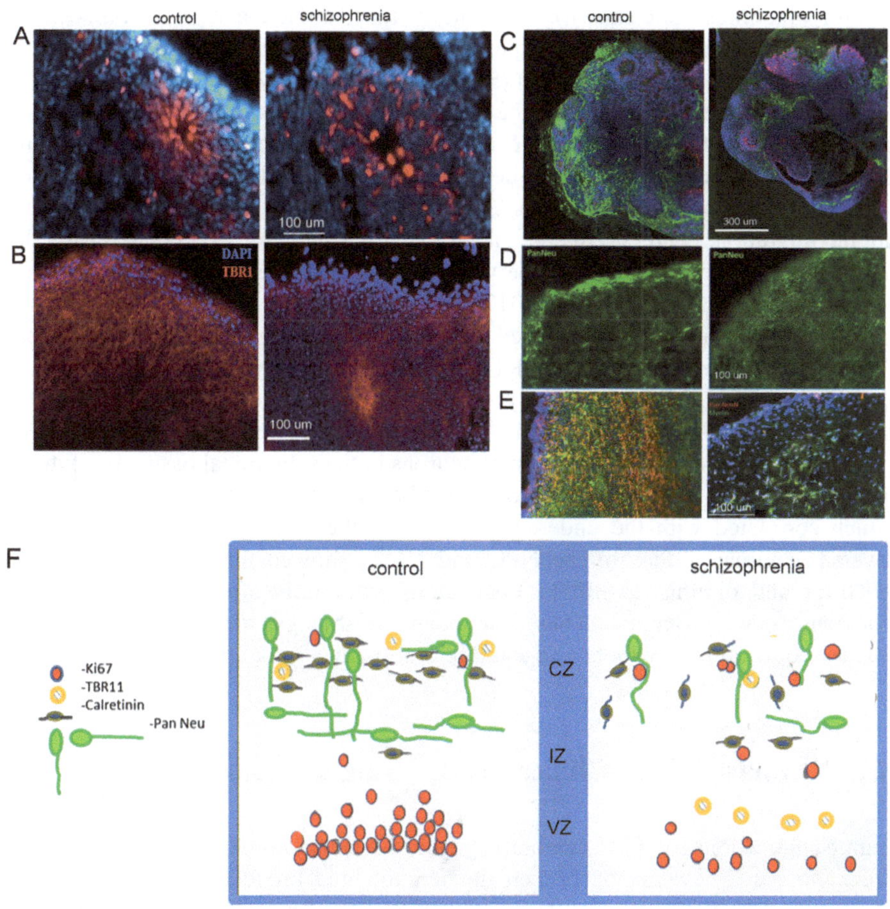

Fig. 8.3 Disorganized migration of proliferating cells and depletion of cortical neurons in schizophrenia 5 weeks iPSC cerebral organoids (Stachowiak et al. 2017). Representative images of control and schizophrenia organoids are shown. (**a**) Organoids were immunostained for Ki67 (red). Nuclei were stained with DAPI (blue) in schizophrenia organoids; note a dispersion of Ki67+ cells into IZ and CZ. (**b**) Decreased nuclear TBR1 (red) expression in the upper cortical zone of 5-week-old schizophrenia organoids. (**c–e**) Reduced density of Pan-Neu+ neurites in basal CZ and the presence of Pan-Neu+ cells with neurites in the IZ are visible. (**e**) Reduction in Pan-Neu-stained neurons (red) and myelinated fibers (green) in the schizophrenia organoid cortex. (**f**) Diagrams show schematic stratification of developing telencephalic-like zones in cerebral organoids—ventricular zone (VZ), intermediate zone (IZ), and cortical zone (CZ). Summary of cortical changes found in iPSC-derived schizophrenia organoids: increased proliferation of Ki67 NPCs and migration from the VZ into the IZ and CZ; reduced cortical accumulation of pioneer TBR1 neurons; reduced formation of Pan-Neu-stained cortical neurons especially those with horizontal neurites; and changes in the orientation of calretinin interneurons (Stachowiak et al. 2017)

control organoids (not shown). Co-staining for Pan-Neu and myelin-associated protein-2 confirmed the reduction neurons and their subcortical projections in the schizophrenia organoids (Fig. 8.3e) (Stachowiak et al. 2017).

Together, our experiments indicate an increased proliferation and migration of the schizophrenia NPCs, a premature development of neurons in the subcortical region, followed by an impaired neuronal development in the cortex of the schizophrenia organoids (Stachowiak et al. 2017). The early changes, increased VZ cell proliferation, migration, and premature neuronal differentiation correlate with the upregulation of gene programs underwriting these functions in early differentiating schizophrenia NPCs (Narla et al. 2017).

An important finding that emerged from this investigation was the role of nuclear form of FGF Receptor 1 (nFGFR1) in the observed dysregulation of cortical development (Fig. 8.3f). At an early stage of the NPC development, nFGFR1 was overexpressed in schizophrenia cells, which appear to correlate with the dispersion of VZ cells and premature formation of neurons in the subcortical tissue. In contrast, at the later stage of cortical development, the nFGFR1 expression was shut off, which correlated with the underdevelopment of the mature cortical neurons and layers. Genomic studies of the NPC and NCCs showed that both an excessive nFGFR1 and diminished nFGFR1 signaling profoundly affected diverse neuro-ontogenic gene programs. Thus, one potential strategy for preventing cortical maldevelopment in schizophrenia could be a normalization of the INFS mechanism.

8.11 "Phase Zero" Clinical Trial Using Organoids

Future applications for iPSC-cerebral organoid-based research in the biomedical field are multifold. They can be broken down into three broad areas of application: (1) organ regeneration and replacement for central nervous system (CNS) injury and degenerative disorders; (2) improved diagnosis of developmental disorders like autism, microcephaly, schizophrenia, and neurodegenerative diseases like Alzheimer or Parkinson disorders; and (3) individualized drug testing and preventive/corrective therapies.

Each of these subject areas has the potential to develop breakthrough discoveries that may not have been predicted or envisioned without the advancement in organoid technology. By sorting through the scientific literature, we can have an idea how proficient basic research is. Yet, translating these discoveries into clinical applications is a costly and long process to say the least (Cerovska et al. 2017). One of the reasons this might be the case is probably the way the drug discovery program is set up. Regulatory agencies require any new drug to be tested first in an animal model or *in vitro* before they can be tested in humans.

Even if the disease may be modeled in mouse, the drug efficacy is often different. So far we have emphasized the important role the genetic profile plays in drug sensitivity testing. For animal and human risk assessments though other aspects need to be taken into consideration. For instance, the response to a drug can vary from one species to another. Notably, some antibiotics are lethal for hamsters and guinea pigs, yet mice and rats can tolerate them very well. In this sense, we are lucky that Alexander Fleming used rats for testing penicillin, had he tested it on a different rodent, history could have taken a different path.

Aspirin is another excellent example of how a drug can have different effects based on the species. Aspirin is known to cause birth defects in mice and rats but not in humans. Can one imagine what could have happened if aspirin would have been tested on mice? Another such drug known to cause birth defects in mice but not in humans is cortisone. In addition, aspartame is known to cause lymphomas in rats, but not in humans. Acetaminophen, within therapeutic range, is well tolerated by humans but is hepatotoxic in mice at low dose. Indomethacin, a drug used to treat rheumatoid arthritis, is well tolerated in humans; however, it causes ulcers in rats and dogs (Matsubara and Bissell 2016). These examples show that a drug can be safe for rodents, but not for humans and vice versa. No wonder, most of the potential new drugs only ever make it to phase 1 of clinical trials.

The iPSC and organoid technologies bring out a possibility of a new "Phase Zero" clinical trial. Established organoids are grown from control and diseased patients' iPSC in a dish, and the drugs or combination of new drugs is applied to the culture to determine their potential toxic and therapeutic effects. Through this approach, researchers would be able evaluate the drug efficacy against the disease of a specific patient.

8.12 New Directions and Challenges for the Organoid Research

The continued use and development of the brain organoid model will extend into many future scientific endeavors and functional studies. For instance, an abnormal brain development and the responsible agents could potentially be counteracted with the existing and new pharmacological agents. In addition, a promising new tool may be developed to control biological development and functions by incorporating light-activated molecular switch proteins into the cells. One upcoming new development will be partnering of the organoid research with optogenetics and optogenomics. Photonic regulation of light-sensitive switches is highly advantageous compared to classical chemical activation methods, due to its ability to precisely activate and inactivate both in space and in time. Especially the use of the new generation of nano-actuators allows the induction of protein–protein interactions among several proteins of interest on a subcellular scale. Recent advancements in nanotechnology provide the engineering community with a new set of new tools to create nanoscale photonic devices with unprecedented functionalities (Feng et al. 2014; Miao et al. 2016; Nafari and Jornet 2017). In our opinion, the plasmonic nano-lasers working in conjunction with nano-antennas can serve as *nano-photo-actuators* of biological processes. Together, networks of nano-actuators and nano-sensors could control development

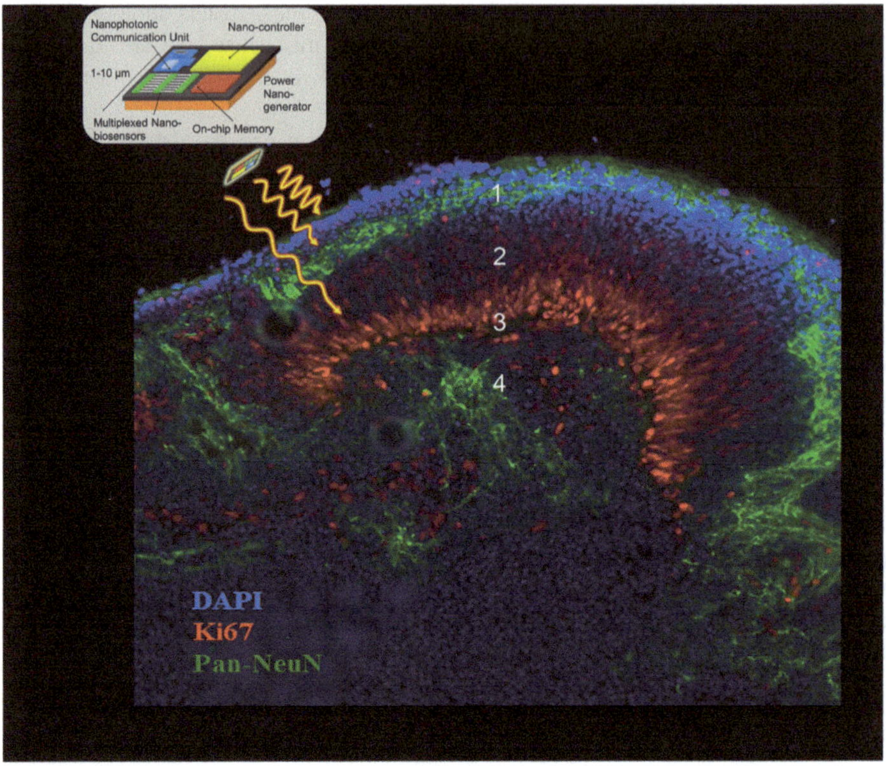

Fig. 8.4 "Chip on the brain"—development of nanophotonic devices for activation and inactivation of molecular switches (protein–protein interactions) to control cell proliferation, migration, and differentiation in brain organoids. Networks of arrayed nano-actuators are developed to control the laser light-sensitive molecular toggle switches. The platform will provide spatial and temporal control of the illuminated area and in addition will contain nanosensors to receive signals generated by the light-emitting cell molecules. The platform diagram was generated by Pedram Johari, the University at Buffalo, Department of Electrical Engineering. The image shows 5-week-old iPSC cerebral organoid: 1—differentiated neurons in CZ; 2—migrating neuroblasts in IZ; 3—proliferating NPC in VZ; 4—neurons in developing deep nuclei

of cells in complex tissues of the developing brain. An important step toward these new technologies will be their testing in the cerebral organoids to see if they can direct and modify organoid development (Fig. 8.4).

It has been proposed that mechanical forces exerted by developing axonal projections influence the surface tension and the shape of the developing human cortex (Hilgetag and Barbas 2006). Cerebral organoids offer an effective tool to test these assumptions and the roles that the changes in surface tension and other vector forces could play in schizophrenia, autism, and other developmental diseases. By combining organoid technology of disease-specific cerebral organoids with the atomic force microscopy (AFM), one could examine for the first time the relationship between tissue elasticity/force and the disease progression (Tan et al. 2017). For instance, in schizophrenia organoids, changes in cortical axonal tracks, i.e., loss of horizontal and maintenance of vertical fibers, and disorientation of horizontal connecting interneurons suggest that the surface tension and the elasticity of cortex may be reduced (Fig. 8.3f) (Stachowiak et al. 2017). How these altered mechanical properties may affect cortical development and its control by the mechanical forces may be addressed using AFM and a novel optogenomics platform applied to the cerebral organoid model.

Further development of the organoid model faces many technical challenges. If the organoids were to be used for developing new drugs, a better way for delivering drugs and removing wastes ought to be devised. A vascular system is needed to enable the flow of nutrients, oxygen, and hormones in the blood, without which the size of grown cerebral organoids is limited to 5–10 mm and their effective integration with the host tissues may not be possible (Munera and Wells 2017). There are already first reports describing vascularization of organoids generated from iPSCs. In one such study, in 2014, a functional human liver with proper structure was generated from liver buds developed from iPSCs (Takebe et al. 2013). A vascular system within the liver bud transplants began to develop with proper vessel connections (Takebe et al. 2013). This is an important first step in developing regenerative medicine protocols for patients with damaged organs.

8.13 Concluding Remarks

We began this chapter by recounting how generous cancer patients helped in the discovery of stem cells in the adult human brain (Eriksson et al. 1998; Kirschenbaum et al. 1994). It is fair that we come full circle and conclude this chapter by recounting how the latest application of stem cells, the organoid system, is now being used to study cancer. Mina Bissell had studied the murine mammary acinus for a while and is now actively studying breast cancer (Peinado et al. 2017; Snijders et al. 2017). Her research has provided important insights on how the microenvironment, the context as she would call it, can influence the development of cancer (Curtin and Heritier 2017; Cancer Genome Atlas 2012).

In parallel to these cancer-related studies, other researchers are using the organoids to study diverse organ systems including the liver, thyroid gland, pituitary gland, intestines, retina, and the brain (Eiraku et al. 2011; Sato et al. 2009; Antonica et al. 2012; Huch et al. 2013; Suga et al. 2011; Koehler et al. 2013; Xia et al. 2013; Takasato et al. 2014; Taguchi et al. 2014). Combining the organoid technology with iPSCs begins to shed light on how certain human diseases affect early brain development and consequently brain functions. This will likely bring important breakthroughs in understanding the underlying pathologies, improve the diagnostics, and lead to new preventive and corrective treatments. With the cerebral organoids, a frequent common goal has been to expand the cultures in time and develop brain-like structures as advanced as may be possible. Such an approach may potentially be beneficial for regenerative medicine. However, one needs to keep in mind that in order to identify the role of genome in shaping up the human brain, it is important to focus on the early stages of the organoid development driven by the inherited genomic programs rather than on highly variable advanced stages produced by diverse technical manipulations. Clearly, the advancements in the field of organoids are far from being over.

References

Abdel-Salam GMH, Abdel-Hamid MS, Mohammad SA et al (2017) Aicardi-Goutières syndrome: unusual neuro-radiological manifestations. Metab Brain Dis 32(3):679–683. https://doi.org/10.1007/s11011-017-9993-4

Altman J, Das GD (1965) Autoradiographic and histological evidence of postnatal hippocampal neurogenesis in rats. J Comp Neurol 124(3):319–335

Anderson S, Mione M, Yun K et al (1999) Differential origins of neocortical projection and local circuit neurons: role of Dlx genes in neocortical interneuronogenesis. Cereb Cortex 9 (6):646–654

Antonica F, Kasprzyk DF, Opitz R et al (2012) Generation of functional thyroid from embryonic stem cells. Nature 491(7422):66–71. https://doi.org/10.1038/nature11525

Arieti S (1975) Child and adolescent psychiatry, sociocultural and community psychiatry. In: American Handbook of Psychiatry, vol 2. Basic Books, New York

Arnold SE, Ruscheinsky DD, Han LY (1997) Further evidence of abnormal cytoarchitecture of the entorhinal cortex in schizophrenia using spatial point pattern analyses. Biol Psychiatry 42 (8):639–647

Barker N, van de Wetering M, Clevers H (2008) The intestinal stem cell. Genes Dev 22 (14):1856–1864. https://doi.org/10.1101/gad.1674008

Bartfeld S, Clevers H (2017) Stem cell-derived organoids and their application for medical research and patient treatment. J Mol Med (Berlin, Germany) 95(7):729–738. https://doi.org/10.1007/s00109-017-1531-7

Brewer GJ (1995) Serum-free B27/neurobasal medium supports differentiated growth of neurons from the striatum, substantia nigra, septum, cerebral cortex, cerebellum, and dentate gyrus. J Neurosci Res 42(5):674–683. https://doi.org/10.1002/jnr.490420510

Brewer GJ, Torricelli JR, Evege EK et al (1993) Optimized survival of hippocampal neurons in B27-supplemented Neurobasal, a new serum-free medium combination. J Neurosci Res 35 (5):567–576. https://doi.org/10.1002/jnr.490350513

Broccoli V, Giannelli SG, Mazzara PG (2014) Modeling physiological and pathological human neurogenesis in the dish. Front Neurosci 8:183. https://doi.org/10.3389/fnins.2014.00183

Cancer Genome Atlas N (2012) Comprehensive molecular characterization of human colon and rectal cancer. Nature 487(7407):330–337. https://doi.org/10.1038/nature11252

Canitano R, Pallagrosi M (2017) Autism spectrum disorders and schizophrenia spectrum disorders: excitation/inhibition imbalance and developmental trajectories. Front Psychiatry 8:69. https://doi.org/10.3389/fpsyt.2017.00069

Cannon TD, Keller MC (2006) Endophenotypes in the genetic analyses of mental disorders. Annu Rev Clin Psychol 2:267–290. https://doi.org/10.1146/annurev.clinpsy.2.022305.095232

Carrel A (1912) Pure cultures of cells. J Exp Med 16(2):165–168

Carrel A, Burrows MT (1911) An addition to the technique of the cultivation of tissues in vitro. J Exp Med 14(3):244–247

Cerovska E, Elsnerova K, Vaclavikova R et al (2017) The role of membrane transporters in ovarian cancer chemoresistance and prognosis. Expert Opin Drug Metab Toxicol 13(7):741–753. https://doi.org/10.1080/17425255.2017.1332179

Choi SH, Kim YH, Quinti L et al (2016) 3D culture models of Alzheimer's disease: a road map to a "cure-in-a-dish". Mol Neurodegener 11(1):75. https://doi.org/10.1186/s13024-016-0139-7
Chuang HC, Huang TN, Hsueh YP (2015) T-Brain-1: a potential master regulator in autism spectrum disorders. Autism Res 8(4):412–426. https://doi.org/10.1002/aur.1456
Clevers H (2009) Searching for adult stem cells in the intestine. EMBO Mol Med 1(5):255–259. https://doi.org/10.1002/emmm.200900034
Clevers H (2013) A gutsy approach to stem cells and signalling: an interview with Hans Clevers. Dis Model Mech 6(5):1053–1056. https://doi.org/10.1242/dmm.013367
Curtin F, Heritier S (2017) The role of adaptive trial designs in drug development. Expert Rev Clin Pharmacol 10(7):727–736. https://doi.org/10.1080/17512433.2017.1321985
DeFelipe J, Alonso-Nanclares L, Arellano JI (2002) Microstructure of the neocortex: comparative aspects. J Neurocytol 31(3-5):299–316
Donovan AP, Basson MA (2017) The neuroanatomy of autism – a developmental perspective. J Anat 230(1):4–15. https://doi.org/10.1111/joa.12542
Dulbecco R (1952) Production of plaques in monolayer tissue cultures by single particles of an animal virus. Proc Natl Acad Sci USA 38(8):747–752
Eckenhoff MF, Rakic P (1988) Nature and fate of proliferative cells in the hippocampal dentate gyrus during the life span of the rhesus monkey. J Neurosci 8(8):2729–2747
Eiraku M, Sasai Y (2012) Self-formation of layered neural structures in three-dimensional culture of ES cells. Curr Opin Neurobiol 22(5):768–777. https://doi.org/10.1016/j.conb.2012.02.005
Eiraku M, Watanabe K, Matsuo-Takasaki M et al (2008) Self-organized formation of polarized cortical tissues from ESCs and its active manipulation by extrinsic signals. Cell Stem Cell 3 (5):519–532. https://doi.org/10.1016/j.stem.2008.09.002
Eiraku M, Takata N, Ishibashi H et al (2011) Self-organizing optic-cup morphogenesis in three-dimensional culture. Nature 472(7341):51–56. https://doi.org/10.1038/nature09941
Eriksson PS, Perfilieva E, Bjork-Eriksson T et al (1998) Neurogenesis in the adult human hippocampus. Nat Med 4(11):1313–1317. https://doi.org/10.1038/3305
Feng L, Wong ZJ, Ma RM et al (2014) Single-mode laser by parity-time symmetry breaking. Science 346(6212):972–975. https://doi.org/10.1126/science.1258479
Foley KE (2017) Organoids: a better in vitro model. Nat Methods 14(6):559–562. https://doi.org/10.1038/nmeth.4307
Freshney RI, Inc e (2010) Culture of animal cells : a manual of basic technique and specialized applications, 6th edn. Wiley-Blackwell, Hoboken, NJ
Garcez PP, Loiola EC, Madeiro da Costa R et al (2016) Zika virus impairs growth in human neurospheres and brain organoids. Science 352(6287):816–818. https://doi.org/10.1126/science.aaf6116
Giandomenico SL, Lancaster MA (2017) Probing human brain evolution and development in organoids. Curr Opin Cell Biol 44:36–43. https://doi.org/10.1016/j.ceb.2017.01.001
Gil-Perotin S, Duran-Moreno M, Cebrian-Silla A et al (2013) Adult neural stem cells from the subventricular zone: a review of the neurosphere assay. Anat Rec (Hoboken) 296 (9):1435–1452. https://doi.org/10.1002/ar.22746
He X, Sanders SJ, Liu L et al (2013) Integrated model of de novo and inherited genetic variants yields greater power to identify risk genes. PLoS Genet 9(8):e1003671. https://doi.org/10.1371/journal.pgen.1003671
Hilgetag CC, Barbas H (2006) Role of mechanical factors in the morphology of the primate cerebral cortex. PLoS Comput Biol 2(3):e22. https://doi.org/10.1371/journal.pcbi.0020022
Howlett AR, Bissell MJ (1990) Regulation of mammary epithelial cell function: a role for stromal and basement membrane matrices. Protoplasma 159(2):85–95. https://doi.org/10.1007/bf01322592
Huch M, Dorrell C, Boj SF et al (2013) In vitro expansion of single $Lgr5^+$ liver stem cells induced by Wnt-driven regeneration. Nature 494(7436):247–250. https://doi.org/10.1038/nature11826
International Schizophrenia Consortium (2008) Rare chromosomal deletions and duplications increase risk of schizophrenia. Nature 455(7210):237–241. https://doi.org/10.1038/nature07239

Jablensky A, Hugler H, Von Cranach M et al (1993) Kraepelin revisited: a reassessment and statistical analysis of dementia praecox and manic-depressive insanity in 1908. Psychol Med 23 (4):843–858

Kaushik V, Yakisich JS, Azad N et al (2017) Anti-tumor effects of cardiac glycosides on human lung cancer cells and lung tumorspheres. J Cell Physiol 232(9):2497–2507. https://doi.org/10.1002/jcp.25611

Kawasaki H, Mizuseki K, Nishikawa S et al (2000) Induction of midbrain dopaminergic neurons from ES cells by stromal cell-derived inducing activity. Neuron 28(1):31–40

Kirov G, Pocklington AJ, Holmans P et al (2012) De novo CNV analysis implicates specific abnormalities of postsynaptic signalling complexes in the pathogenesis of schizophrenia. Mol Psychiatry 17(2):142–153. https://doi.org/10.1038/mp.2011.154

Kirschenbaum B, Nedergaard M, Preuss A et al (1994) In vitro neuronal production and differentiation by precursor cells derived from the adult human forebrain. Cereb Cortex 4(6):576–589

Kneeland RE, Fatemi SH (2013) Viral infection, inflammation and schizophrenia. Prog Neuro-psychopharmacol Biol Psychiatry 42:35–48. https://doi.org/10.1016/j.pnpbp.2012.02.001

Koehler KR, Mikosz AM, Molosh AI et al (2013) Generation of inner ear sensory epithelia from pluripotent stem cells in 3D culture. Nature 500(7461):217–221. https://doi.org/10.1038/nature12298

Kolk SM, Whitman MC, ME Y et al (2006) A unique subpopulation of Tbr1-expressing deep layer neurons in the developing cerebral cortex. Mol Cell Neurosci 32(1-2):200–214

Kriegstein AR, Noctor SC (2004) Patterns of neuronal migration in the embryonic cortex. Trends Neurosci 27(7):392–399. https://doi.org/10.1016/j.tins.2004.05.001

Laflamme MA, Chen KY, Naumova AV et al (2007) Cardiomyocytes derived from human embryonic stem cells in pro-survival factors enhance function of infarcted rat hearts. Nat Biotechnol 25(9):1015–1024. https://doi.org/10.1038/nbt1327

Lancaster MA, Knoblich JA (2014) Generation of cerebral organoids from human pluripotent stem cells. Nat Protoc 9(10):2329–2340. https://doi.org/10.1038/nprot.2014.158

Lancaster MA, Renner M, Martin CA et al (2013) Cerebral organoids model human brain development and microcephaly. Nature 501(7467):373–379. https://doi.org/10.1038/nature12517

Lazar RM, Mohr JP (2011) Revisiting the contributions of Paul Broca to the study of aphasia. Neuropsychol Rev 21(3):236–239. https://doi.org/10.1007/s11065-011-9176-8

Lee AS, Tang C, Rao MS et al (2013) Tumorigenicity as a clinical hurdle for pluripotent stem cell therapies. Nature Med 19(8):998–1004. https://doi.org/10.1038/nm.3267

Lowenthal J, Gerecht S (2016) Stem cell-derived vasculature: a potent and multidimensional technology for basic research, disease modeling, and tissue engineering. Biochem Biophys Res Commun 473(3):733–742. https://doi.org/10.1016/j.bbrc.2015.09.127

Malhotra D, McCarthy S, Michaelson JJ et al (2011) High frequencies of de novo CNVs in bipolar disorder and schizophrenia. Neuron 72(6):951–963. https://doi.org/10.1016/j.neuron.2011.11.007

Mariani J, Coppola G, Zhang P et al (2015) FOXG1-dependent dysregulation of GABA/glutamate neuron differentiation in autism spectrum disorders. Cell 162(2):375–390. https://doi.org/10.1016/j.cell.2015.06.034

Marin O, Yaron A, Bagri A et al (2001) Sorting of striatal and cortical interneurons regulated by semaphorin-neuropilin interactions. Science 293(5531):872–875. https://doi.org/10.1126/science.1061891

Matsubara M, Bissell MJ (2016) Inhibitors of Rho kinase (ROCK) signaling revert the malignant phenotype of breast cancer cells in 3D context. Oncotarget 7(22):31602–31622. https://doi.org/10.18632/oncotarget.9395

McBride AA (2017) Mechanisms and strategies of papillomavirus replication. Biol Chem 398 (8):919–927. https://doi.org/10.1515/hsz-2017-0113

Miao P, Zhang Z, Sun J et al (2016) Orbital angular momentum microlaser. Science 353 (6298):464–467. https://doi.org/10.1126/science.aaf8533

Munera JO, Wells JM (2017) Generation of gastrointestinal organoids from human pluripotent stem cells. Methods Mol Biol (Clifton, NJ) 1597:167–177. https://doi.org/10.1007/978-1-4939-6949-4_12
Muotri AR (2016) The human model: changing focus on autism research. Biol Psychiatry 79 (8):642–649. https://doi.org/10.1016/j.biopsych.2015.03.012
Muotri AR, Nakashima K, Toni N et al (2005) Development of functional human embryonic stem cell-derived neurons in mouse brain. Proc Natl Acad Sci USA 102(51):18644–18648. https://doi.org/10.1073/pnas.0509315102
Nafari M, Jornet JM (2017) Modeling and performance analysis of metallic plasmonic nano-antennas for wireless optical communication in nanonetworks. IEEE Access 5:6389–6398. https://doi.org/10.1109/Access.2017.2690990
Narla ST, Lee YW, Benson CA et al (2017) Common developmental genome deprogramming in schizophrenia—role of integrative nuclear FGFR1 signaling (INFS). Schizophr Res 185:17–32. https://doi.org/10.1016/j.schres.2016.12.012
Need AC, Attix DK, McEvoy JM et al (2009) A genome-wide study of common SNPs and CNVs in cognitive performance in the CANTAB. Hum Mol Genet 18(23):4650–4661. https://doi.org/10.1093/hmg/ddp413
Nottebohm F (1985) Neuronal replacement in adulthood. Ann N Y Acad Sci 457:143–161
Nuechterlein KH, Dawson ME (1984) Information processing and attentional functioning in the developmental course of schizophrenic disorders. Schizophr Bull 10(2):160–203
Peinado H, Zhang H, Matei IR et al (2017) Pre-metastatic niches: organ-specific homes for metastases. Nature Rev Cancer 17(5):302–317. https://doi.org/10.1038/nrc.2017.6
Peters A, Palay SL (1996) The morphology of synapses. J Neurocytol 25(12):687–700
Peters A, Palay SL, Hd W (1991) The fine structure of the nervous system : neurons and their supporting cells, 3rd edn. Oxford University Press, New York
Quadrato G, Brown J, Arlotta P (2016) The promises and challenges of human brain organoids as models of neuropsychiatric disease. Nature Med 22(11):1220–1228. https://doi.org/10.1038/nm.4214
Quiñones-Hinojosa A, Chaichana K (2007) The human subventricular zone: a source of new cells and a potential source of brain tumors. Exp Neurol 205(2):313–324. https://doi.org/10.1016/j.expneurol.2007.03.016
Raja WK, Mungenast AE, Lin YT et al (2016) Self-organizing 3D human neural tissue derived from induced pluripotent stem cells recapitulate Alzheimer's disease phenotypes. PLoS One 11(9): e0161969. https://doi.org/10.1371/journal.pone.0161969
Rakic P (1985) Limits of neurogenesis in primates. Science 227(4690):1054–1056
Renner M, Lancaster MA, Bian S et al (2017) Self-organized developmental patterning and differentiation in cerebral organoids. EMBO J 36(10):1316–1329. https://doi.org/10.15252/embj.201694700
Ruzzo EK, Geschwind DH (2016) Schizophrenia genetics complements its mechanistic understanding. Nat Neurosci 19(4):523–525. https://doi.org/10.1038/nn.4277
Sadeghipour S, Mathias RA (2017) Herpesviruses hijack host exosomes for viral pathogenesis. Semin Cell Dev Biol 67:91–100. https://doi.org/10.1016/j.semcdb.2017.03.005
Sanders SJ, Murtha MT, Gupta AR et al (2012) De novo mutations revealed by whole-exome sequencing are strongly associated with autism. Nature 485(7397):237–241. https://doi.org/10.1038/nature10945
Sasai Y (2013) Next-generation regenerative medicine: organogenesis from stem cells in 3D culture. Cell Stem Cell 12(5):520–530. https://doi.org/10.1016/j.stem.2013.04.009
Sato T, Vries RG, Snippert HJ et al (2009) Single Lgr5 stem cells build crypt-villus structures in vitro without a mesenchymal niche. Nature 459(7244):262–265. https://doi.org/10.1038/nature07935
Seok J, Warren HS, Cuenca AG et al (2013) Genomic responses in mouse models poorly mimic human inflammatory diseases. Proc Natl Acad Sci USA 110(9):3507–3512. https://doi.org/10.1073/pnas.1222878110

Snijders AM, Lee SY, Hang B et al (2017) FAM83 family oncogenes are broadly involved in human cancers: an integrative multi-omics approach. Mol Oncol 11(2):167–179. https://doi.org/10.1002/1878-0261.12016

Stachowiak EK, Benson CA, Narla ST et al (2017) Cerebral organoids reveal early cortical maldevelopment in schizophrenia—computational anatomy and genomics, role of FGFR1. Translational Psychiatry 7:6

Stahnisch FW, Nitsch R (2002) Santiago Ramon y Cajal's concept of neuronal plasticity: the ambiguity lives on. Trends Neurosci 25(11):589–591

Stefansson H, Rujescu D, Cichon S et al (2008) Large recurrent microdeletions associated with schizophrenia. Nature 455(7210):232–236. https://doi.org/10.1038/nature07229

Suga H, Kadoshima T, Minaguchi M et al (2011) Self-formation of functional adenohypophysis in three-dimensional culture. Nature 480(7375):57–62. https://doi.org/10.1038/nature10637

Sutcliffe M, Lancaster MA (2017) A simple method of generating 3D brain organoids using standard laboratory equipment. Methods Mol Biol (Clifton, NJ). https://doi.org/10.1007/7651_2017_2

Taguchi A, Kaku Y, Ohmori T et al (2014) Redefining the in vivo origin of metanephric nephron progenitors enables generation of complex kidney structures from pluripotent stem cells. Cell Stem Cell 14(1):53–67. https://doi.org/10.1016/j.stem.2013.11.010

Takahashi K, Yamanaka S (2006) Induction of pluripotent stem cells from mouse embryonic and adult fibroblast cultures by defined factors. Cell 126(4):663–676. https://doi.org/10.1016/j.cell.2006.07.024

Takasato M, Er PX, Becroft M et al (2014) Directing human embryonic stem cell differentiation towards a renal lineage generates a self-organizing kidney. Nat Cell Biol 16(1):118–126. https://doi.org/10.1038/ncb2894

Takebe T, Sekine K, Enomura M et al (2013) Vascularized and functional human liver from an iPSC-derived organ bud transplant. Nature 499(7459):481–484. https://doi.org/10.1038/nature12271

Tan SS, Kalloniatis M, Sturm K et al (1998) Separate progenitors for radial and tangential cell dispersion during development of the cerebral neocortex. Neuron 21(2):295–304

Tan Q, Choi KM, Sicard D et al (2017) Human airway organoid engineering as a step toward lung regeneration and disease modeling. Biomaterials 113:118–132. https://doi.org/10.1016/j.biomaterials.2016.10.046

Thomas CA, Tejwani L, Trujillo CA et al. (2017) Modeling of TREX1-dependent autoimmune disease using human stem cells highlights L1 accumulation as a source of neuroinflammation. Cell Stem Cell 21(3):319–331.e318. doi:https://doi.org/10.1016/j.stem.2017.07.009

Wataya T, Muguruma K, Sasai Y (2008) Human pluripotent stem cell and neural differentiation. Brain Nerve/Shinkei kenkyu no shinpo 60(10):1165–1172

Welter D, MacArthur J, Morales J et al (2014) The NHGRI GWAS catalog, a curated resource of SNP-trait associations. Nucleic Acids Res 42(Database issue):D1001–D1006. https://doi.org/10.1093/nar/gkt1229

Xia Y, Nivet E, Sancho-Martinez I et al (2013) Directed differentiation of human pluripotent cells to ureteric bud kidney progenitor-like cells. Nat Cell Biol 15(12):1507–1515. https://doi.org/10.1038/ncb2872

Xu B, Roos JL, Levy S et al (2008) Strong association of de novo copy number mutations with sporadic schizophrenia. Nat Genet 40(7):880–885. https://doi.org/10.1038/ng.162

Yang Y, Yu L, Li J et al (2017) Autophagy regulates the stemness of cervical cancer stem cells. Biologics 11:71–79. https://doi.org/10.2147/BTT.S134920

Zielinski BA, Prigge MB, Nielsen JA et al (2014) Longitudinal changes in cortical thickness in autism and typical development. Brain 137. (Pt 6:1799–1812. https://doi.org/10.1093/brain/awu083

Chapter 9
Bioengineering of the Human Neural Stem Cell Niche: A Regulatory Environment for Cell Fate and Potential Target for Neurotoxicity

Leonora Buzanska, Marzena Zychowicz, and Agnieszka Kinsner-Ovaskainen

Abstract Human neural stem/progenitor cells of the developing and adult organisms are surrounded by the microenvironment, so-called neurogenic niche. The developmental processes of stem cells, such as survival, proliferation, differentiation, and fate decisions, are controlled by the mutual interactions between cells and the niche components. Such interactions are tissue specific and determined by the biochemical and biophysical properties of the niche constituencies and the presence of other cell types. This dynamic approach of the stem cell niche, when translated into in vitro settings, requires building up "biomimetic" microenvironments resembling natural conditions, where the stem/progenitor cell is provided with diverse extracellular signals exerted by soluble and structural cues, mimicking those found in vivo. The neural stem cell niche is characterized by a unique composition of soluble components including neurotransmitters and trophic factors as well as insoluble extracellular matrix proteins and proteoglycans. Biotechnological innovations provide tools such as a new generation of tunable biomaterials capable of releasing specific signals in a spatially and temporally controlled manner, thus creating in vitro nature-like conditions and, when combined with stem cell-derived tissue specific progenitors, producing differentiated neuronal tissue structures. In addition, substantial progress has been made on the protocols to obtain stem cell-derived cell aggregates such as neurospheres and self-assembled organoids.

L. Buzanska (✉) · M. Zychowicz
Stem Cell Bioengineering Unit, Mossakowski Medical Research Centre Polish Academy of Sciences, Warsaw, Poland
e-mail: buzanska@imdik.pan.pl; mzychowicz@imdik.pan.pl

A. Kinsner-Ovaskainen
European Commission, Joint Research Centre, Directorate for Health Consumers and Reference Materials, Ispra, Italy
e-mail: agnieszka.kinsner-ovaskainen@ec.europa.eu

L. Buzanska (ed.), *Human Neural Stem Cells*, Results and Problems in Cell Differentiation 66, https://doi.org/10.1007/978-3-319-93485-3_9

In this chapter, we have assessed the application of bioengineered human neural stem cell microenvironments to produce in vitro models of different levels of biological complexity for the efficient control of stem cell fate. Examples of biomaterial-supported two-dimensional and three-dimensional (2D and 3D) complex culture systems that provide artificial neural stem cell niches are discussed in the context of their application for basic research and neurotoxicity testing.

Keywords Bioengineering · Neural stem cells · Stem cell niche · Neurotoxicity

Abbreviations

CNS	Central nervous system
DNT	Developmental neurotoxicity testing
ECM	Extracellular matrix
ESC	Embryonic stem cells
Fn	Fibronectin
GFAP	Glial fibrillary acidic protein
hESC	Human embryonic stem cells
HUCB-NSC	Human umbilical cord blood-derived neural stem cells
iPSC	Induced pluripotent stem cells
Map-2	Microtubule-associated protein-2
MAPK	Mitogen-activated protein kinase
mdDA	Midbrain dopaminergic
MEA	Multielectrode array
MeHgCl	Methylmercury chloride
NFA	Network formation assay
NSC	Neural stem cells
PDMS	Polydimethylsiloxane
PEO-like	Poly(ethylene) oxide-like
PI-3K	Phosphoinositide-3-kinase
SGZ	Subgranular zone
SVZ	Subventricular zone
TH	Tyrosine hydroxylase

9.1 Introduction

Human stem cells have the capacity for unlimited in vitro expansion and differentiation into essentially all cell types of the organism. They therefore represent a highly promising cell source for a wide range of biomedical applications and are increasingly being exploited in studies of human diseases, pharmacology, and toxicology. Within the realm of human diseases, personalized medicine, and therapeutic interventions, the target evaluation of stem cells obtained from defined

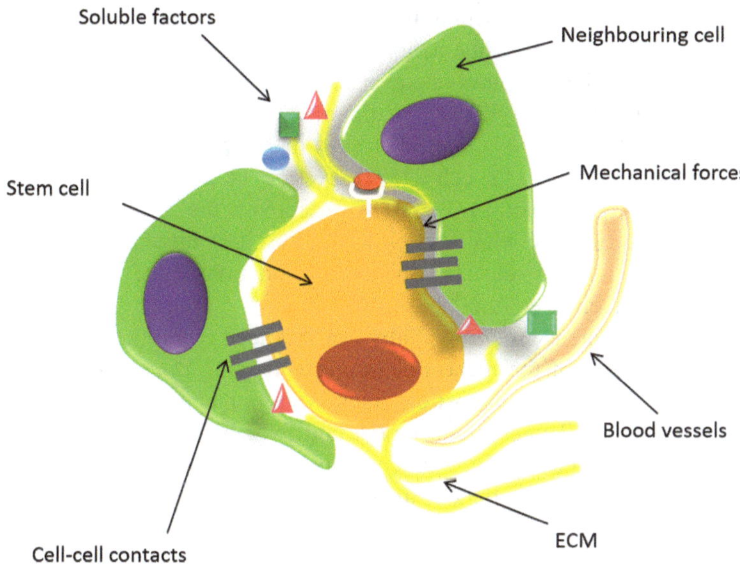

Fig. 9.1 Components of the stem cell niche: structural (extracellular matrix proteins and proteoglycans, blood vessels, and intercellular contacts) and soluble factors (trophic molecules, hormones, neurotransmitters, and signaling proteins). Different factors, such as interactions between the stem cell and surrounding cells, ligand–receptor interactions, and access to trophic factors, determine proper functioning of the stem cells in their microenvironments. Based on and reprinted with permission from Buzanska et al. (2013)

populations with genetic risk or ongoing disease has proved to be useful in advancing our understanding of cellular processes.

9.1.1 The Complexity of the Stem Cell Niche as the Challenge for the Establishment of Reliable In Vitro Stem Cell Based Models

In the developing and adult organism, stem and progenitor cells exist in the specific microenvironment, called stem cell niche, which determines crucial features of stem cell biology: maintenance, proliferation, differentiation, apoptosis, migration, and integration into the tissue architecture. The three-dimensional (3D) complex microenvironment of the niche comprises 1) structural elements, such as cells (stem cell progeny and other type of cells) (Chen et al. 2013), blood vessels, and extracellular matrix (ECM) proteins and proteoglycans (Gattazzo et al. 2014), and 2) soluble factors, including growth factors, cytokines, cell adhesion molecules, hormones, and neurotransmitters (Brizzi et al. 2012) (Fig. 9.1, Buzanska et al. 2013). Mutual, time- and spatial-dependent cell–cell and cell–ECM proteins/proteoglycans interactions have been shown to play an essential role in cell fate decisions (Guilak et al. 2009; Stabenfeldt et al. 2010). These interactions also influence binding of soluble signaling molecules to cellular receptors (Kerever et al. 2007). Anchorage of stem cells to the basal lamina or their morphological alignment with neighboring cells plays an essential role in determining symmetric or asymmetric planes of cell division—the crucial aspect contributing to the cell fates (Kerever et al. 2007; Dityatev et al. 2010; Lathia et al. 2008; Bjornsson et al. 2015). Moreover, the basal lamina of blood vessels, containing ECM proteins and proteoglycans, can influence the niche components by collecting and releasing the active and non-active forms of specific soluble factors (Riquelme et al. 2008).

Although knowledge is growing rapidly, we still need to unravel the mechanisms underlying the control of the niche functioning, enabling the stem cells to differentiate and establish tissue homeostasis. Increasing our understanding of the components of the stem cell microenvironment and their histoarchitecture, as well as current advances in biotechnology and nanobiotechnology, would allow investigators to reproduce the stem cell niche under in vitro conditions (Ranga et al. 2014b).

9.1.2 Neural Stem Cell Niche of the Developing and Adult Central Nervous System in Vertebrates

Adult neurogenic niches develop as the remnants of embryonic signaling centers. The subgranular zone (SGZ) is generated from the dentate neuroepithelium close to the cortical hem, while subventricular zone (SVZ) is the continuation of embryonic ventricular zone (Sugiyama et al. 2013). Similar to their place of origin, they are source of instructive signals for neural stem/progenitor cell fate decisions (Urbán and Guillemot 2014).

The earliest event of vertebrate CNS development is the establishment of a pseudostratified neuroepithelial tube composed of one layer of neuroepithelial cells (NECs) tightly adhered by lateral connections (through adherent and tight junctions). NECs give rise to radial glial cells (RGCs) which extend processes to the outer surface of the developing brain. Lineage-traced RGCs of the neonatal ventricular zone have been shown to serve as neural stem cells (NSC), dividing symmetrically and asymmetrically, generating neural progenitor cells (NPC) which give rise to neurons, astrocytes, and oligodendrocytes in a time-dependent and spatially defined manner (Merkle et al. 2004; Götz and Huttner 2005). In the human developing brain, as compared to the rodent brain, the number of proliferating progenitor cells, including radial glia (stem cells) and transient-amplifying progenitor cells, is largely increased, contributing to the formation of the additional layer of the developing neocortex such as outer subventricular zone (OSVZ) (Hansen et al. 2010). During aging, there is a change in the balance of new cells generated from NSCs: while in the young brain the NSCs generate neurons and glial cells, in the aged brain gliogenesis (mostly oligodendrogenesis) is maintained and neurogenesis is diminished (Capilla-Gonzalez et al. 2015).

The architecture of the human SVZ niche is "spatially organized." It is composed of four layers: Layer I composed of ependymal cells; Layer II consisting of the nearly acellular space built up by cytoplasmic extensions from GFAP+ cells; Layer III consisting of densely packed cell bodies of neural stem cells that are GFAP positive; and finally Layer IV, the transitional zone between the other layers and the brain parenchyma (Capilla-Gonzalez et al. 2015). Migration of neuroblasts generated in the SVZ differs between rodents and humans. While migration in the rodent brain is limited to a rostral migratory stream to olfactory bulb, recent findings suggest that in the adult human brain, the striatum is the place where migrating neuroblasts are being found (Alvarez-Buylla and Garcia-Verdugo 2002; Ernst et al. 2014; Ernst and Frisén 2015).

In the SGZ of the hippocampus, which is located between the granular layer and hilus of the dentate gyrus, the neural stem cells produce only one type of cell, granule neurons, but they are produced during the entire life span (Eriksson et al. 1998). The number of neurons generated and their turnover is much higher in human than rodent SGZ (Ernst et al. 2014). The composition of the surrounding cells is different than in SVZ: NSCs are positioned in between the neuroblasts, astrocytes, and transit-amplifying/highly dividing precursors, where ependymal cells are not present. After division they migrate a short distance to the nearby granule cell layer. However, the NSCs are not anatomically separated from the other structural components of the brain niche and are in close contact with their progeny and vascular endothelial cells (Bjornsson et al. 2015).

To summarize, the adult neural stem cells in their brain niche proliferate, acquire specific fates, migrate to their place of destination, and differentiate into specified neuronal or glial cells (Suh et al. 2007; Aimone et al. 2014). Moreover, the complex cellular and extracellular signals provided by the entire context of the niche, such as ECM physicochemical and mechanical properties (elasticity, cross-linking, specified topography of niche elements), govern the fate of neural stem cells giving unique abilities to promote and sustain neurogenesis (Conway and Schaffer 2012; Massirer et al. 2011) and can influence cell responses to neurotoxicants (Reilly and Engler 2010; Lee et al. 2011). In addition, the neurogenesis that takes place in the NSC niches during development and adulthood is influenced by the factors that come from the changing microenvironment. It is important to note that the SVZ and SGZ niches, though sharing many common features, differ not only in the cellular content and architecture, but also regarding their content of soluble factors. In the adult SGZ, these include molecules involved in Wnt and IGF signaling for stimulation of neurogenesis, while BMP, GABA, and Notch and auto-secreted VEGF are required for NSC maintenance (Urbán and Guillemot 2014; Kirby et al. 2015). The signaling molecules of the SVZ, the source of which are choroid plexus, cerebrospinal fluid (Zappaterra and Lehtinen 2012), vascular endothelial cells (Delgado et al. 2014), and microglia (Harry 2013), are different between embryonic and adult niches and have different effects on NSC maintenance, self-renewal, proliferation, neural migration, or differentiation (Bjornsson et al. 2015). Such microenvironmental cues are relatively stable in adult neurogenic niches in contrast to the developing brain, which experiences a continuously changing environment.

9.1.3 The Advantage of Human Stem Cell Models for Developmental Neurotoxicity Testing

The use of human stem cell-based approaches in neurotoxicology, including developmental neurotoxicity, has several advantages. Firstly, it is debatable the extent to which the animal in vivo models provide accurate insights into human neurotoxicity due to species differences. For the same reason, although in vitro models based on rodent cells are available and have been extensively investigated, increasing evidence shows that there are significant differences in response to neurotoxicants (Baumann et al. 2016); thus, models based on human cells are preferred for predicting hazards to the human CNS linked to exposure to environmental contaminants and drugs (Hou et al. 2013). However, until recently, the only sources of human neuronal cells were tumor tissues, which have altered genetics and cellular mechanisms due to their diseased status. Stem cell technology allows working with cells of human origin which are not transformed, are genetically stable, and can be differentiated at the same time into neuronal and glial cells. Moreover, the application of in vitro stem cell models in toxicology helps to reduce the use of animals in testing and has the potential to markedly speed up and increase the throughput of toxicity screenings (Berthuy et al. 2016; Bal-Price et al. 2010).

Several human stem cell models for neurotoxicity testing have been derived. In this context, the neuronal lineages derived from human embryonic stem cells (hESCs), (He et al. 2012), fetal human neuronal precursors (neuroprogenitor cells) (Breier et al. 2010), and human cord blood-derived neural stem cells (HUCB-NSC) (Buzanska et al. 2009a, b) have been proposed as valuable in vitro models of the central nervous system. Neurotoxicity testing in human ESC-based neuronal differentiation has proven successful in identifying known neurotoxicants such as methylmercury (He et al. 2012). However, the use of ESC raises ethical concerns. The discovery of induced pluripotent stem cell (iPSC) technology helped to overcome this problem. The use of iPSC technique also offers the possibility to assess personalized toxicological responses and to predict individual susceptibility to specific environmental agents (Kumar et al. 2012; Snyder 2017). Nowadays, we are able to recapitulate in vivo developmental milestones of human corticogenesis in in vitro conditions using human pluripotent stem cells, reconstructing the human SVZ niche with developmental morphogenes and their specific inhibitors in 3D culture settings (van den Ameele et al. 2014).

In the following sections, we review the research strategies for micro/nano bioengineering to construct in vitro biomimetic cell culture systems resembling the neurogenic niche. Examples of the cell/biomaterial 2D and 3D complex systems that have been used for applications in neurotoxicity testing are discussed with regard to their strengths and weaknesses. Ideas for future directions are presented.

9.2 Bioengineering of Two-Dimensional Niches: Bioactive Domains

Nowadays, the definition of bioengineering, which historically encompassed the biological or medical applications of engineering equipment or principles, has expanded to include engineering at the molecular and cellular levels. In many laboratories, biomolecular engineering is combined with stem cell biology to resolve the question of how the components of tissue-specific microenvironments influence the behavior of stem cells and control their development. One of the strategies would be a reductionist approach, allowing stem cells to interact in 2D culture with defined microenvironmental factors.

9.2.1 Significance of 2D Bioengineered Surfaces for Neurotoxicity Screening

Bioengineered 2D culture systems, while structurally simpler than 3D model systems in terms of mutual cell/niche interactions, are suitable to probe cellular responses to specific microenvironmental stimuli (the presence of ECM proteins, growth factors, or other signaling molecules) and allow investigation of molecular mechanisms underlying these responses in the presence and absence of neurotoxicants. Basic research conducted on 2D bioengineered cell growth platforms offers the opportunity to follow molecular pathways in healthy and diseased cells, dissecting the specific interactions and cellular responses to gradually added environmental stimuli. Based on the functionality of matrix proteins for replicating stem cell niches, our laboratory and others have developed miniaturized systems with micropatterned bioactive surfaces (domains) for directing and monitoring of neural stem cell developmental processes and for assessment of vulnerability to different tested compounds (Soen et al. 2006; Ruiz et al. 2008a, b; Ceriotti et al. 2009). Besides identifying the molecular mechanisms governing neural stem cell fate decisions (Buzanska et al. 2010; Lutolf et al. 2009), such systems can be used to generate different cell phenotypes by modifying physical properties and/or geometry of the bioactive domains (Lutolf et al. 2009). Thus, various arrays of signaling microenvironments have been shown to influence neural stem cell developmental processes (Soen et al. 2006; Zychowicz et al. 2011, 2012). The functionalization of the cell growth surface with active biomolecules and the miniaturization of these cell growth platforms by micropatterning are the main goals of current research to achieve a controllable cell microenvironment for the "microscale" 2D approach testing systems.

The advanced technology platforms with ECM components and soluble signaling molecules immobilized on the surface developed to experimentally probe the influence on the stem cell fate were described previously (Buzanska et al. 2013; Soen et al. 2006; Lutolf et al. 2009). The classic cell culture is characterized by a homogeneous adhesive surface providing the same individual signals to adherent cells where the type of homogeneous adhesive surface with defined physico-mechanical properties (substrate stiffness, i.e., elasticity of biomaterials) can determine cell shape. Printed signal domains composed of the same biomaterial but with different geometry of the domain influence cell shape by spatial limitations or, on the other hand, printed domains composed of more than one protein or ECM derivative, i.e., multi-compositional domains, influence cell-specific responses mainly through the ligand/receptor interactions.

9.2.2 *Micro/Nano Technologies to Create 2D Patterns of Bioactive Compounds*

To form clusters (microarrays) of cells within a small, defined area of a 2D substrate, microcontact printing or piezoelectric (noncontact) microspotting is used to deposit biomolecules on the cell growth platforms (Brétagnol et al. 2006; Ruiz and Chen 2008; Ceriotti et al. 2009; Ruiz et al. 2013). While both techniques are suitable for fabricating cell growth platforms important for controlling stem cell fate decisions, they have different merits. Microspotting permits patterning of different microenvironments on a single platform. Microcontact printing allows for the deposition of only one type of microenvironment, but it allows for fine-tuning of the size and shape of the patterns. It does this by creating homogenous areas with different geometry and physicochemical properties permisive for cell attachment. As a first step, a soft lithography technique is used to fabricate a polymeric stamp (e.g., polydimethylsiloxane, PDMS) with defined pattern of domains. Following immersion of this stamp in the biomolecule suspension (e.g., fibronectin, laminin, or poly-L-lysine), the array of bioactive material is transferred onto a cell-repellent surface (e.g., poly(ethylene) oxide (PEO)-like, plasma-deposited films) (Ruiz et al. 2007, 2008a, b, 2009). This method allows establishing a variety of patterns with microscale domains up to a single-cell resolution (Fig. 9.2) (Buzanska et al. 2013; Ruiz et al. 2008a, b, 2013; Zychowicz et al. 2012).

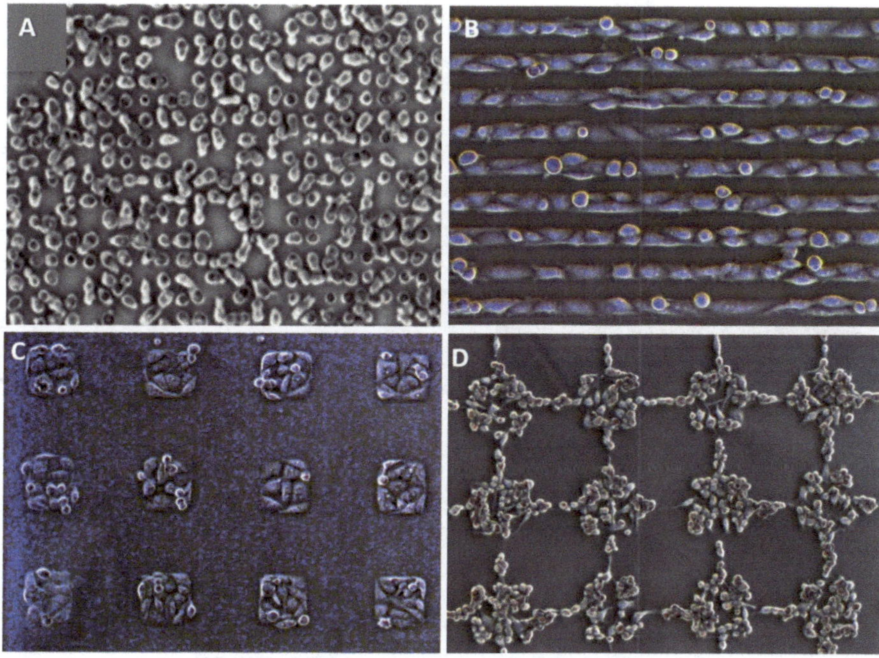

Fig. 9.2 Neural stem cells positioned on fibronectin-microprinted biofunctional domains: (**a**) 10 × 10 μm single-cell posts; (**b**) 20 μm width lines; (**c**) 100 μm side squares; (**d**) 120 μm side squares with interconnecting lines. Based on and reprinted with permission from Buzanska et al. (2013)

9.2.3 Neural Stem Cell Growth and Differentiation on the Arrays of Various 2D Microenvironments

Studies examining HUCB-NSC cells cultured on both microcontact-printed (Ruiz et al. 2008a, b; 2009; Buzanska et al. 2009a, b) or microspotted (Ceriotti et al. 2009; Buzanska et al. 2010) bioactive surfaces show that a domain's geometry and composition have a significant impact on cell adhesion, proliferation, and differentiation (Fig. 9.3), which are key developmental stages of these cells (Buzanska et al. 2010; Zychowicz et al. 2011, 2012; Ruiz et al. 2009, 2013). Characteristics of the domains can differentially influence the HUCB-NSC phenotype. When cultured on thin, 10-μm-wide lines that are microcontact printed with fibronectin (ECM protein allowing specific integrin receptor-driven interactions with the domains), the neuronal phenotype is promoted, as demonstrated by β-tubulin III and Map-2 immunoreactivity in elongated, thin cells (Zychowicz et al. 2012). Single cells positioned on smaller domains (10 x 10 μm), fibronectin or poly-L-lysine (PLL) posts, maintain an undifferentiated phenotype, but also show immunoreactivity for nestin, the marker of early neural commitment (Ruiz et al. 2009). In contrast, staining for connexin-43,

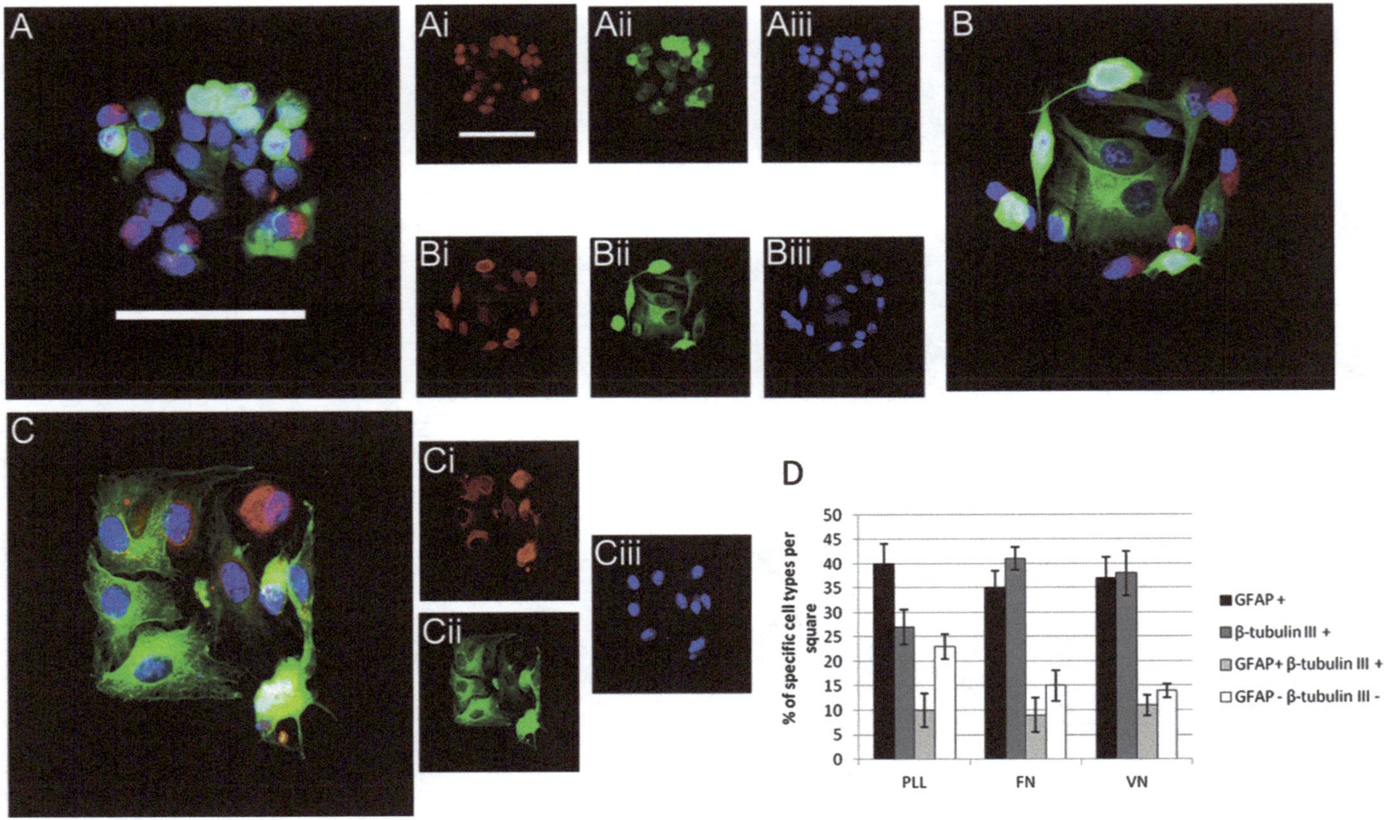

Fig. 9.3 Images of HUCB-NSCs cultured in control conditions on patterns printed with poly-L-lysine (**a**), fibronectin (**b**), and vitronectin (**c**), immunostained against GFAP (red, **a–c** i), b-tubulin III (green, **a–c** ii), and counterstained with Hoechst 33258 (blue, **a–c** iii). Scale bar 100 μm. (**d**) Distribution of the

the main gap junction protein, was not detected on PLL, but was detected on fibronectin posts of the same size (Zychowicz et al. 2011). The above findings are important because the development of gap junctions is the key marker for the stem cell/progenitor stage commitment, and cell–cell communication via gap junctions is one of the fundamental targets in assessing the potential toxicity of chemicals (Kang and Trosko 2011). The influence of the strength of intercellular connections/interactions and forces generated by diverse geometry of the domain was also shown to be crucial in other tissues, such as for the control of differentiation of mesenchymal stem cells into adipose and bone tissues (Ruiz and Chen 2008). In addition, experiments conducted on 2D micropatterned surface have demonstrated that two niche determinates—the composition and the size of domains—are powerful tools to regulate human ESC self-renewal and differentiation (Peerani et al. 2007; Reilly and Engler 2010). Such determinants and their dynamic changes within the niche should be considered while planning experiments to test stem cell responses to toxic factors.

Another technique used for stem cell niche bioengineering is piezoelectric (non-contact) microspotting that allows generating domains composed of a variety of ECM components, growth factors, and signaling proteins immobilized to the culture surface. The advantage of this technique is that the resulting microarrays support investigation of different substances used for the microspotting within the same screening platform. In addition, each microspotted domain can be separately functionalized with regard to growth factors and signaling molecules, giving the possibility for parallel investigation of a large number of different microenvironments. The first spotted technology platform to experimentally probe the influence of ECM components and soluble growth factors on stem cell cellular differentiation was established by Flaim et al. (2005). In their study, the influence of 32 different combinations of five ECM proteins (collagen I, collagen III, collagen IV, laminin, and fibronectin) on mouse embryonic stem cell differentiation was examined. These investigators later studied the developmental response of human embryonic stem cells to combinations of different ECM proteins, such as fibronectin, laminin, collagen I, collagen III, and collagen IV, resulting in 240 unique signaling environments (Flaim et al. 2008).

More complicated arrays of combinatorial signaling microenvironments deposited on cell growth platforms have been implemented by our laboratory and others

Fig. 9.3 (continued) phenotypes of non-treated HUCB-NSC at different stages of differentiation regarding to the expression of GFAP and β-tubulin III in cells growing on bioactive domains. Average percentage of HUCB-NSC: committed to astroglial lineage (GFAP+); committed to neuronal lineage (β-tubulin III+); non-committed progenitors, expressing neural markers (GFAP+/β-tubulin III+), and undifferentiated cells non-expressing neural markers (GFAP− and β-tubulin III−) growing on the patterns of poly-L-lysine (PLL), fibronectin (FN), and vitronectin domains. Error bars indicate standard error. Results represent three independent experiments, for each experimental endpoint readouts are based on 8 separate domains. Reprinted from Buzanska et al. (2010) with permission from Elsevier

for studying embryonic (Flaim et al. 2005, 2008) as well as human neural stem cell fate (Soen et al. 2006; Buzanska et al. 2010). For that purpose, ECM proteins were immobilized to the surface together with small signaling molecules, influencing selected intracellular pathways involved in neural stem cell differentiation. Our results indicate that microdomains of fibronectin functionalized with molecules of Wnt, Shh, CNTF, Jagged, or Notch type resulted in various fate decisions of HUCB-NSC to either the neuronal or astroglial lineage (Buzanska et al. 2010, 2017). Signaling molecules immobilized on fibronectin domains exerted a significant influence on cell fate decisions as compared to the control fibronectin domain. Jagged and CNTF directed HUCB-NSC into the astrocytic lineage, while Wnt-3 and Shh supported proliferation and self-renewal, but also neuronal commitment of HUCB-NSC. The signaling molecule Dkk1 supported neuronal differentiation. This data revealed that the response of HUCB-NSCs to the signaling molecules, which were immobilized to the surface together with fibronectin, was in accordance with their intracellular "mode of action" (Buzanska et al. 2010, 2017). The microarrays of "smart" microenvironments, besides being used to control and probe neural stem cell fates, were also implemented for toxicity screening (Zychowicz et al. 2014).

9.2.4 Neurotoxicity Screening on the Patterns of Functional Substrates

Such arrays of various microenvironments could be easily adapted for toxicity screening. The array platforms with microspotted domains served as a tool for the screening of HUCB-NSC's susceptibility to MeHgCl when attached to different biomolecules (poly-L-lysine, fibronectin, and vitronectin) (Zychowicz et al. 2014). We have observed that HUCB-NSC cells growing on domains microspotted with extracellular matrix proteins, such as fibronectin and vitronectin, were more resistant to MeHgCl treatment as compared to cells grown on classical polystyrene culture dish, indicating that receptor-mediated interactions between cells and ECM proteins can be protective. The cell response to MeHgCl was also found to be dependent on their developmental stage (cell type). Phenotypic distribution of cells on the microspotted domains indicated that undifferentiated cells were the most sensitive, as compared to differentiated progenitors, regardless of the type of the domain. However, the type of the biofunctional domain was important for differentiated cells: while neural stem cells committed to astrocytic lineage (GFAP+) revealed the highest sensitivity to MeHgCl on poly-L-lysine domains, neuronally committed β-Tubulin III expressing cells were most sensitive on fibronectin domains. Notably, this is correlated with electrostatic nonspecific versus specific, integrin receptor-mediated cell/surface domain interactions (Fig. 9.4) (Zychowicz et al. 2014).

HUCB-NSCs have been previously shown to be susceptible to the set of tested compounds: sodium tellurite, methylmercury chloride, cadmium chloride, chlorpyrifos, and L-glutamate in a developmentally dependent manner as revealed in 96 well plates 2D culture tests (Buzanska et al. 2009a, b). Further, HUCB-NSCs have been

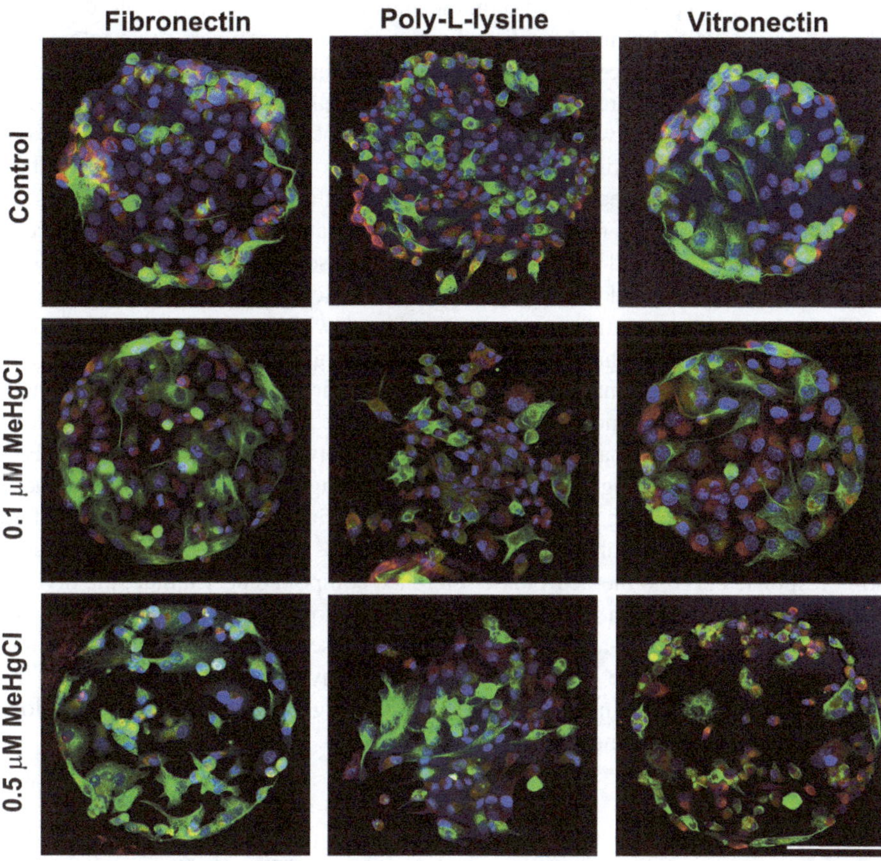

Fig. 9.4 Differentiation of HUCB-NSC cultured on biofunctional domains microspotted with poly-L-lysine, fibronectin, or vitronectin in the presence of MeHgCl at different concentrations for 48 h. Immunocytochemical analysis of neuronal marker b-tubulin III (green) and astrocytic marker GFAP (red). Cell nuclei are contrastained with Hoechst (blue). Scale bar 100 μm. Reprinted from Buzanska et al. (2010) with permission from Elsevier

also investigated for their vulnerability to MeHgCl treatment while immobilized onto poly-L-lysine microcontact-printed patterns (Buzanska et al. 2010). We have shown that concentrations higher than 0.5 μM MeHgCl inhibit free movement of cells between bioactive domains and disrupt well-confined patterns, including the pattern of single-cell resolution (10 × 10 μm) as well as the patterns of squares (150 × 150 μm) separated or interconnected by lines (Fig. 9.2) (Buzanska et al. 2010).

In another study from our group, different types of biofunctional surfaces created by microprinting/microspotting of ECM proteins (fibronectin and vitronectin) or poly-L-lysine have been used to investigate the sensitivity of HUCB-NSC's developmental processes (viability, proliferation, and differentiation) upon exposure to MeHgCl (Zychowicz et al. 2014). The studies on the microcontact-printed patterned domains demonstrated that developmental sensitivity of HUCB-NSC to MeHgCl depend upon the type of adhesive biomolecules and the geometry of bio-domains. The cells in spatially limited bioactive domains (of μm size) were more tolerant to MeHgCl treatment than cells in non-patterned larger areas of the same adhesive substrate. After MeHgCl treatment of HUCB-NSC cells on microspotted domains of the same size, but coated with different ECM proteins or poly-L-Lysine, the greater decreases in cell proliferation and cell viability were observed on poly-L-lysine than on fibronectin and vitronectin domains. The distribution of cellular phenotypes on printed bioactive domains revealed that nondifferentiated cells, as compared to cells committed in neuronal and glial lineage, are the most sensitive to MeHgCl treatment, regardless of the type of domain (Fig. 9.4) (Zychowicz et al. 2014).

The microcontact printing approach combined with the patterning of defined groups of neuronal cells has been proposed as a rapid and sensitive neurotoxicity testing platform by Frimat et al. (2010). In their elegant study, the authors bioengineered a cellular microenvironment for directed growth of axonal protrusions from the groups of neuronal cells and designed a "network formation assay" (NFA) for neurotoxicity screening. This neurotoxicity testing platform resembled pseudo-3D conditions by forcing neuronal cells to grow in aggregates and respond to microenvironmental stimuli as interconnected groups of cells. After establishing a dose–response curve to the neurotoxic reference compound acrylamide, the investigators showed a reduction in network formation at non-cytotoxic concentrations of inhibitors of mitogen-activated protein kinase (MAPK) ERK1/2 and phosphoinositide-3-kinase (PI-3K) signaling pathways, thereby demonstrating the potential of the NFA testing platform for high-throughput applications in developmental neurotoxicity testing (Frimat et al. 2010).

9.3 Bioengineering of Three-Dimensional (3D) Culture Systems: For Human Neural Stem Cell Fate Control and Toxicity Studies

In vivo the stem cells reside in 3D microenvironments. Capturing these 3D characteristics of the stem cell niche ex vivo is considered essential to precisely control stem cell fate and mimic the complex morphology of neuronal tissue.

The 3D microenvironment has been shown in our laboratory to be essential for neural stem cells derived from human cord blood to acquire electrical activity. We have shown that HUCB-NSC differentiated into neuronal lineage and revealed spontaneous electrical activity on multielectrode array (MEA) chips only when cells were cultured on 3D scaffolds, while in 2D conditions such an activity was not observed (Jurga et al. 2009). Also sensitivity to toxic compounds is changed in 3D as compared to 2D cultures. Recently, it was documented by BoNT/A1 treatment of iPSC-NSCs cultured on TCP (tissue culture polystyrene) or PEG (poly(ethylene glycol) in 3D hydrogel surfaces (Pellett et al. 2015). The cells grown in 3D conditions, which formed neuronal clusters interconnected with thick axonal bundles, were much more susceptible to BoNT/A1 than 2D polystyrene grown cells.

9.3.1 *Encapsulating Stem cells in 3D Microenvironment*

At least four aspects of the mutual relationship between stem cells and other niche components have to be taken into consideration when trying to reconstruct the in vivo environment in an in vitro model encapsulating cells in 3D biomaterial-based structures. These include (1) structural elements comprising the 3D niche architecture (cell–cell and cell–ECM proteins interaction), (2) physical matrix characteristics (elasticity, stiffness), (3) nutritional status (O_2 level, nutrients), and (4) known soluble signaling molecules responsible for molecular cell responses (cytokines, growth factors, neurotransmitters). Controlled delivery of such factors and their regulated presence is one of the main challenges for tissue engineering research (Lutolf et al. 2009; Buzanska et al. 2013), for example, different types of 3D microenvironments and their "fine-tuning" in order to influence cell fates "on demand." To provide a uniform distribution of signaling molecules, one approach has been to surround the cells with a homogeneously cross-linked polymer or the gradient of cross-linked polymer may provide differential signaling within the niche. Manipulation of the biopolymer structure (cross-linking of the polymer) and function (release of bioactive factors) can be carried out using temperature, UV-light, or electromagnetic field (Buzanska et al. 2013).

9.3.2 *Self-assembly-Based Cellular "Micro-tissues" for Neurotoxicity Testing*

One of the approaches commonly used to reconstitute three-dimensional neuronal tissues in vitro is the formation of 3D cell aggregates called neurospheres. The classic "neurosphere" test for DNT screening established in Elaine Fritsche laboratory (Baumann et al. 2016) is based on human neural progenitor cells (hNPCs) from gestational week 16 to 20, grown as neurospheres in proliferation medium. Recently, a high-content image analysis (HCA) software "Omnisphero" was applied by this group for phenotypic DNT screenings of these 3D structures (Schmuck et al. 2017). Hogberg et al. (2013) established from human iPSCs derived from healthy or diseased donors the model of neurospheres which have differentiated into neural progenitor cells and then neuronal and astroglial lineages. This step-wise protocol involves formation of embryoid bodies, dissociation and selection of neural precursor cells, and further culture of these precursors under constant agitation in stirred culture systems for up to 8 weeks. This cell culture model was exposed to reference chemicals with well-described neurotoxic effects, and the predictivity was assessed by gene expression, immunohistochemistry, and calcium signaling (functional endpoint) measurements (Pamies et al. 2014). Effects were also compared to human data from the literature and to previously obtained data from the 3D rat aggregating brain cell culture model. The model was further developed into so-called 3D brain microphysiological system (BMPS) that reproduce neuronal-glial interactions and connectivity. Since as much as 40% overall myelination can be achieved, this model allows assessing oligodendroglia function and vulnerability to neurotoxic compounds (Pamies et al. 2017). This approach gives broader information on gene/environmental interactions than previous studies with less complex cell culture systems.

Emerging studies of several groups have resulted in more developmentally advanced than BMPS, neural tissue-like self-assembling structures called "organoids" (Sasai et al. 2012; Lancaster and Knoblich 2014). In this methodology, ESC/iPSC neurosphere formation was followed by tissue-like self-assembly into neuroectodermal structures within 3D hydrogel-based microenvironment. The polarized neuronal structures such as cerebral cortex (Lancaster et al. 2013) and retina (Sasai et al. 2012) were developed. The spontaneous aggregation and complicated arrangement of retina cells was achieved in suspension in the presence of medium suitable for retinal differentiation and a low concentration of Matrigel (Sasai et al. 2012). Derivation of a 3D organoid culture system resembling stratified layers of developing cortex was enabled by a dense hydrogel matrix and bioreactor steering conditions (Lancaster et al. 2013). Finally, a human pluripotent stem cell-derived 3D CNS-like organoid culture system was shown to recapitulate features of human cortical development, but also succeeded in modeling microcephaly disease (Lancaster et al. 2013) or schizophrenia (Narla et al. 2017). Such "self-assembling" organoids are predominantly used for developmental studies and disease modeling; however, they are important to recall in this review in the context that they give future possibility to test neurotoxicity in the organ-like structure in the context of neurodevelopment.

9.3.3 Bioengineered, Scaffold-Based Neural Tissue-like Constructs for Neurotoxicity Testing

While the abovementioned neural "tissue-like" 3D structures were obtained entirely by cellular "self-assembly" processes in the ECM-based microenvironment, the bioengineering approach using scaffolds combines the advantages of 3D cultures with biotechnologies for plating different cell types and biomaterials to obtain environmental niche patterning.

The most spectacular example of neural tissue-like construct successfully applied in high-throughput toxicology was recently obtained in John Thomson's Laboratory (Schwartz et al. 2015). The human neuronal constructs were obtained by sequential plating of four different cell types: neural progenitor cells (NPCs), endothelial cells

(ECs), mesenchymal stem cells (MSCs), and microglia/macrophage precursors (MG) in tunable, photodegradable poly(ethylene glycol) hydrogel. The synthetic hydrogel was bioengineered to present peptides that allow the cells to attach and degrade the matrix, thus enabling plated cell types to grow naturally and self-assemble into a complex network of tissues. The further step was to design experimental settings to test toxicity of library of toxicological compounds (Hou et al. 2013; Schwartz et al. 2015). For that purpose RNA sequencing data was collected from 240 neural tissue constructs that were individually exposed to 60 different "training" chemicals (controls which were safe compounds and known toxins). In silico computer-based learning was applied to build up a predictive model based on these results. At the end, the model correctly classified nine out of ten chemicals in a blind trial.

The presented above tissue construct is an example of how recent advances in biotechnology made it possible to better reproduce the natural conditions for stem cells in vitro and to provide 3D culture environments that resemble the complexity of the in vivo stem cell microenvironment. This combination of methodologies may allow for establishing neuronal circuits and supporting niche elements (e.g., microvessels) as well as gaining functionality in tested stem cell populations.

The main types of 3D scaffolds used to support stem cell cultures include hydrogels (Jang et al. 2015; Ylä-Outinen et al. 2014), macroporous alginate scaffolds (Bozza et al. 2014; Kim et al. 2013), fibrin gel (Navaei-Nigjeh et al. 2013), interpenetrating networks such as collagen matrix (Li et al. 2014; Pietrucha et al. 2017), and nanofiber scaffolds (Ni et al. 2013). The biomaterials used for fabrication of three-dimensional scaffolds, which support tissue development, are expected to be nontoxic, biodegradable, non-immunogenic, and with a porous structure enabling 3D sensing of the seeded cells (Ranga et al. 2014a, b). Each scaffold type has its own advantages over others in supporting ex vivo expansion and differentiation of stem cells, and the choice of scaffolds for tissue engineering and screening applications depends upon the type of the cell or tissue used.

The benefit of hydrogels is the high water content, which is very close to what prevails in tissues, ensuring that the scaffold is soft, elastic, and supportive for cell growth. Ylä-Outinen et al. (2014) have shown that hESC cultured inside the hydrogel could be differentiated and matured into neurons, astrocytes, and oligodendrocytes. Importantly, neuronal cells were able to form electrically active connections that were verified using microelectrode arrays.

One of the important factors used for stem cells differentiation in the 3D hydrogel scaffold is the orientation of the ECM components that influences the development and alignment of astrocytes and neurons. Jang et al. (2015) used a microfluidic device which generated a continuous flow across the Matrigel matrix as it gelled. Due to the flow, more than 70% of ECM components were oriented along the direction of flow, compared with randomly cross-linked Matrigel. Guided by the oriented ECM components, primary rat cortical neurons and mouse neural stem cells showed oriented outgrowth of neuronal processes within the 3D Matrigel matrix. This and other experiments show that, when bioengineering the stem cell niche, it is crucial to take into consideration diverse aspects of mechanical properties of the niche environments, such as shear stress created by flowing fluids, as well as rigidity and stiffness of the matrix. Such "oriented" outgrowth of neuronal processes was also observed in 3D microenvironment for human neural stem cells. The example of directional control and enhanced neurite outgrowth of human neuronal cells in 3D microenvironment was reported by Richard J McMurtrey (2014), who has implemented the patterned and functionalized nanofiber scaffolds in three-dimensional hydrogel constructs.

Compared to hydrogels, which must encapsulate cells in situ during gelation, it is much easier to seed cells into macroporous scaffolds such as alginate beads. Bozza et al. (2014) tested whether encapsulation of mouse embryonic stem cells (mESCs) within alginate beads, with or without modification by fibronectin or hyaluronic acid (HA), could support and/or enhance in vitro neural differentiation with respect to two-dimensional cultures. Gene expression and immunohistochemistry analyses indicated that mESCs grown in alginate and alginate-HA supports increased differentiation toward neural lineages (expression of synaptic markers and markers of different neuronal subtypes) when compared to the two-dimensional control and the Fn group. Alginate-based supports were also used successfully to culture and differentiate human ESC into neurons. Kim et al. (2013) differentiated hESCs grown on alginate microcapsules into midbrain dopaminergic (mdDA) neurons. Gene and protein expression analysis during neuronal differentiation showed an increased expression of various specific DA neuronal markers (e.g., tyrosine hydroxylase) as compared to cells differentiated on a conventional two-dimensional (2D) platform. The encapsulated TH (+) cells also secreted higher dopamine levels when induced. The authors conclude that the 3D platform allows for an early onset of DA neuronal generation compared to the 2D system and is a very useful model to study the proliferation and directed differentiation of hESCs to various lineages, and it allows the separation of feeder cells from hESCs during the process of differentiation. This model could be used for screening of neurodevelopmental toxicants influencing development of human nervous tissue.

Another type of 3D support consists of fibrous meshes that can recapitulate important structural and topographical aspects of the extracellular matrix found in vivo. Li et al. (2014) used a highly ordered collagen fibril matrix to study cell–matrix interaction. With high-resolution imaging, they have shown that stem cells interact with the matrix by deforming the cell shape, harvesting the nearby collagen fibrils, and reorganizing the fibrils around the cell body to transform a 2D matrix to a localized 3D matrix. Such a unique 3D matrix prompted high expression of β-1 integrin around the cell body that mediates and facilitates stem cell differentiation toward neural cells. Ni et al. (2013) cultured mouse ESCs and iPSCs on a self-assembling peptide made from natural amino acids, which has the property of generating a true 3D environment for dopaminergic differentiation. The resulting nanofiber scaffolds led to a significant increase in dopaminergic differentiation compared to either a laminin-coated 2D culture or Matrigel-encapsulated 3D culture.

A breakthrough in studying the effects of microenvironmental cues on cellular responses was proposed by Ranga et al. (2014a, b) who generated large-scale libraries of 3D microenvironments and used them to assess the combined effects of matrix properties and signaling proteins on mouse embryonic stem cells. A subsequent systems-level analysis allowed a better understanding of the multi-factorial 3D cell–matrix interactions involved. The same team provided very recent technological elaboration on 3D microenvironment to mimic in vitro neural tube development (Ranga et al. 2016); however, it was still done with mice neural stem cells. We may speculate that the combining of adverse outcome pathway (AOP) development program (Bal-Price and Meek 2017) with such systems as proposed by Ranga et al. (2014a, b, 2016) but based on human neural stem cells may increase mechanistic understanding of the cellular processes and pathophysiological pathways underlying responses of human organism to the toxic compound.

While dimensionality is of high importance for the proper sensing and interactions between neural stem cells and the structural components of their niche, in order to create proper biomimetic niche in vitro one should also consider the proper, similar to endogenous level of oxygen. Lowering of the oxygen level from 21 to 5% (Ivanovic 2009) was beneficial in experiments of our group, using human neural stem cells for both reprogramming of NSCs to induced pluripotent stem cells and differentiation of neural progenitors into neuronal lineage (Szablowska-Gadomska et al. 2011, 2012).

Yang et al. (2013) used a microfluidic device to investigate hNSC self-renewal and differentiation in the biomimetic, similar to in vivo microenvironment. The implemented NSC niche conditions included 3D extracellular matrices and low oxygen tension. In these experiments, 3D microenvironments with 5% oxygen tension allowed the maintenance of hNSC self-renewal capacity and direct neuronal commitment during hNSC differentiation. The proper reconstitution of the biomimetic microenvironment in the microfluidic array in a combinatorial manner allowed for the quantifying the effects of the biomimetic conditions on hNSC self-renewal and differentiation with quantitative real-time polymerase chain reaction in the high-throughput manner (Yang et al. 2013).

Although the bioengineered 3D stem cell niche is more biologically relevant than the standard 2D culture conditions, the deficiency of reliable analytical tools to measure the endpoints of toxicity pathways is a great impediment for an efficient use of these culture models in toxicology (Simão et al. 2015; Meli et al. 2014). Thus, the application of 3D bioengineered microenvironments for neurotoxicity testing is still very limited. The alginate hydrogel encapsulation system designed by Kim et al. (2013) and described above appears to be the most biologically relevant current 3D model for high-throughput toxicity screening, but it does not fully replicate conditions in vivo. The main current and future challenge will be focused on closer recapitulation of the complexity of the niche structure. In that respect, the advanced systems such as those described by Yang et al. (2013), Ranga et al. (2014a, b, 2016), and Barthes et al. (2014) hold promise for providing physical, chemical, and biological control of the cell microenvironment and possibility for unlimited experimental combinations within one microarray, in order to gain information on the sensitivity of certain molecular pathways to selected neurotoxicants.

9.4 Conclusions

This chapter provides the review of the critical features of human neural stem cell microenvironments and methods applied for the bioengineering of such an environment in vitro. In order to allow for a proper stem cell differentiation in as close to the in vivo environment as possible, the accurate replication of the stem cell niche in the in vitro conditions must be considered. To better predict the toxicity and efficacy of potential drugs in physiologically relevant conditions, several screening systems at different levels of biological complexity have been proposed.

Bioengineered 2D domains do not closely replicate the natural existing three-dimensional microenvironment; however, they are important for mechanistic studies as well as high-content and high-throughput screening. Micropatterned cell growth platforms have been applied successfully to study the adverse effects of toxic compounds on neural stem cell fate decisions and advanced differentiation (e.g., directed axonal outgrowth) and can be used to gain information on the sensitivity of certain molecular pathways to selected neurotoxicants.

The biomimetic approach of experimentation requires more precise replication of 3D natural conditions than does the 2D modeling approach. Traditional 3D culture models, such as self-assembling neurospheres (human neural stem/progenitor cell aggregates cultured on non-adhesive surface or in agitation conditions), were recently advanced into 3D brain microphysiological system (BMPS) that reproduces neuronal–glial interactions and shows elements of neuronal function. Such systems are useful to recapitulate the developmental program and are a good model for developmental neurotoxicity testing.

2D cultures and traditional 3D models used for mechanistic, proof-of-concept studies are being recently upgraded into human pluripotent stem cell-derived self-assembling organoids and bioengineered tissue-like constructs, mimicking multilevel in vivo interactions. While derived from patient-specific pluripotent stem cells, they are reliable tools for personalized investigations for drug discovery and toxicity screening. Bioengineered tissue-like neuronal constructs with stem cell-derived neural progenitor cells, vascular cells, and microglia encapsulated in engineered hydrogels were recently integrated into protocols for toxicity screening to establish highly consistent and relevant drug toxicity prediction model. Such combined strategy implementing advancement in human neural stem cell biology, tissue engineering, bioinformatics, and machine learning successfully proved the value of human cell-based assays with the importance of microenvironmental control.

Acknowledgments This work was sponsored by statutory funds to Mossakowski Medical Research Centre and Wroclaw Research Centre EIT+ under the project "Biotechnologies and advanced medical technologies"—BioMed (POIG.01.01.02-02-003/08) financed from the European Regional Development Fund (Operational Programme Innovative Economy, 1.1.2).

References

Aimone JB, Li Y, Lee SW et al (2014) Regulation and function of adult neurogenesis: from genes to cognition. Physiol Rev 94(4):991–1026

Alvarez-Buylla A, Garcia-Verdugo JM (2002) Neurogenesis in adult subventricular zone. J Neurosci 22(3):629–634

Bal-Price A, Meek MEB (2017) Adverse outcome pathways: application to enhance mechanistic understanding of neurotoxicity. Pharmacol Ther 179:84–95

Bal-Price AK, Hogberg HT, Buzanska L et al (2010) In vitro developmental neurotoxicity (DNT) testing: relevant models and endpoints. Neurotoxicology 31:545–554

Barthes J, Özçelik H, Hindié M et al (2014) Cell microenvironment engineering and monitoring for tissue engineering and regenerative medicine: the recent advances. Biomed Res Int 2014: 921905

Baumann J, Gassmann K, Masjosthusmann S et al (2016) Comparative human and rat neurospheres reveal species differences in chemical effects on neurodevelopmental key events. Arch Toxicol 90(6):1415–1427

Berthuy OI, Blum LJ, Marquette CA (2016) Cells on chip for multiplex screening. Biosens Bioelectron 15(76):29–37

Bjornsson C, Apostolopoulou TY et al (2015) It takes a village: constructing the neurogenic niche. Dev Cell 32(4):435–446

Bozza A, Coates EE, Incitti T et al (2014) Neural differentiation of pluripotent cells in 3D alginate-based cultures. Biomaterials 35(16):4636–4645

Breier JM, Gassmann K, Kayser R et al (2010) Neural progenitor cells as models for high-throughput screens of developmental neurotoxicity: state of the science. Neurotoxicol Teratol 32(1):4–15

Brétagnol F, Lejeune M, Papadopoulou-Bouraoui A et al (2006) Fouling and non-fouling surfaces produced by plasma polymerization of ethylene oxide monomer. Acta Biomater 2(2):165–172

Brizzi MF, Tarone G, Defilippi P (2012) Extracellular matrix, integrins, and growth factors as tailors of the stem cell niche. Curr Opin Cell Biol 24(5):645–651

Buzanska L, Ruiz A, Zychowicz M et al (2009a) Patterned growth and differentiation of human cord blood-derived neural stem cells on bio-functionalized surfaces. Acta Neurobiol Exp 69(1):24–36

Buzanska L, Sypecka J, Nerini-Molteni S et al (2009b) A human stem cell-based model for identifying adverse effects of organic and inorganic chemicals on the developing nervous system. Stem Cells 27(10):2591–2601

Buzanska L, Zychowicz M, Ruiz A et al (2010) Neural stem cells from human cord blood on bioengineered surfaces—novel approach to multiparameter bio-tests. Toxicology 270(1):35–42

Buzanska L, Zychowicz M, Sarnowska A et al (2013) Bioengineering of neural stem cell niche. Postepy Biochem 59(2):175–186

Buzanska L, Zychowicz M, Ruiz A et al (2017) Neural stem cell fate control on micropatterned substrates. In: Srivastava AK, Snyder EY, Teng YD (eds). Springer ProtocolsStem cell technologies in neuroscience, Neuromethods, vol 126, pp 19–45

Capilla-Gonzalez V, Herranz-Pérez V, García-Verdugo JM (2015) The aged brain: genesis and fate of residual progenitor cells in the subventricular zone. Front Cell Neurosci 24(9):365

Ceriotti L, Buzanska L, Rauscher H et al (2009) Fabrication and characterization of protein arrays for stem cells patterning. Soft Matters 5:1406–1416

Chen S, Lewallen M, Xie T (2013) Adhesion in the stem cell niche: biological roles and regulation. Development 140(2):255–265

Conway A, Schaffer DV (2012) Biophysical regulation of stem cell behavior within the niche. Stem Cell Res Ther 3(6):50

Delgado AC, Ferro´ n, SR, Vicente D et al (2014) Endothelial NT-3 delivered by vasculature and CSF promotes quiescence of subependymal neural stem cells through nitric oxide induction. Neuron 83, 572–585

Dityatev A, Seidenbecher CI, Schachner M (2010) Compartmentalization from the outside: the extracellular matrix and functional microdomains in the brain. Trends Neurosci 33:503–512
Eriksson PS, Perfilieva E, Björk-Eriksson T et al (1998) Neurogenesis in the adult human hippocampus. Nat Med 4:1313–1317
Ernst A, Frisén J (2015) Adult neurogenesis in humans-common and unique traits in mammals. PLoS Biol 26, 13(1)
Ernst A, Alkass K, Bernard S et al (2014) Neurogenesis in the striatum of the adult human brain. Cell 156(5):1072–1083
Flaim CJ, Chien S, Bhatia SN (2005) An extracellular matrix microarray for probing cellular differentiation. Nat Methods 2(2):119–125
Flaim CJ, Teng D, Chien S et al (2008) Combinatorial signaling microenvironments for studying stem cell fate. Stem Cells Dev 17:29–39
Frimat JP, Sisnaiske J, Subbiah S et al (2010) The network formation assay: a spatially standardized neurite outgrowth analytical display for neurotoxicity screening. Lab Chip 10(6):701–709
Gattazzo F, Urciuolo A, Bonaldo P (2014) Extracellular matrix: a dynamic microenvironment for stem cell niche. Biochim Biophys Acta 1840(8):2506–2519
Götz M, Huttner WB (2005) The cell biology of neurogenesis. Nat Rev Mol Cell Biol 6(10): 777–788
Guilak F, Cohen DM, Estes BT et al (2009) Control of stem cell fate by physical interactions with the extracellular matrix. Cell Stem Cell 5(1):17–26
Hansen DV, Lui JH, Parker PR et al (2010) Neurogenic radial glia in the outer subventricular zone of human neocortex. Nature 464(7288):554–561
Harry GJ (2013) Microglia during development and aging. Pharmacol Ther 139:313–326
He X, Imanishi S, Sone H et al (2012) Effects of methylmercury exposure on neuronal differentiation of mouse and human embryonic stem cells. Toxicol Lett 212(1):1–10
Hogberg HT, Bressler J, Christian KM et al (2013) Toward a 3D model of human brain development for studying gene/environment interactions. Stem Cell Res Ther 4(Suppl 1):S4
Hou Z, Zhang J, Schwartz MP et al (2013) A human pluripotent stem cell platform for assessing developmental neural toxicity screening. Stem Cell Res Ther (Suppl 1):S12
Ivanovic Z (2009) Hypoxia or in situ normoxia: the stem cell paradigm. J Cell Physiol 219:271–275
Jang JM, Tran SH, Na SC et al (2015) Engineering controllable architecture in matrigel for 3D cell alignment. ACS Appl Mater Interfaces 7:2183–2188
Jurga M, Lipkowski AW, Lukomska B et al (2009) Generation of functional neural artificial tissue from human umbilical cord blood stem cells. Tissue Eng Part C Methods 15(3):365–372
Kang KS, Trosko JE (2011) Stem cells in toxicology: fundamental biology and practical considerations. Toxicol Sci 120(Suppl 1):S269–S289
Kerever A, Schnack J, Vellinga D et al (2007) Novel extracellular matrix structures in the neural stem cell niche capture the neurogenic factor fibroblast growth factor 2 from the extracellular milieu. Stem Cells 25:2146–2157
Kim J, Sachdev P, Sidhu K (2013) Alginate microcapsule as a 3D platform for the efficient differentiation of human embryonic stem cells to dopamine neurons. Stem Cell Res 11(3):978–989
Kirby ED, Kuwahara AA, Messer RL et al (2015) Adult hippocampal neural stem and progenitor cells regulate the neurogenic niche by secreting VEGF. Proc Natl Acad Sci USA 112(13): 4128–4133
Kumar KK, Aboud AA, Bowman AB (2012) The potential of induced pluripotent stem cells as a translational model for neurotoxicological risk. Neurotoxicology 33(3):518–529
Lancaster MA, Knoblich JA (2014) Generation of cerebral organoids from human pluripotent stem cells. Nat Protoc 9(10):2329–2340
Lancaster MA, Renner M, Martin CA et al (2013) Cerebral organoids model human brain development and microcephaly. Nature 501(7467):373–379
Lathia JD, Mattson MP, Cheng A (2008) Notch: from neural development to neurological disorders. J Neurochem 107(6):1471–1481

Lee DA, Knight MM, Campbell JJ et al (2011) Stem cell mechanobiology. J Cell Biochem 112(1): 1–9

Li W, Zhu B, Strakova Z et al (2014) Two-way regulation between cells and aligned collagen fibrils: local 3D matrix formation and accelerated neural differentiation of human decidua parietalis placental stem cells. Biochem Biophys Res Commun 450(4):1377–1382

Lutolf MP, Gilbert PM, Blau HM (2009) Designing materials to direct stem-cell fate. Nature 462(7272):433–441

Massirer KB, Carromeu C, Griesi-Oliveira K et al (2011) Maintenance and differentiation of neural stem cells. Wiley Interdiscip Rev Syst Biol Med 3:107–114

McMurtrey RJ (2014) Patterned and functionalized nanofiber scaffolds in three-dimensional hydrogel constructs enhance neurite outgrowth and directional control. J Neural Eng 11(6):066009

Meli L, Barbosa HS, Hickey AM et al (2014) Three dimensional cellular microarray platform for human neural stem cell differentiation and toxicology. Stem Cell Res 13(1):36–47

Merkle FT, Tramontin AD, García-Verdugo JM et al (2004) Radial glia give rise to adult neural stem cells in the subventricular zone. Proc Natl Acad Sci USA 101(50):17528–17532

Narla ST, Lee YW, Benson CA et al (2017) Common developmental genome deprogramming in schizophrenia: role of Integrative Nuclear FGFR1 Signaling (INFS). Schizophr Res 185:17–32

Navaei-Nigjeh M, Amoabedini G, Noroozi A et al (2013) Enhancing neuronal growth from human endometrial stem cells derived neuron-like cells in three-dimensional fibrin gel for nerve tissue engineering. J Biomed Mater Res A 102(8):2533–2543

Ni N, Hu Y, Ren H et al (2013) Self-assembling peptide nanofiber scaffolds enhance dopaminergic differentiation of mouse pluripotent stem cells in 3-dimensional culture. PLoS One 8(12): e84504

Pamies D, Hartung T, Hogberg HT (2014) Biological and medical applications of a brain-on-a-chip. Exp Biol Med 239(9):1096–1107

Pamies D, Barreras P, Block K et al (2017) A human brain microphysiological system derived from induced pluripotent stem cells to study neurological diseases and toxicity. Altex 34(3):362–376

Peerani R, Rao BM, Bauwens C et al (2007) Niche-mediated control of human embryonic stem cell self-renewal and differentiation. EMBO J 26:4744–4755

Pellett S, Schwartz MP, Tepp WH et al (2015) A human induced pluripotent stem cell derived neuronal cells cultured on chemically-defined hydrogels for sensitive in vitro detection of botulinum neurotoxin. Sci Rep 5:14566

Pietrucha K, Zychowicz M, Podobinska M et al (2017) Functional properties of different collagen scaffolds to create a biomimetic niche for neurally committed human induced pluripotent stem cells (iPSC). Folia Neuropathol 55(2):110–123

Ranga A, Gjorevski N, Lutolf MP (2014a) Drug discovery through stem cell-based organoid models. Adv Drug Deliv Rev 69–70:19–28

Ranga A, Gobaa S, Okawa Y et al (2014b) 3D niche microarrays for systems-level analyses of cell fate. Nat Commun 5:4324

Ranga A, Girgin M, Meinhardt A, Eberle D et al (2016) Neural tube morphogenesis in synthetic 3D microenvironments. Proc Natl Acad Sci USA 113(44):E6831–E6839

Reilly GC, Engler AJ (2010) Intrinsic extracellular matrix properties regulate stem cell differentiation. J Biomech 43:55–62

Riquelme PA, Drapeau E, Doetsch F (2008) Brain micro-ecologies: neural stem cell niches in the adult mammalian brain. Philos Trans R Soc Lond B Biol Sci 12:123–137

Ruiz SA, Chen CS (2008) Emergence of patterned stem cell differentiation within multicellular structures. Stem Cells 26:2921–2927

Ruiz A, Valsesia A, Ceccone G et al (2007) Fabrication and characterization of plasma processed surfaces with tuned wettability. Langmuir 23(26):12984–12989

Ruiz A, Buzanska L, Ceriotti L et al (2008a) Stem-cell culture on patterned bio-functional surfaces. J Biomater Sci Polym Ed 19(12):1649–1657

Ruiz A, Buzanska L, Gilliland D et al (2008b) Micro-stamped surfaces for the patterned growth of neural stem cells. Biomaterials 29(36):4766–4774

Ruiz A, Zychowicz M, Buzanska L et al (2009) Single stem cell positioning on polylysine and fibronectin microarrays. Micro Nanosyst 1:50–56

Ruiz A, Zychowicz M, Ceriotti L et al (2013) Microcontact printing and microspotting as methods for direct protein patterning on plasma deposited polyethylene oxide: application to stem cell patterning. Biomed Microdevices 15(3):495–507

Sasai Y, Eiraku M, Suga H (2012) In vitro organogenesis in three dimensions: self-organising stem cells. Development 139(22):4111–4121

Schmuck MR, Temme T, Dach K et al (2017) Omnisphero: a high-content image analysis (HCA) approach for phenotypic developmental neurotoxicity (DNT) screenings of organoid neurosphere cultures in vitro. Arch Toxicol 91(4):2017–2028

Schwartz MP, Hou Z, Propson NE et al (2015) Human pluripotent stem cell-derived neural constructs for predicting neural toxicity. Proc Natl Acad Sci USA 112:12516–12521

Simão D, Pinto C, Piersanti S et al (2015) Modeling human neural functionality in vitro: three-dimensional culture for dopaminergic differentiation. Tissue Eng Part A 21(3–4):654–668

Snyder EY (2017) Finding a new purpose for old drugs. Science 357(6354):869–870

Soen Y, Mori A, Palmer TD et al (2006) Exploring the regulation of human neural precursor cell differentiation using arrays of signalling microenvironments. Mol Syst Biol 2:37

Stabenfeldt E, Brown AC, Barker TH (2010) Engineering ECM complexity into biomaterials for directing cell fate. Stud Mechanobiol Tissue Eng Biomater 2:1–18

Sugiyama T, Osumi N, Katsuyama Y (2013) The germinal matrices in the developing dentate gyrus are composed of neuronal progenitors at distinct differentiation stages. Dev Dyn 242(12): 1442–1453

Suh H, Consiglio A, Ray J et al (2007) In vivo fate analysis reveals the multipotent and self-renewal capacities of Sox2+ neural stem cells in the adult hippocampus. Stem Cell 1:515–528

Szablowska-Gadomska I, Zayat V, Buzanska L (2011) Influence of low oxygen tensions on expression of pluripotency genes in stem cells. Acta Neurobiol Exp (Wars) 71(1):86–93

Szablowska-Gadomska I, Sypecka J, Zayat V et al (2012) Treatment with small molecules is an important milestone towards the induction of pluripotency in neural stem cells derived from human cord blood. Acta Neurobiol Exp (Wars) 72(4):337–350

Urbán N, Guillemot F (2014) Neurogenesis in the embryonic and adult brain: same regulators, different roles. Front Cell Neurosci 27(8):396

van den Ameele J, Tiberi L, Vanderhaeghen P et al (2014) Thinking out of the dish: what to learn about cortical development using pluripotent stem cells. Trends Neurosci 37(6):334–342

Yang K, Han S, Shin Y et al (2013) A microfluidic array for quantitative analysis of human neural stem cell self-renewal and differentiation in three-dimensional hypoxic microenvironment. Biomaterials 34(28):6607–6614

Ylä-Outinen L, Joki T, Varjola M et al (2014) Three-dimensional growth matrix for human embryonic stem cell-derived neuronal cells. J Tissue Eng Regen Med 8(3):186–194

Zappaterra MW, Lehtinen MK (2012) The cerebrospinal fluid: regulator of neurogenesis, behavior, and beyond. Cell Mol Life Sci 69:2863–2878

Zychowicz M, Mehn D, Ana Ruiz A et al (2011) Proliferation capacity of cord blood derived neural stem cell line on different micro-scale biofunctional domains. Acta Neurobiol Exp 71:12–23

Zychowicz M, Mehn D, Ruiz A et al (2012) Patterning of human cord blood-derived stem cells on single cell posts and lines: implications for neural commitment. Acta Neurobiol Exp 72: 325–336

Zychowicz M, Dziedzicka D, Mehn D et al (2014) Developmental stage dependent neural stem cells sensitivity to methylmercury chloride on different biofunctional surfaces. Toxicol In Vitro 28(1):76–87

Part III
Therapeutic Potential of Human Neural Stem Cells

Chapter 10
Updates on Human Neural Stem Cells: From Generation, Maintenance, and Differentiation to Applications in Spinal Cord Injury Research

Yang D. Teng, Lei Wang, Xiang Zeng, Liquan Wu, Zafer Toktas, Serdar Kabatas, and Ross D. Zafonte

Abstract Human neural stem cells (hNSCs) and human induced pluripotent stem cells (hiPSCs) have been the primary focuses in basic science and translational research as well as in investigative clinical applications. Therefore, the capability to perform reliable derivation, effective expansion, and long-term maintenance of uncommitted hNSCs and hiPSCs and their targeted phenotypic differentiations through applying chemically and biologically defined medium in vitro is essential for expanding and enriching the fundamental and technological capacities of stem cell biology and regenerative medicine. In this chapter, we systematically summarized a set of protocols and unique procedures that have been developed in the laboratories of Prof. Teng and his collaborators. These regimens have been, over the years, reproducibly and productively used to derive, propagate, maintain, and differentiate hNSCs, including those derived from hiPSCs. We emphasize the multimodal methodologies that were pioneered and established in our laboratories

Y. D. Teng (✉)
Department of Physical Medicine and Rehabilitation, Harvard Medical School and Spaulding Rehabilitation Hospital, Boston, MA, USA

Department of Neurosurgery, Harvard Medical School and Brigham & Women's Hospital, Boston, MA, USA

Division of SCI Research, VA Boston Healthcare System, Boston, MA, USA
e-mail: yang_teng@hms.harvard.edu

L. Wang · X. Zeng · L. Wu · Z. Toktas · S. Kabatas
Department of Physical Medicine and Rehabilitation, Harvard Medical School and Spaulding Rehabilitation Hospital, Boston, MA, USA

Department of Neurosurgery, Harvard Medical School and Brigham & Women's Hospital, Boston, MA, USA

R. D. Zafonte
Department of Physical Medicine and Rehabilitation, Harvard Medical School and Spaulding Rehabilitation Hospital, Boston, MA, USA

L. Buzanska (ed.), *Human Neural Stem Cells*, Results and Problems in Cell Differentiation 66, https://doi.org/10.1007/978-3-319-93485-3_10

for characterizing functional multipotency of stem cells and its value in basic science as well as translational biomedical studies.

10.1 Methods for Working with Human Induced Pluripotent Stem Cell-Derived Neural Stem Cells

Background Induced pluripotent stem cells (iPSCs) are cells that were genetically reprogrammed from adult somatic cells (e.g., dermal fibroblast cells), which possess restored embryonic stem cell (ESC)-like pluripotency and, by definition, can be differentiated into all cell phenotypes of the human body (Takahashi and Yamanaka 2006; Takahashi et al. 2007; Yu et al. 2007). In 2006, Yamanaka and his colleagues reversed cell developmental fate of murine embryonic and adult fibroblasts and established the first set of cell lines of mouse iPSCs (Takahashi and Yamanaka 2006). In the next year, hiPSCs were effectively derived by the same team from human dermal fibroblasts via retroviral transduction of four transcriptional factors: Oct3/4, Sox2, Klf4, and c-Myc (Takahashi et al. 2007). Thompson and his team, using a different combination of Oct4, Sox2, Lin28, and Nanog, were also able to generate hiPSCs (Yu et al. 2007). Human iPSCs (hiPSCs) can develop into any cell type of all three germ layers and thus share common pluripotent characteristics with ESCs in morphology, proliferation, surface antigens, gene expression, global epigenetic status of pluripotent cell-specific genes, tissue/organ genesis participation, and telomerase activity (Takahashi and Yamanaka 2006; Takahashi et al. 2007; Yu et al. 2007).

The establishment of hiPSCs has provided a powerful platform for the fields of stem cell biology, developmental biology, genetic engineering, tissue engineering, and regenerative medicine. Since hiPSCs can be derived from patient tissue specimens in a disease-specific manner and then be guided to differentiate into target cell types, they offer technical and pathophysiological feasibilities to model different pathogenic conditions in vitro (e.g., controlled cell culture, iPSC-derived organoids, etc.) and to develop personalized therapeutic strategies in order to design precision medicine, in addition to generation of autologous cells for transplantation therapies. These properties of hiPSCs not only lower the risk of immune rejection but also mitigate common ethical, religious, and legal challenges that are associated with using human ESCs, overcoming major barriers to demands of scaling up cell numbers for tissue or organ repair. Conversely, there is persistent concern over possible consequences of intuitional teratogenicity and genetic editing following transplantation of hiPSCs and their derivative cell types, respectively, in experimental and putative clinical settings. Efforts have been given to reduce such risks by devising more definitive phenotypic differentiation and cell type purification technologies for hiPSC lines and by validating formulas of using virus-free and transgene-free reprogramming tactics besides inventing in situ direct reprogramming and conditional reprogramming approaches (Kele et al. 2016; Kim et al. 2011; Parsons et al. 2011a, b).

The following sections contain standard methods for common hiPSC culture and neural cell induction procedures that have been used in the laboratories of Prof. Teng at HMS/SRH/BWH and VABHS. Note: all cell culture medium, tropic factors, cytokines, genetic molecules, and other experimental reagents can be ordered from well-established manufacturers and vendors including American Type Culture Collection (Manassas, VA, USA), Atlanta Biologicals (Flowery Branch, GA, USA), EMD Millipore-Life Science (Billerica, MA, USA), Innovative Cell Technologies, Inc. (San Diego, CA, USA), Life Technologies (Grand Island, NY, USA), Invitrogen (Carlsbad, CA, USA), Thermo Fisher Scientific Inc. (Hudson, NH, USA), Sigma-Aldrich Corporation (St. Louis, MO, USA), Promega (Madison, WI, USA), Corning Inc. (Corning, NY, USA), STEMCELL Technologies Inc. (Cambridge, MA, USA), etc.

10.1.1 Maintenance and Propagation of Human iPSC in Feeder-Free (FF) and Serum-Free (SF) Culture

Researchers initially used murine tissue-derived feeder cells and serum-containing medium for the culture of hiPSCs (Takahashi et al. 2007; Yu et al. 2007). The required time and effort for preparing feeder cells and biological hazard possibility regarding cross-species exposure of animal-derived products have imposed a serious roadblock to the endeavors of investigating these hiPSCs for translational and clinical applications (Kele et al. 2016). Therefore, in our laboratories, we mainly use feeder-free and serum-free formulas to maintain and propagate hiPSCs.

One of the hiPSC lines that we have used was reprogrammed by retroviral vectors containing OCT4, SOX2, KLF4, and c-Myc from skin fibroblasts of a 64-year-old male healthy donor (ND41864, Coriell Institute, Camden, NJ). Mycoplasma and common viruses are checked periodically to ensure the quality of the cell stocks.

Standard Protocols All procedures are performed under sterile conditions with investigators exercising antiseptic tissue culture approaches and using individual biosafety protection gears for handling laminar flow hood, CO_2 incubator, and cell/tissue culture glassware or plasticware.

10.1.1.1. Corning® Matrigel® Matrix Coating: Prepare Corning® Matrigel® Matrix in 0.5 mg/mL with cold mTeSR™1 medium (STEMCELL Technologies Inc., Cambridge, MA) and coat the dish evenly. Incubate coated dishes at 37 °C for at least 1 h before use and ensure the dishes are adequately moisturized. Discard the excess solution, wash the dish with mTeSR™1 three times, and add pre-warm mTeSR™1 medium.

10.1.1.2. Thawing Cells: Pick up the cell stock from designated liquid nitrogen cryopreservation tank or −150 °C freezer and quickly thaw the cells in a 37 °C water bath. Mix warm mTeSR™1 with the existing cryopreserved protective medium (e.g., CryoStor® CS2 from STEMCELL Technologies Inc. or cryopreservation medium made in the laboratory, see Sect. 10.1.1.6). Centrifuge and discard the mixed medium while leave the cell pellets intact. Resuspend the cell pellets in warm mTeSR™1 medium and seed the cells in Corning® Matrigel®-coated dish for recovery.

10.1.1.3. Maintenance: hiPSCs can be maintained in mTeSR™1 medium on Corning® Matrigel® Matrix with daily evaluation and medium refreshment, based on parameters of cell viability and growth rate.

10.1.1.4. Passage: Passage hiPSCs when the colonies approach borders of any adjacent colony (usually 5–7 days). Cell colonies are normally detached by Dispase® (1 U/mL; STEMCELL Technologies Inc.). Split ratios range from 1:4 to 1:8, pending on cell density/numbers. Note: immediately after thawing and passaging, adding a dose of 2–10 μM Y27632, a specific inhibitor of p160ROCK, is recommend for enhancing survival rate of hiPSCs (Watanabe et al. 2007). After 24 h, the medium should be replaced with normal mTeSR™1 medium.

10.1.1.5. hiPSC Quality: hiPSC quality can be verified by immunocytochemical staining of pluripotent stem cell markers (e.g., SSEA3, SSEA4, TRA-1-81, CD30, etc.). After cultured hiPSCs form embryoid bodies, their differentiation potential and degrees can be evaluated by quantitative RT-PCR analysis of relative expression levels of genes specific for ectoderm (e.g., Fgf5, Otx2, Sox1, and Pax6), mesoderm (e.g., Hand1, Brachyury, Twist2, FoxA2, and Mixl1), and endoderm (e.g., Gata4, Gata6, and Sox17), compared to control hiPSCs.

10.1.1.6. Storage and Freezing: Prepare "Freezing Medium A" (50% of Final Volume): 50% DMEM/F12 50% knockout serum replacement (KSR) and "Freezing Medium B" (50% of final volume): 80% DMEM/F12 20% DMSO (final composition: 65% DMEM/F12, 25% KSR, and 10% DMSO) (Wagner and Welch 2010).

1. Pre-warm the required volume of Dispase® in a 37 °C water bath (estimated volume: 1 mL/60 mm dish). Pre-warm the needed volume of KSR-FF (1 mL/vial) in a 37 °C water bath for 10–15 min.
2. Remove the spent medium from the culture vessel using a pipette and rinse the cells twice with Dulbecco's phosphate-buffered saline (D-PBS; ~4 mL/per 60 mm dish cells).
3. Smoothly mix pre-warmed Dispase® solution to the culture vessel (~1 mL of Dispase® solution per 60 mm dish). Swirl the culture vessel to immerse the entire cell surface and incubate the culture vessel at 37 °C for 3–5 min before removing the Dispase® solution and gently wash the cells with D-PBS.
4. Gently scrape the cells off the surface of the culture dish using a cell scraper and transfer the cells to a sterile 15 mL centrifuge tube. Rinse (2×) the culture dish with KSR-FF and pool the rinse medium with the cells collected in the 15 mL tube.
5. Centrifuge the tube at 200 × *g* for 5 min at room temperature to pellet the cells. Discard the supernatant without disturbing the cell pellet. Prepare the needed number of cryovials.

6. Gently and fully dislodge the cell pellet from the tube bottom and resuspend the cells in Freezing Medium A by carefully pipetting up and down. After reaching uniform suspension of cells, add equal volume of Freezing Medium B to the tube, using a dropwise manner along swirling the cell suspension for mixing. Note: at this point, cells are in contact with DMSO, they should be aliquoted and frozen within 2–3 min.
7. Aliquot 1 mL of the cell suspension into each cryovial and swiftly place the vials in a Mr. Frosty™ freezing container (Thermo Fisher Scientific) to rapidly freeze and transfer it to −80 °C overnight followed by transferring to a liquid nitrogen tank for long-term storage. Alternatively, Dispase® (1 U/mL)-dissociated FF and SF hiPSC pellets can be resuspended in the solution containing 50% mTeSR™1, 40% knockout serum replacement, 10% DMSO, and 10 μM Y-27632 for cryopreservation. The total number of ~90% confluent hiPSCs in a 60 mm dish can be packed into two vials with each containing 1 mL cryopreservation media such as FreSR™-S (STEMCELL Technologies Inc.). Freeze cell aggregates using standard slow rate controlled cooling protocol that reduces temperatures at approximately 1 °C/min, followed by −80 °C overnight and long-term storage in liquid nitrogen, or using an isopropanol freezing container for storing at liquid nitrogen temperature (−195.79 °C). Importantly, do not keep vital cells for long-term storage at −80 °C.

10.2 Human Neural Stem Cells (hNSCs)

Stable clones of hNSCs have been established following isolation from the fetal telencephalon and induction from hiPSCs, respectively (Eriksson et al. 1998; Flax et al. 1998; Redmond et al. 2007; Roy et al. 2000, 2007; Vescovi et al. 1999). These self-renewing clones in vitro give rise to all three neural lineages (i.e., neuron, astrocyte, and oligodendrocyte) under proper culture conditions. Following transplantation into the brain or the spinal cord of newborn rodents, they participate in aspects of normal development, including migration along established migratory pathways to disseminated CNS regions, differentiation into multiple developmentally and regionally appropriate cell types, and nondisruptive interspersion with host progenitors and their progeny. Prototype or genetically engineered ex vivo, hNSCs are capable of expressing innate genes or foreign transgenes in vitro and in vivo in these disseminated locations (Eriksson et al. 1998; Flax et al. 1998; Roy et al. 2007). The soluble products (e.g., trophic factors, cytokines, exosomes, etc.) secreted or shed from and gap junctions formed by these hNSCs can cross-correct an inherited genetic metabolic defect in neurons and glial cells in vitro or in vivo, further enriching their cell therapy and gene therapy capabilities. Finally, hNSCs have been demonstrated to have potential to protect and/or replace specific deficient neuronal and neural cell populations in the adult rodent, nonhuman primate, and human nervous systems (Brustle et al. 1998; Eriksson et al. 1998; Flax et al. 1998; Redmond et al. 2007; Roy et al. 2007; Teng et al. 2012).

For research and clinical applications, hNSCs have been studied and found feasible for cryopreservation. Moreover, these cells can be propagated by both epigenetic and genetic means that are comparably safe and effective. The outcomes have encouraged investigators to study hNSC transplantation for a range of neurological disorders. To date, we have tested various culture conditions and genetic manipulations to develop formulas for optimizing the continuous, efficient expansion and passaging of prototype hNSCs (Redmond et al. 2007; Teng et al. 2012; Yu et al. 2009). For genetic engineering modifications, v-*myc* (i.e., the viral homolog of c-Myc that is usually a p110 gag-myc fusion protein derived from the avian retroviral genome) seems to be one of the most effective genes (see the following sections for more details). In either case, we have also identified a strict requirement for the presence of mitogens (FGF2 and EGF) in the growth medium, in effect constituting a conditional perpetuality or immortalization (note: LIF has been effective for blunting potential senescence) (Haragopal et al. 2015; Yu et al. 2009; Redmond et al. 2007; Roy et al. 2000).

10.2.1 hNSC Induction and Selection Protocols

10.2.1.1. Induction of hNSCs from hiPSCs In Vitro

1. After thawing, culture hiPSCs on Corning® Matrigel® Matrix coated dishes with mTeSR™1. The process usually takes 7–10 days.
2. Stage 1: when cells reach 30–40% confluency, remove mTeSR™1 and replace it with "Induction Medium": DMEM/F12 medium with 1% N2 supplement, 1% B27, 10–20 μg/mL FGF2, 10–20 μg/mL EGF, and 2–5 ng/mL heparin for NSC and neural progenitor cell (NPC) induction. Change the media fully every day till spheres start forming. It normally takes 5–9 days for the cells to show typical sphere morphology. Collect the spheres under microscope in the cell culture hood by using microdissection tools and dissociate the cells by trypsin-EDTA (0.05%) or Accutase®.
3. Stage 2: gently resuspend the cell pellets in the 2nd phase Induction Medium (i.e., Induction Medium plus 10 ng/mL LIF) and seed the cells in dishes pre-coated with recombinant human (rh) fibronectin and poly-D-lysine (1–5 μg/mL; 1:1) or rh-laminin (5–10 μg/mL). Partially change (~50%) the medium every 2–3 days for 2–4 weeks. Note: 10 ng/mL EGF is continuously applied only during the first 3 days.
4. Stage 3: when the cells reach 80% confluency, detach the cells by trypsin-EDTA (0.05%) or Accutase®. Resuspend the cell pellets in maintenance medium (i.e., Induction Medium without LIF and EGF is optional) and passage the cells by 1:3–1:5 ratio. Partially change (~50%) the medium every 2–3 days to maintain NSC differentiation and phenotype. The cells should be regularly checked for their NSC marker expression and proliferation rate in order to pick the most proper cells for research use.

10.2.1.2. Induction, Selection, and Establishment of Prototype hNSC Lines In Vitro

1. A suspension of primary dissociated neural cells (5 × 10^5 cells/mL), prepared from the telencephalon (particularly the periventricular region) of an early second-trimester human fetus (e.g., 13 weeks or 15 weeks) made available for medical reasons under formal institutional approval (Flax et al. 1998; Vescovi et al. 1999; Redmond et al. 2007; Teng et al. 2012), is plated on uncoated tissue culture dishes (Corning®) first in serum-containing medium for ~24–48 h and then in the following "growth medium": DMEM/F12 medium supplemented with 1% N2, to which FGF2 (10–20 μg/mL), heparin (8 μg/mL), and EGF (10–20 μg/mL) and LIF (10 ng/mL) are added.
2. Cultures are then put through the following "growth factor" selection process: cells are transferred to FGF2-containing serum-free medium alone for 2–3 weeks; then they are cultured in EGF-containing serum-free medium alone for another 2–3 weeks; consequently, they are returned to FGF2-containing serum-free medium alone for 2–3 weeks.
3. Finally, they are maintained in serum-free medium containing FGF2 plus LIF (EGF is optional). Medium is changed every 5–7 days and cells are passaged by trypsin-EDTA (0.05%) or Accutase® when >10–20 cell diameters in cluster or sphere sizes are formed. They should be subsequently replated in the maintenance medium at 5 × 10^5 cells/mL.

Note: at each stage of selection or induction, large numbers of cells would die or fail to survive passaging. What would be left following the above selection process are passageable, immature, and proliferative cells with qualities that are essential and sufficient, in our assessment, for meeting the standard operational definition of NSCs. In our experience, inclusion of 1% penicillin/streptomycin in the medium could be optional in the culture medium in induction, selection, and maintenance.

10.2.1.3. Polyclonal populations of the hNSCs are then separated into single clonal lines either by serial dilution alone (i.e., one cell per well—the process often has poor yields for prototypic hNSCs) or by first infecting the cells with a retrovirus to insert an engineered mitotic gene and/or molecular marker of clonality (i.e., the proviral integration site) and then performing serial dilution. The genes transduced via retrovirus are either (non-transforming) propagation enhancement genes, such as *v-myc* or human telomerase reverse transcriptase (hTERT) and/or purely reporter genes such as *lacZ* (Flax et al. 1998; Roy et al. 2007).

10.2.1.4. For retrovirus-mediated gene transfer, two xenotropic, replication-incompetent retroviral vectors have been used to infect hNSCs. A vector encoding *lacZ* is similar to BAG (Snyder et al. 1992) except for bearing a PG13 xenotropic envelope. An amphotropic vector encoding *v-myc* was generated using the ecotropic vector described for generating murine NSCs clone C17.2 (Ryder et al. 1990) to infect the GP + envAM12 amphotropic retroviral packaging cell line (Markowitz et al. 1988). No helper virus needs to be produced. Infection of FGF2-maintained growing human neural cells with either vector (titer, 4 × 10^5 CFUs) can follow similar, previously detailed procedures (Markowitz et al. 1988; Snyder et al. 1992).

10.2.1.5. For cloning of hNSCs, cells are dissociated as above, diluted to 1 cell/15 μL, and plated at 15 μL/well of a Terasaki or 96-well dish. Wells with single cells were noted immediately. Single-cell clones can be expanded and maintained in FGF2-containing growth medium. Single cells grow best when conditioned medium from dense hNSC cultures is included at a ratio between 20 and 50% of the growth medium.

10.2.1.6. Monoclonality is confirmed, as an example, by identifying in all progeny a single and identical genomic insertion site on Southern analysis for either the *lacZ*- or the *v-myc*-encoding provirus as previously detailed (Ryder et al. 1990). The *v-myc* probe is generated by nick translation labeling with 32P dCTP; a probe to the neo-sequence of the *lac*Z-encoding vector is generated by PCR utilizing 32P dCTP.

10.2.1.7. Lastly, cryopreservation of hNSCs is done by resuspending post-trypsinized or accutased human cells in a freezing solution composed of growth medium containing 10% DMSO, 50% FBS, and 40% FGF2. Afterward, the temperature of cells is brought down slowly first to 4 °C for an hour and then to −80 °C for 24 h and then to −140 °C in a freezer or a liquid nitrogen tank for longer-term storage (Flax et al. 1998).

10.2.2 hNSC Differentiation Protocols

To verify differentiation potential, some dissociated hNSCs were plated on poly-L-lysine (PLL)-coated slides (Nunc) in DMEM (devoid of growth factors) + 10% fetal bovine serum (FBS; one may add 2 mM L-glutamine and 1% penicillin/streptomycin) and processed weekly for immunocytochemistry (ICC). In most cases, differentiation occurred spontaneously or under enhanced promotion of selected trophic factors (Haragopal et al. 2015; Teng et al. 2012).

10.2.2.1. In general, our monoclonal or prototype, nestin-positive, hNSC lines perpetuated in this way divide every ~40 h and stop proliferation upon mitogen removal, undergoing spontaneous morphological differentiation and upregulating markers of the three fundamental lineages in the CNS (neurons, astrocytes, and oligodendrocytes) (Flax et al. 1998; Yu et al. 2009; Teng et al. 2012; Haragopal et al. 2015). Specified recipes can be used to augment a particular type of neural cell induction. As examples, for astrocytic maturation, clones can be co-cultured with primary dissociated embryonic CD-1 mouse brain (Ryder et al. 1990). In a newly described approach, neural cell (neuron, in particular) differentiation can be induced by culturing hNSCs in nanofibers that are fabricated in specially designed patterns (Jia et al. 2014).

10.2.3 *In Vivo Validation of hNSCs*

By using the methods described above, hNSC lines can be maintained or transfected and transduced via a variety of procedures and genes encoding proteins for labeling purposes (e.g., lacZ) and of therapeutic interest (e.g., BDNF: brain-derived neurotrophic factor). Such cell lines (e.g., HFB2050) can retain basic features of epigenetically expandable hNSCs (Redmond et al. 2007; Teng et al. 2012; Wakeman et al. 2014; Haragopal et al. 2015). Clonal analysis confirmed the stability, multipotency, and self-renewability of the cell lines (Kim et al. 2011; Snyder et al. 1992, 1995). Ultimately, all hNSC lines derived in vitro should be validated for their developmental, physiological, immunogenic, and therapeutic properties using well-established preclinical in vivo models for investigations of developmental neurobiology, neurotrauma, and neurological disorders.

10.3 Differentiation of Specific Phenotypes of Neurons from hiPSCs: Protocols for Deriving Serotonergic Neurons

Serotonergic neurons, despite their relatively limited numbers and highly selective locations in the brain, have a profound and widespread impact on sensorimotor, respiratory, mood, feeding, sleep, and sexual functions as well as on brain development. With advancement of stem cell technology, in vitro differentiation of hiPSCs or direct reprogramming of human fibroblast cells into serotonergic neurons could reach ~25–50% yielding rate in a time frame of ~30–63 days (Lu et al. 2016; Vadodaria et al. 2016; Xu et al. 2016). Recently, we reported that enhanced integrity of submidbrain neural circuits including serotonergic reticulospinal innervation at spinal cord levels below T9-10 lesion by scaffolded hMSCs implantation treatment promoted hind limb locomotion recovery in a rat model of spinal cord injury (Ropper et al. 2017). Therefore, it is desirable to devise effective recipes of serotonergic neuron differentiation for basic science research, transplantation study, translational investigation, and possible clinical application in the future.

10.3.1 *Differentiation of Serotonergic Neuron from hiPSCs*

1. After thawing, culture hiPSCs on Corning® Matrigel® Matrix coated dishes with mTeSR™1 for 7–10 days.
2. Dissociate hiPSC pellets with Dispase® (1 U/mL) and directly seed the cells onto 5 µg/mL laminin-coated dishes filled with mTeSR™1 medium.
3. When hiPSCs reach approximately 20% confluence (usually 1 day after passaging), completely change the medium and culture the cells in the "serotonin induction medium" (SIM) for 1 week: 50% DMEM/F12, 50% neurobasal, 1%

N2 supplement, 1% B27, 1% nonessential amino acids, 1% GlutaMAX®, 2–10 μM SB431542, 2 μM DMH1, and 1.4 μM CHIR99021 (Lu et al. 2016; Xu et al. 2016).
4. After 1 week of differentiation, cells are passed mechanically in the same SIM medium with 200–1000 ng/mL SHH C25II or 2.5 μg/mL purmorphamine at the ratio of 1:3–1:6 onto laminin-coated plates.
5. In the third week of differentiation, 10 ng/mL FGF4 or 100 ng/mL FGF8 along with 200–1000 ng/mL SHH C25II or 2.5 μg/mL purmorphamine will be added into SIM medium (Lu et al. 2016; Vadodaria et al. 2016).
6. From the 4th week, the serotonergic progenitor cells are seeded onto laminin-coated glass coverslips and cultured in the medium of 100% neurobasal plus 1% N2, 1% B27, and 1% NEAA supplemented with 1 μg/mL laminin, 0.2 mM vitamin C, 2.5 μM DAPT, 10–20 ng/mL glial cell line-derived neurotrophic factor (GDNF), 10–20 ng/mL BDNF, 10 ng/mL insulin-like growth factor-I (IGF-I), and 1 ng/mL transforming growth factor β3 (TGF-β3) (Lu et al. 2016; Vadodaria et al. 2016).
7. Continue culturing the cells for another 2–4 weeks in order to develop largely matured serotonergic neurons. Verification of serotonin neurons can be done by ICC staining of representative cell markers including serotonin (5HT: 5-hydroxytryptophan), aromatic L-amino acid decarboxylase (AADC; also known as DOPA decarboxylase, tryptophan decarboxylase), and TPH2 (tryptophan hydroxylase 2), as well as by performing serotonin/5HT release assays.

10.4 Preparation of hNSCs In Vitro for In Vivo Transplantation Studies of Experimental Spinal Cord Injury (SCI)

For preparation of cells for transplantation in the injured spinal cord of rodents, here is a succinct description of the bench workflow.

10.4.1. Rodent Spinal Cord Injury Models. Young adult or adult Sprague-Dawley (or a different strain) rats or selected strains of mice (body weight: 225–250 g and 20–30 g, respectively) are randomly assigned for receiving different pre- and/or post-injury treatments with group size carrying adequate statistical power (Teng et al. 2002a, b, 2004; Choi et al. 2005; Ropper et al. 2015, 2017). All in vivo study protocols should be in accordance with the principal investigator's Institutional Animal Care and Use Committee (IACUC) guidelines and policies issued by federal and local regulatory agencies.

1. At a predetermined SCI site, midline contusion (Teng et al. 2004) or compression (Ropper et al. 2015) or hemicontusion (Choi et al. 2005) or segmental hemisection (Teng et al. 2002a, b) is created using a standardized injury device or a size #11 surgical scalpel, under IACUC-approved sufficient anesthesia.

Hemostasis is usually achieved by using Gelfoam (Pfizer, New York, NY) plus room temperature saline washes of the surgical area.

2. An independent observer who is blinded for the experimental group design should confirm the adequacy and consistency of the contusion or compression severity or length and breadth of the surgical lesion. Only at that time is the surgeon informed of a particular treatment (e.g., cell suspension solution injection, implanting of NSC-seeded polymer construct, etc.) to be administered locally or systematically.
3. Following either the experimental or control treatment, the musculature is sutured, skin closed, and the rat or mouse recovers in a cage with clean bedding materials on a heating pad till it fully awakes. Ringer's lactate solution (0.5–1.0 mL/mouse, 5.0–10.0 mL/rat) should be given daily for 5–7 days post-operation. When injury affects spontaneous micturition reflex, the bladder should be gently evacuated twice daily by a trained investigator or research facility staff member until a so-called "reflex bladder" function is established.

10.4.2. Cultured hNSCs grow as a combination of adherent cells and floating clusters within T25 flasks or 10 mm dishes. All cells must be collected and be well dispersed into a suspension of individual cells in order for them to engraft well.

10.4.3. To accomplish this, all medium and cells (including those adherents which are mechanically dislodged) are transferred into a 15-mL centrifuge tube and centrifuged for 3 min at 1000 rpm in common desktop centrifuges. Following removal of the supernatant, 0.7 mL of trypsin/EDTA (0.05%) or Accutase®, typically 2.5–5 mL for a T25 flask depending upon the degree of confluency and density of the cell culture, is added to the centrifuge tube, and the cells are again triturated briefly before a 3–5 min incubation at 37 °C to facilitate dissociation of cells from each other.

Note: further gentle triturating is required to break up pellets and reach a true single-cell suspension status (although trypsin can also be added to the original flask in order to retrieve cells that may have still been adherent to the flask, it must be used with caution as over digestion could trigger irreversible cell clunking). Trypsinization is then terminated by adding 0.7 mL trypsin inhibitor (0.25 mg/mL in PBS) into the tube (or the flask) and triturating the mixture thoroughly. After a 3-min centrifuge at 1000 rpm (setting: radius of rotor, 100 mm; RCF (relative centrifugal force) = 112 × *g*), and removal of supernatant, cells are washed 1–3 times by resuspending them in 10 mL PBS.

10.4.4. Procedures for additional labeling of cells with trackers such as DiI or Hoechst can be done at this stage according to protocols suggested by manufacturers. Then cells can be resuspended with small volume of PBS (and/or with adding a sufficient amount of trypan blue, e.g., ~0.05% w/v, to permit viability assay or localization of the injected suspension).

10.4.5. An ideal injection concentration of hNSCs or mNSCs can be achieved by cell counting using a hemocytometer for final volume adjustments (i.e., $5–10 \times 10^4$ cells/μL). We use finely drawn glass micropipettes produced by Flaming/Brown Micropipette Puller (Model: P-97; Sutter Instrument Co. Novato, CA, USA). One can also use a Hamilton syringe, though we recommend connection of a glass micropipette instead of a conventional metal needle, with it the cell injection is done.

Note: it is important to minimize cell settlement out of suspension (which can occur rather quickly if the suspension solution is not prepared correctly) and to ensure that there is no clump formation that is usually detected by sharply increased resistance to injection advancement. Thus, we maintain the cells on ice and gently triturate them regularly. One should be careful not to fall into the trap of implanting cell carrying vehicle from which the cells have settled out yet believing that one has implanted cells. Interpretation of data, as one might imagine, will be entirely erroneous.

10.5 hNSCs-Based Investigative and Translational Approaches in SCI Research

Per discussions of the previous sections, transplantation of hNSCs into areas of injury can be useful for cell replacement and/or for delivery of therapeutic genes and their products. Some of the most impressive examples of this, in platform technology establishing studies, were observed in rat models of traumatic SCI (Teng et al. 2002a, b). Although NSCs appear to have the potential to repopulate severely injured spinal cord, their ability to survive, reconstitute neural tissue, and reform neural connections is often limited by the vast amount of parenchyma loss and the consequent secondary injury cellular and molecular milieu. Since the epicenter of the primary lesion changes rapidly into a necrotic syrinx (the so-called cystic cavity, lesion cavity, syrinx or lesion volume), even the most vital NSC may need an organization template that partially serves as extracellular matrix (ECM) skeletons to support survival and guide restructuring. In addition, large volumes of cells will not survive if located greater than a few hundreds of pm from the nearest capillary (Park et al. 2002). Thereby, we first hypothesized that three-dimensional highly porous (or purposely "structurally patterned") "scaffolds" composed of biodegradable (natural or synthetic) copolymers (or polymers) such as poly(lactic-*co*-glycolic acid) (PLGA) coated with poly-L-lysine or other molecules that also support live biological systems, if pre-seeded with NSCs (or other types of cells) for co-transplantation into the lesion cavity or space in solid organs, might facilitate donor cell survival, migration, differentiation, and their functional multipotency to promote structural repair and neural circuit activation (Teng et al. 2009, 2011). Since PLGA or PGA is a synthetic biodegradable polymer with Food and Drug Administration (FDA) approval for a variety of clinical applications, this approach has since become a primary platform technology for devising research and clinical applications of cells, especially stem cells. Highly hydrophilic, PLGA or PGA loses its mechanical strength rapidly over 2–4 weeks in the body; the scaffold can initially provide a matrix to guide cellular organization and growth, allowing diffusion of oxygen/nutrients to the transplanted cells, become vascularized, before being hydrolyzed and removed, obviating concerns over long-term biocompatibility.

To test our hypothesis for spinal cord repair, a multicomponent, biodegradable, synthetic PLGA scaffold of specified architecture and seeded with mNSCs was designed to support and structure neural repair, including possibilities of neural regeneration, direct cell replacement, impediment of reactive gliosis (glial scar) formation and penetration, stabilization of blood-spinal cord barrier, and mitigation of secondary injury events (Teng et al. 2002a, 2009, 2011). The implantation of the scaffold seeded with mNSCs in an adult rat segmental hemisection model of SCI (T9-T10; initial lesion length: 4 mm) led to robust long-term improvement in hind limb function relative to the lesion only or mNSC alone control groups. At 70 days post injury, the scaffolded mNSC treatment group exhibited coordinated weight-bearing stepping as compared with limited movement of two to three hind limb joints in the two control groups. Importantly, transplantation of scaffold alone also showed significantly discernible benefit for locomotion improvement, with the treated rats demonstrating body weight-bearing stepping in the hind limbs. In contrast with conventional data, neural pathway tracing revealed no corticospinal tract (CST) axons passing through the injury epicenter to the caudal side of the cord; there were, however, very few BDA (biotinylated dextran amine)-labeled CST fibers spotted along the interface between the injured side and the contralateral side of the spinal cord, suggesting that they were likely spared by the mNSC-seeded polymer or polymer-only treatment. Histological and ICC analyses including antibodies against donor-specific or general NSC markers (e.g., nestin), neurofilament (NF-H, M, and L), GFAP, and GAP-43 revealed that functional recovery observed in the study was the product of a group of combinatorial effects that comprise marked mitigation of secondary tissue loss via neuroprotection, reduction of reactive gliosis scale, and intensity in and around the lesion epicenter, mitigation of inflammatory responses, and promotion of beneficial neural plasticity (e.g., increased sprouting of serotonergic neurites) in the spinal cord. Lastly, we observed no donor-derived neuronal replacement in the lesion epicenter. The data suggested that comparing to the brain, the adult mammalian spinal cord shows an overall suppressive and inhibitory environment for neurogenesis. Our conclusion was additionally corroborated by the fact that there were only very few long-term (>10 weeks after implantation) surviving donor cells (~1–3%) that remained to be progenitors showing nestin-positive immunoreactivity. Therefore, we for the first time concluded that NSCs, in addition to their neural lineage differentiation capacity, produce neurotrophic and anti-inflammatory factors (e.g., BDNF, GDNF, a variety of cytokines) in response to lesion niche demands to promote tissue and functional repair through rebuilding homeostasis (Teng et al. 2002a, b). This evidence, together with the finding that NSCs can nurture functioning neural network via forming gap junctions with their surrounding cells, enabled us to engender a novel concept of stem cell biology—the functional multipotency of the stem cell (Teng et al. 2009, 2011).

These groundbreaking findings demonstrate that multimodal applications of stem cells (e.g., using polymer scaffolding to promote cell survival and guide cell fate, building "biological reactors", etc.) can further augment effectiveness of applying stem cells not only as investigation entities but also as multifunctional therapeutic devices. These innovative approaches elevate the repair potential, function, and differentiation specificity of donor and host NSCs, which jointly enable activation of neural circuits intrinsic to the spinal cord and the "reptilian brain" to reanimate locomotion pattern generator network. Together, the studies clearly suggest that prototype or iPSC-derived NSCs can play a key role in neural repair via mechanistically oriented strategies of neurobiological investigation (Teng et al. 2002a, b, 2009, 2011, 2012; Park et al. 2002; Redmond et al. 2007), tissue engineering (Jia et al. 2014), controlled drug release (Yu et al. 2009), conditional reprogramming (Kim et al. 2011), cell replacement (Park et al. 2002; Redmond et al. 2007), in vitro disease modeling (Tobe et al. 2017), and gene-directed enzyme prodrug therapy (GDEPT) (Ropper et al. 2016). Indeed, the data jointly reinforce the idea that for neural repair, NSCs can serve as powerful individual investigative or therapeutic vehicles; furthermore, they can act as an anchor that holds concomitant approaches together: molecular and cell interaction, gene therapy, targeted drug delivery, biomaterial, tissue engineering, niche modification, and cell replacement.

Acknowledgments Research work in Prof. Teng's Laboratories has been supported by Project ALS, NIH, VA, DoD, BSF, The Cele H. and William B. Rubin Family Fund, Inc., The Gordon Program to Treat and Cure Clinical Paralysis, and CASIS-NASA. The authors hold no COI for the paper.

References

Brustle O, Choudhary K, Karram K, Huttner A, Murray K, Dubois-Dalcq M, McKay RD (1998) Chimeric brains generated by intraventricular transplantation of fetal human brain cells into embryonic rats. Nat Biotechnol 16:1040–1044

Choi H, Liao WL, Newton KM, Onario RC, King AM, Desilets FC, Woodard EJ, Eichler ME, Frontera WR, Sabharwal S, Teng YD (2005) Respiratory abnormalities resulting from midcervical spinal cord injury and their reversal by serotonin 1A agonists in conscious rats. J Neurosci 25:4550–4559

Eriksson PS, Perfilieva E, Bjork-Eriksson T, Alborn AM, Nordborg C, Peterson DA, Gage FH (1998) Neurogenesis in the adult human hippocampus. Nat Med 4:1313–1317

Flax JD, Aurora S, Yang C, Simonin C, Wills AM, Billinghurst LL, Jendoubi M, Sidman RL, Wolfe JH, Kim SU, Snyder EY (1998) Engraftable human neural stem cells respond to developmental cues, replace neurons, and express foreign genes. Nat Biotechnol 16:1033–1039

Haragopal H, Yu D, Zeng X, Kim SW, Han IB, Ropper AE, Anderson JE, Teng YD (2015) Stemness enhancement of human neural stem cells following bone marrow MSC coculture. Cell Transplant 24:645–659

Jia C, Yu D, Lamarre M, Leopold PL, Teng YD, Wang H (2014) Patterned electrospun nanofiber matrices via localized dissolution: potential for guided tissue formation. Adv Mater 26:8192–8197

Kele M, Day K, Ronnholm H, Schuster J, Dahl N, Falk A (2016) Generation of human iPS cell line CTL07-II from human fibroblasts, under defined and xeno-free conditions. Stem Cell Res 17:474–478

Kim KS, Lee HJ, Jeong HS, Li J, Teng YD, Sidman RL, Snyder EY, Kim SU (2011) Self-renewal induced efficiently, safely, and effective therapeutically with one regulatable gene in a human somatic progenitor cell. Proc Natl Acad Sci USA 108:4876–4881

Lu J, Zhong X, Liu H, Hao L, Huang CT, Sherafat MA, Jones J, Ayala M, Li L, Zhang SC (2016) Generation of serotonin neurons from human pluripotent stem cells. Nat Biotechnol 34:89–94

Markowitz D, Goff S, Bank A (1988) Construction and use of a safe and efficient amphotropic packaging cell line. Virology 167:400–406

Park KI, Teng YD, Snyder EY (2002) The injured brain interacts reciprocally with neural stem cells supported by scaffolds to reconstitute lost tissue. Nat Biotechnol 20:1111–1117

Parsons XH, Teng YD, Parsons JF, Snyder EY, Smotrich DB, Moore DA (2011a) Efficient derivation of human neuronal progenitors and neurons from pluripotent human embryonic stem cells with small molecule induction. J Vis Exp 56:e3273

Parsons XH, Teng YD, Parsons JF, Snyder EY, Smotrich DB, Moore DA (2011b) Efficient derivation of human cardiac precursors and cardiomyocytes from pluripotent human embryonic stem cells with small molecule induction. J Vis Exp 57:e3274

Redmond DE Jr, Bjugstad KB, Teng YD, Ourednik V, Ourednik J, Wakeman DR, Parsons XH, Gonzalez R, Blanchard BC, Kim SU, Gu Z, Lipton SA, Markakis EA, Roth RH, Elsworth JD, Sladek JR Jr, Sidman RL, Snyder EY (2007) Behavioral improvement in a primate Parkinson's model is associated with multiple homeostatic effects of human neural stem cells. Proc Natl Acad Sci USA 104:12175–12180

Ropper AE, Zeng X, Anderson JE, Yu D, Han I, Haragopal H, Teng YD (2015) An efficient device to experimentally model compression injury of mammalian spinal cord. Exp Neurol 271:515–523

Ropper AE, Zeng X, Haragopal H, Anderson JE, Aljuboori Z, Han I, Abd-El-Barr M, Lee HJ, Sidman RL, Snyder EY, Viapiano MS, Kim SU, Chi JH, Teng YD (2016) Targeted treatment of experimental spinal cord glioma with dual gene-engineered human neural stem cells. Neurosurgery 79:481–491

Ropper AE, Thakor DK, Han I, Yu D, Zeng X, Anderson JE, Aljuboori Z, Kim SW, Wang H, Sidman RL, Zafonte RD, Teng YD (2017) Defining recovery neurobiology of injured spinal cord by synthetic matrix-assisted hMSC implantation. Proc Natl Acad Sci USA 114:820–829

Roy NS, Wang S, Jiang L, Kang J, Benraiss A, Harrison-Restelli C, Fraser RA, Couldwell WT, Kawaguchi A, Okano H, Nedergaard M, Goldman SA (2000) In vitro neurogenesis by progenitor cells isolated from the adult human hippocampus. Nat Med 6:271–277

Roy NS, Chandler-Militello D, Lu G, Wang S, Goldman SA (2007) Retrovirally mediated telomerase immortalization of neural progenitor cells. Nat Protoc 2:2815–2825

Ryder EF, Snyder EY, Cepko CL (1990) Establishment and characterization of multipotent neural cell lines using retrovirus vector-mediated oncogene transfer. J Neurobiol 21:356–375

Snyder EY, Deitcher DL, Walsh C, Arnold-Aldea S, Hartwieg EA, Cepko CL (1992) Multipotent neural cell lines can engraft and participate in development of mouse cerebellum. Cell 68:33–51

Snyder EY, Taylor RM, Wolfe JH (1995) Neural progenitor cell engraftment corrects lysosomal storage throughout the MPS VII mouse brain. Nature 374:367–370

Takahashi K, Yamanaka S (2006) Induction of pluripotent stem cells from mouse embryonic and adult fibroblast cultures by defined factors. Cell 126:663–676

Takahashi K, Tanabe K, Ohnuki M, Narita M, Ichisaka T, Tomoda K, Yamanaka S (2007) Induction of pluripotent stem cells from adult human fibroblasts by defined factors. Cell 131:861–872

Teng YD, Lavik EB, Qu X, Park KI, Ourednik J, Zurakowski D, Langer R, Snyder EY (2002a) Functional recovery following traumatic spinal cord injury mediated by a unique polymer scaffold seeded with neural stem cells. Proc Natl Acad Sci USA 99:3024–3029

Teng YD, Park KI, Larvik EB, Langer R, Synder EY (2002b) Stem cell culture: neural stem cells. In: Atala A, Lanzer RP (eds) Methods of tissue engineering. Academic, San Diego, pp 421–437

Teng YD, Choi H, Onario RC, Zhu S, Desilets FC, Lan S, Woodard EJ, Snyder EY, Eichler ME, Friedlander RM (2004) Minocycline inhibits contusion-triggered mitochondrial cytochrome c release and mitigates functional deficits after spinal cord injury. Proc Natl Acad Sci USA 101:3071–3076

Teng YD, Kabatas S, Wakeman DR, Li J, Synder EY, Sidman RL (2009) Functional multipotency of neural stem cells and its therapeutic implications. In: Ulrich H (ed) Perspectives of stem cells: from tools for studying mechanisms of neuronal differentiation towards therapy. Springer, San Diego, pp 255–270

Teng YD, Yu D, Ropper AE, Li J, Kabatas S, Wakeman DR, Wang J, Sullivan MP, Redmond DE Jr, Langer R, Snyder EY, Sidman RL (2011) Functional multipotency of stem cells: a conceptual review of neurotrophic factor-based evidence and its role in translational research. Curr Neuropharmacol 9:574–585

Teng YD, Benn SC, Kalkanis SN, Shefner JM, Onario RC, Cheng B, Lachyankar MB, Marconi M, Li J, Yu D, Han I, Maragakis NJ, Llado J, Erkmen K, Redmond DE Jr, Sidman RL, Przedborski S, Rothstein JD, Brown RH Jr, Snyder EY (2012) Multimodal actions of neural stem cells in a mouse model of ALS: a meta-analysis. Sci Transl Med 4:165ra164

Tobe BTD, Crain AM, Winquist AM, Calabrese B, Makihara H, Zhao WN, Lalonde J, Nakamura H, Konopaske G, Sidor M, Pernia CD, Yamashita N, Wada M, Inoue Y, Nakamura F, Sheridan SD, Logan RW, Brandel M, Wu D, Hunsberger J, Dorsett L, Duerr C, Basa RCB, McCarthy MJ, Udeshi ND, Mertins P, Carr SA, Rouleau GA, Mastrangelo L, Li J, Gutierrez GJ, Brill LM, Venizelos N, Chen G, Nye JS, Manji H, Price JH, McClung CA, Akiskal HS, Alda M, Chuang DM, Coyle JT, Liu Y, Teng YD, Ohshima T, Mikoshiba K, Sidman RL, Halpain S, Haggarty SJ, Goshima Y, Snyder EY (2017) Probing the lithium-response pathway in hiPSCs implicates the phosphoregulatory set-point for a cytoskeletal modulator in bipolar pathogenesis. Proc Natl Acad Sci USA 114:4462–4471

Vadodaria KC, Mertens J, Paquola A, Bardy C, Li X, Jappelli R, Fung L, Marchetto MC, Hamm M, Gorris M, Koch P, Gage FH (2016) Generation of functional human serotonergic neurons from fibroblasts. Mol Psychiatry 21:49–61

Vescovi AL, Parati EA, Gritti A, Poulin P, Ferrario M, Wanke E, Frolichsthal-Schoeller P, Cova L, Arcellana-Panlilio M, Colombo A, Galli R (1999) Isolation and cloning of multipotential stem cells from the embryonic human CNS and establishment of transplantable human neural stem cell lines by epigenetic stimulation. Exp Neurol 156:71–83

Wagner K, Welch D (2010) Cryopreserving and recovering of human iPS cells using complete knockout serum replacement feeder-free medium. J Vis Exp 41:e2237

Wakeman DR, Redmond DE Jr, Dodiya HB, Sladek JR Jr, Leranth C, Teng YD, Samulski RJ, Snyder EY (2014) Human neural stem cells survive long term in the midbrain of dopamine-depleted monkeys after GDNF overexpression and project neurites toward an appropriate target. Stem Cells Transl Med 3:692–701

Watanabe K, Ueno M, Kamiya D, Nishiyama A, Matsumura M, Wataya T, Takahashi JB, Nishikawa S, Nishikawa S, Muguruma K, Sasai Y (2007) A ROCK inhibitor permits survival of dissociated human embryonic stem cells. Nat Biotechnol 25:681–686

Xu Z, Jiang H, Zhong P, Yan Z, Chen S, Feng J (2016) Direct conversion of human fibroblasts to induced serotonergic neurons. Mol Psychiatry 21:62–70

Yu J, Vodyanik MA, Smuga-Otto K, Antosiewicz-Bourget J, Frane JL, Tian S, Nie J, Jonsdottir GA, Ruotti V, Stewart R, Slukvin II, Thomson JA (2007) Induced pluripotent stem cell lines derived from human somatic cells. Science 318:1917–1920

Yu D, Neeley WL, Pritchard CD, Slotkin JR, Woodard EJ, Langer R, Teng YD (2009) Blockade of peroxynitrite-induced neural stem cell death in the acutely injured spinal cord by drug-releasing polymer. Stem Cells 27:1212–1222

Chapter 11
Human Neural Stem Cells for Ischemic Stroke Treatment

Zaal Kokaia and Vladimer Darsalia

Abstract Ischemic stroke is the second most common cause of death worldwide and a major cause of disability. It takes place when the brain does not receive sufficient blood supply due to the blood clot in the vessels or narrowing of vessels' inner space due to accumulation of fat products. Apart from thrombolysis (dissolving of blood clot) and thrombectomy (surgical removal of blood clot or widening of vessel inner area) during the first hours after an ischemic stroke, no effective treatment to improve functional recovery exists in the post-ischemic phase. Due to their narrow therapeutic time window, thrombolysis and thrombectomy are unavailable to more than 80% of stroke patients.

Many experimental studies carried out in animal models of stroke have demonstrated that stem cell transplantation may become a new therapeutic strategy in stroke. Transplantation of stem cells of different origin and stage of development has been shown to lead to improvement in experimental models of stroke through several mechanisms including neuronal replacement, modulation of cellular and synaptic plasticity and inflammation, neuroprotection and stimulation of angiogenesis. Several clinical studies and trials based on stem cell delivery in stroke patients are in progress with goal of improvements of functional recovery through mechanisms other than neuronal replacement. These approaches may provide therapeutic benefit, but generation of specific neurons for reconstruction of stroke-injured neural circuitry remains ultimate challenge. For this purpose, neural stem cells could be developed from multiple sources and fated to adopt required neuronal phenotype.

Z. Kokaia (✉)
Laboratory of Stem Cells and Restorative Neurology, Lund Stem Cell Center, University Hospital, Lund, Sweden
e-mail: Zaal.Kokaia@med.lu.se

V. Darsalia
Department of Clinical Science and Education, Södersjukhuset, Internal Medicine, Karolinska Institutet, Stockholm, Sweden

L. Buzanska (ed.), *Human Neural Stem Cells*, Results and Problems in Cell Differentiation 66, https://doi.org/10.1007/978-3-319-93485-3_11

11.1 Sources of Human Neural Stem Cells

11.1.1 Neural Stem Cells Derived from Embryonic Stem Cells

Embryonic stem cells (ESCs) are pluripotent stem cells that can propagate and differentiate into any cell type (Thomson et al. 1998). Human ESCs (hESCs) can be derived from the inner cell mass of the blastocyst at 4–5 days after fertilization. Isolation of hESCs requires the destruction of a blastocyst, thus raising ethical issues. However, the alternative methods of establishing ESC lines from single-cell embryo biopsy without interfering with the developmental potential of embryos have been proposed (Chung et al. 2006). The common practice of obtaining hESCs is the use of leftover fertilized embryos from in vitro fertilization that were not implanted and were discarded. Human ESCs derived from the inner cell mass of the blastocyst can be easily maintained as undifferentiated pluripotent cell lines in the in vitro culture conditions. Because of their strong capacity to self-renew and propagate, transplanted ESCs can form tumours in the host (Erdo et al. 2003; Pomper et al. 2009); thus, the risk of adverse outcome after transplantation is high.

To reduce the tumour-forming potential, hESCs can be pre-differentiated in vitro in committed precursor cells or neural precursor cells (NPCs). Several in vitro techniques have been developed to derive NPCs from hESCs, notably retinoic acid (RA) induction pathway (Schuldiner et al. 2001), inhibition of bone morphogenetic protein (BMP) signalling by noggin (Pera et al. 2004; Gerrard et al. 2005) and fibroblast growth factor (FGF) signalling (Carpenter et al. 2001; Dhara et al. 2008). Human ESCs can also be used to generate a stable population of long-term self-renewing neuroepithelial stem cells (lt-NES cells) (Koch et al. 2009), which are easily expandable and give rise to high numbers of neurons in vitro and after intracerebral transplantation (Steinbeck et al. 2012).

NPCs derived from ESCs maintain self-renewal capacity while at the same time are restricted to generation of neural cells (neurons and glia) upon differentiation. Human ESC-derived NPCs have been successfully implanted in rodents after cerebral ischemia, and neural differentiation and improved functional recovery have been shown (Hara et al. 2007; Daadi et al. 2008; Seminatore et al. 2010; Jiang et al. 2013). However, due to the large population of proliferating cells within the grafts, the overgrowth from hESCs-derived NPC grafts remains a potential hazard (Seminatore et al. 2010).

11.1.2 Neural Stem Cells Derived from Foetal Brain

Foetal neural stem cells can be isolated from aborted died human foetus at gestational age of 6–20 weeks. The age of foetus at which the human NSCs (hNSCs) are usually isolated are limited by the statutory law. The hNSCs can be isolated from almost any part of the foetal human CNS. Isolated hNSCs can be maintained for long periods of time in in vitro conditions where they can be expanded and passaged numerous times without losing their self-renewing and neurogenic capacity (Kallur et al. 2006; Darsalia et al. 2007).

The main drawback of using foetal hNSCs as the source of the transplantation material is their limited availability. As these NSCs are derived from aborted foetuses, it is unpredictable when, where, or in what condition the source material will be obtained. It is also undetermined for how long foetal NSCs can be maintained in vitro without transforming them into cell lines, although successful maintenance for few months or even up to 2 years and through several freeze/thaw cycles, as well as successful survival and neuronal differentiation after consequent transplantation has been reported (Darsalia et al. 2007; Anderson et al. 2007).

Preclinical studies have also reported that hNSCs isolated from different parts of foetal CNS maintain their region specificity after transplantation (Kallur et al. 2006; Tailor et al. 2013), indicating that the structural origin of hNSCs should be considered when grafting hNSCs for treating neurological disorders linked to specific brain regions or cell types.

11.1.3 Neural Stem Cells Derived from Adult Brain

Adult mammalian brain including human brain maintains the niches of stem cells that could serve as a source of new neurons for the adult brain. The new neurons might contribute to new circuit formation (learning, memory) or cell replacement after brain damage. For reviews, see Alunni and Bally-Cuif (2016), Kuhn et al. (2016) and Kempermann (2015).

Unlike NSCs from other sources, prospective therapeutic application of adult neural stem cells for treating ischemic stroke could be accomplished through two potential routes: stimulation of de novo neurogenesis from endogenous NSCs from brain's own neurogenic niches and transplantation of previously isolated and expanded adult NSCs.

Significant stimulation of adult NSCs after ischemic stroke and generation of new neurons has been first reported in rodents (Arvidsson et al. 2002; Parent et al. 2002). However, later studies have confirmed similar occurrence in primates (Tonchev et al. 2003) and also possibly in human (Jin et al. 2006; Minger et al. 2007; Nakayama et al. 2010). The functional significance of stroke-generated new neurons for the recovery after stroke is not clear. However, some data indicate that the possible involvement of endogenously derived neural progenitors can contribute to the recovery of post-stroke behavioural deficits (Jin et al. 2010; Wang et al. 2012).

The possibility to stimulate neurogenesis from endogenous NSCs after ischemic stroke to such extent which will have therapeutic significance through cell replacement remains unclear. In humans, endogenous NSCs would be required to migrate from NSC niches relatively long distance to the injury site, which could be another limiting factor. However, recent studies indicate on possibility that subventricular zone adult NPCs might exert "bystander" effect and act as regulators of neuronal homeostasis for glutamatergic excitotoxicity in stroke-lesioned striatum (Butti et al. 2012).

Transplantation of NSCs isolated from adult brain and expanded in vitro could overcome the limitations of endogenous neurogenesis. However, surgical samples are usually very small, and adult hNSCs have limited capacity of proliferation. Therefore, it is difficult task not only to isolate but also to expand them in vitro and develop adult hNSCs as the reliable source for neural transplantation with therapeutic significance (Murrell et al. 2013; Joo et al. 2013).

11.1.4 Neural Stem Cells Derived from Induced Pluripotent Stem Cells

In 2006, a publication by Takahashi and Yamanaka (2006) regarding reprogramming of somatic cells to pluripotent stem cells opened new opportunities for generation of NPCs with much less ethical, social, and tumorigenic concerns as compared to foetal- or embryonic stem cell-derived NSCs. The iPSCs have a powerful capacity for proliferation and differentiation similar to ESCs which make them very attractive for application in cell therapy approaches. Similar to ESCs, iPSCs could be used to generate NPCs or lt-NESCs with very high neurogenic potential and virtually no tumorigenicity after intracerebral transplantation. Up to now, several studies have been carried out in rodent model of stroke using intracerebral transplantation of these cells. Several studies used mouse fibroblast-derived iPSCs for transplantation in stroke-damaged brain of mice (Kawai et al. 2010; Liu et al. 2014) and rat (Chau et al. 2014; Chen et al. 2010) with goal to improve various functions impaired by this insult. However, the majority of studies have been carried out using human-derived iPSCs. In only few studies, human iPSCs have been transplanted directly without prior fating or pre-differentiation (Jiang et al. 2011; Qin et al. 2015). More often, before grafting in stroke-damaged brain, iPSCs are treated to adopt NPC phenotype (Chang et al. 2013; Eckert et al. 2015; Jensen et al. 2013; Lam et al. 2014; Mohamad et al. 2013; Yuan et al. 2013) or are transformed into lt-NESCs (Oki et al. 2012; Tatarishvili et al. 2014; Tornero et al. 2013, 2017).

Pretreatment of iPSCs ensures that after intracerebral transplantation, iPSCs will develop neural phenotype, and they will lose tumorigenic capacity. The fate mapping of grafted human iPSCs could be achieved by means of human-specific antibodies. Alternatively, iPSCs could be pre-labelled with GFP-containing lentivirus and later traced with immune fluorescence microscopy. Pre-labelling of cells with GFP also allows to carry out electrophysiological studies from acute slice of brain tissue subjected to stroke and intracerebral transplantation. Transplanted iPSCs have been detected in the rodent brain 4–5 months after transplantation with variable survival rate depending on host strain and species as well as immunosuppression mode (Majid et al. 2000; Sauter and Rudin 1995; Braeuninger and Kleinschnitz 2009; Oki et al. 2012). All published studies clearly indicate that when human iPSCs are transformed into iPSC-NPCs or iPSC-lt-NESCs after transplantation in the stroke-damaged brain, they become prone to develop into cells with neuronal phenotype. At early time points, cells transplanted in ischemic brain extensively express either immature or young neuronal markers such as nestin (Chang et al. 2013; Eckert et al. 2015; Yuan et al. 2013; Jiang et al. 2011; Jensen et al. 2013), DCX (Oki et al. 2012; Tatarishvili et al. 2014; Tornero et al. 2013) and βIII tubulin (Eckert et al. 2015; Jensen et al. 2013; Yuan et al. 2013). In studies with long-term survival, human iPSC-derived cells revealed expression of generic mature neuronal markers NeuN (Chang et al. 2013; Mohamad et al. 2013; Tornero et al. 2013), MAP 2 (Chang et al. 2013; Jensen et al. 2013), and HuD (Oki et al. 2012; Tatarishvili et al. 2014; Tornero et al. 2013). In several studies, more specific neuronal phenotype was also revealed with markers such as GABA/GAD65 (inhibitory neurons) (Chang et al. 2013; Tornero et al. 2013), kidney-type glutamate (KGA, glutamatergic neurons) (Tornero et al. 2013), TH (dopaminergic neurons) (Chang et al. 2013), and DARPP-32 (striatal projection neurons) (Polentes et al. 2012; Chang et al. 2013). As shown by Tornero and colleagues (2013), it is also possible to fate iPSC-derived cells towards specific neuronal phenotype. Fating of iPSC-lt-NESCs towards cortical phenotype before intracerebral transplantation in stroke-subjected rats, at 2 months after transplantation, led to efficient generation to mature neurons (increased number of HuD+, Fox3+, and KGA+ cells) with pyramidal morphological and immunohistochemical marker such as Tbr1 as well as higher axonal projection density. Only in a few studies, grafted iPSC-derived cells differentiated into GFAP+ (Chang et al. 2013; Jensen et al. 2013; Tatarishvili et al. 2014; Yuan et al. 2013) or s100β+ (Eckert et al. 2015) astrocytes. Cells expressing markers of oligodendroglia or oligodendrocyte precursor such as CNPase (Buhnemann et al. 2006), NG2 (Hicks et al. 2009) or A2B5, and GalC (Kim et al. 2007) were also detected among grafted cells. However, the number of glial cells was always very low as compared to cells with neuronal properties.

11.1.5 Highlight Summary Table of Sources of Human Neural Stem Cells (Table 11.1)

Table 11.1 Human neural stem cells

Origin	Concept	Advantages	Disadvantages
Embryo, usually left over after IVF	Derived from the inner cell mass of the blastocyst	Strong capacity to self-renew and easily expandable. Can be easily maintained in vitro or transformed into cell lines. Can be used for derivation of multiple cell types	Continued proliferative activity after transplantation. High risk of tumour formation. Ethically controversial

(continued)

Table 11.1 (continued)

Origin	Concept	Advantages	Disadvantages
Foetal, aborted between gestational weeks 6–20	Isolated from almost any part of the foetal human CNS	Committed towards neural phenotype. Can be expanded and passaged numerous times without losing self-renewing and neurogenic capacity	Limited availability. Ethically controversial
Adult brain	Isolated from neurogenic niches of adult brain or stimulated after injury to migrate and promote regeneration	Committed towards neural phenotype	Low viability in vitro after isolation, difficult to expand and maintain
Induced pluripotent stem cells	Reprogramming of somatic cells to pluripotent stem cells	Readily available sources. Powerful capacity for proliferation and differentiation in vitro. Ethically noncontroversial	Continued proliferative activity after transplantation. High risk of tumour formation

11.2 Routes of hNSC Transplantation

One of the key components for developing NSCs transplantation for clinical application is the choice of the graft delivery route. The graft should be able to survive the transplantation procedure (cell suspension injected via a syringe) and rebuild damaged neuronal circuits by replacing lost neurons. Over the past decades, varieties of cell types and delivery routes have been tested in animal models of ischemic stroke. The following section will exclusively focus on studies using hNSC delivered through a variety of routes in stroke-subjected animals.

Intraparenchymal transplantation implies injecting cell suspension directly into the brain parenchyma in proximity to injury site. The main advantage of this route of graft delivery is the ability to precisely control the graft placement. Several studies have used this route of hNSC graft delivery in rats after focal ischemia. These studies have shown that graft had survived for up to 2 months after intrastriatal or intracortical transplantation, and the functional impairments had been significantly reduced compared to controls (Saporta et al. 1999; Hara et al. 2007; Jeong et al. 2003; Hicks et al. 2009). The drawback of using this delivery route is its invasiveness. Depending on the localization of injury site, the trauma from implantation surgery may compound the tissue damage and could potentially offset the benefits of transplantation and even worsen the functional outcome. Additionally, the path of implanter needle may pass through healthy parts of the brain, thus potentially affecting the normal functioning of these brain regions.

Intraparenchymal transplantation of hNSCs has been tested in a limited number of stroke patients in several clinical trials. These studies have shown motor and cognitive improvements in grafted patients, although the scale of the studies was limited to up to 30 patients (Nelson et al. 2002; Kondziolka et al. 2005; Stilley et al. 2004; Kalladka et al. 2016). Recent clinical trial immortalized human neural stem cell line was used in an open-label, single-site, dose-escalation clinical study. Men aged 60 years or older with stable disability 6–60 months after ischemic stroke were implanted. Single intracerebral doses up to 20 million cells induced no adverse events and were associated with improved neurological function (Kalladka et al. 2016).

The efficacy of intraventricular rodent NSC implantation has been examined by some preclinical studies (Doeppner et al. 2015; Jin et al. 2005). Although cells injected in the lateral ventricle reached the stroke-lesioned site, they seemed to be less efficient for the recovery as compared to intracerebral implantation. Human-derived NSCs have never been tested for the intraventricular transplantation in stroke-damaged brain.

Intravascular route of cell delivery has been mostly used to study the efficacy of mesenchymal stem cells, although limited trials using hNSCs have also been performed. The main advantage of intravascular delivery is the ability to introduce greater number of cells through the least invasive route. Studies performed in rodents show that intravenously injected hNSCs survive and migrate into the ischemia-damaged regions of the brain where they differentiate into neurons and glia and ameliorate functional deficits, although the number of injected cells that had migrated to the brain was significantly smaller than amount of the number of injected cells. Distribution of injected cells in internal organs was also observed (Chu et al. 2003, 2004). Using SPECT imaging, Lappalainen et al. have shown the significant accumulation of hNSCs in internal organs (liver, spleen, and kidneys) after intravenous injection (Lappalainen et al. 2008). These data show the main drawback of intravascular injection route, which is the risk of accumulation of injected cells in internal organs and tumour formation. Interestingly, the attempts have been made to direct the migration of injected hNSCs to the brain. Song et al. have used neodymium magnets to target the delivery of hNSCs to ischemic brain. This method increased the number of hNSCs that reached the ischemic brain, although the accumulation of injected cells in internal organs was not determined (Song et al. 2010). Without first developing such techniques that would limit the unwanted cell accumulation outside the targeted brain regions, thus ensuring the safety of intravascular cell delivery, it is unlikely that this approach will develop clinically.

11.3 Influence of Host Environment on the Graft

The microenvironment of uninjured host brain plays an important role in the survival and fate determination of the grafted NSCs. Several studies have shown that the same neural stem cell line grafted in different brain structures gave rise to region-specific neurons (Lundberg et al. 1996; Onifer et al. 1993; Shihabuddin et al. 1996). These studies point towards the possibility that perhaps it is unnecessary to develop brain region-specific cell lines for human application. If it is true (has not yet been tested using hNSCs), this will dramatically simplify the development of clinically usable hNSC lines.

Host immunity will also potentially affect the graft survival. It is unknown to what extent the hNSCs will stimulate the immune response after allogenic grafting (i.e. in human brain). However, in vitro experiments have shown that hNSC can induce immune response of allogenic natural killer cells and T lymphocytes (Preynat-Seauve et al. 2009). The potential immune response to allogenic NSC transplants could likely be detrimental factor for hNSC survival and an additional hurdle to overcome for successful clinical application.

The microenvironment of the brain after ischemic stroke will likely have a strong influence on the graft fate. Ischemic damage activates microglia and astrocytes, as well as induces infiltration of leukocytes from the blood. Increased inflammatory state can persist for several weeks or even months and finally leads to the formation of glial scar by reactive astrocytes around the injury site. This creates a kind of barrier that excludes the damaged region from the axonal regrowth and regenerative processes. Inflammation after ischemic injury is a dynamic process than can first exacerbate the injury and later promote tissue regeneration (for a recent review, see Kim et al. 2016; Guruswamy and ElAli 2017). Therefore, it is of great importance to have a clear understanding of the brain's inflammatory state after ischemic stroke to successfully apply stem cell therapy. For example, several studies have demonstrated that the timing of transplantation (in relation to stroke onset) and the state of neuroinflammation affect hNSCs survival, differentiation, and integration (Darsalia et al. 2011).

11.4 hNSC Graft-Induced Functional Recovery: Trophic Effects

It has been long believed that to improve neurological function, grafted NSC would need to differentiation into functional neurons and integrate into the host neural circuitry. However, experimental evidence began to accumulate showing noticeable functional improvements shortly after NSCs transplantation without apparent neuronal differentiation or with the limited number of graft-derived mature neurons that could not explain significant functional improvement. From prior experiments using marrow stromal cells or other nonneural cells, it has been known that such graft can reduce injury, dampen the inflammation, and promote neuronal regrowth (Daadi et al. 2010; Li et al. 2010; Ma et al. 2013; Schwarting et al. 2008; Shichinohe et al. 2006). Similar effects have been observed after hNSC transplantation in rodents. Transplanted hNSC reduced cell death near the graft (Skardelly et al. 2011) and reduced inflammation (Lee et al. 2008). Furthermore, transplanted hNSCs have been shown to promote proliferation and neuronal differentiation of endogenous NSCs from subventricular zone and hippocampal subgranular zone (Ryu et al. 2016).

11.5 hNSC Graft-Induced Functional Recovery: Neuronal Replacement

Neuronal differentiation and neuronal replacement still remain one of the main focuses of hNSCs development for therapy for stroke. Several studies using NSC transplantation strategy for treatment of stroke-damaged brain clearly demonstrated the capacity of these cells to survive intracerebral transplantation in lesioned brain and morphologically and electrophysiogically differentiate into mature neurons. However, we still lack direct evidence of their integration in host neuronal circuitry and contribution to post-stroke recovery through restoration of damaged neuronal network.

Demonstration that grafted neurons exhibit spontaneous excitatory postsynaptic currents indicated on possibility that host neurons establish synaptic connection with grafted neurons (Oki et al. 2012; Buhnemann et al. 2006; Daadi et al. 2009). In these studies, grafted NSCs exhibited spontaneous postsynaptic currents mostly excitatory in their nature. However, whether the synaptic inputs observed in these studies were coming from host neurons or reflected synaptic connections established between grafted cells was not clear.

Further evidence on possible functional integration of grafted neural cells was obtained by Oki and co-workers (2012). Injection of retrograde tracer Fluorogold in the ipsilateral globus pallidus in the stroke-subjected mouse 9 weeks after they received intrastriatal transplantation of hNSCs revealed that some graft-derived neurons incorporated the tracer. This indicated that grafted neurons from the striatum extended their axons to the globus pallidus which is the natural projection area for the striatal neurons. Similarly, when the monkey ESC-derived NPCs were implanted in the ischemic striatum (Hayashi et al. 2006), local injection of Fluorogold in ipsilateral anterior thalamus and substantia nigra led to the labelling of grafted cells. All these histological data speaks in favour of neuronal integration of grafted NSCs in host brain.

Additional morphological data supporting functional integration of grafted neurons is high density of projections from transplanted NSCs in the striatum towards globus pallidus, corpus callosum, and substantia nigra (Polentes et al. 2012). In agreement, the study by Tornero et al. also demonstrated axons of cortically fated iPSC-lt-NESCs extending from the site of engraftment in the stroke-lesioned cortex over the corpus callosum towards contralateral cortex (Tornero et al. 2013). In follow-up study, Tornero and co-workers (2017) presented convincing data that transplanted human iPSC-derived neurons receive synaptic inputs from stroke-injured host brain. This study took advantage of rabies virus-based trans-synaptic tracing technology and demonstrated that grafted cortically fated neurons in stroke-lesioned cortex receive direct synaptic inputs from neurons anatomically properly located in different structures of host brain. Immuno-electron microscopy confirmed that neurons of host brain establish excitatory axo-dendritic synaptic contacts with the grafted human cortical neurons (Tornero et al. 2017). Speaking in favour of functional integration of grafted neurons, mechanical stimulation of the nose and forepaw in live animals altered electrical activity in the grafted neurons. Moreover, electrophysiological recordings of GFP-labelled grafted neurons and photostimulation of virally transfected, channelrhodopsin-2-expressing thalamo-cortical axons in acute brain slices clearly demonstrated that exons of host thalamic neurons form functional synapses on the grafted human cortical neurons derived from iPSCs (Tornero et al. 2017). Despite of all these encouraging data, it still remains to demonstrate that restoration of host neuronal network takes place, and it contributes to restoration of stroke-impaired functions.

11.6 Clinical Perspective

Although significant headways have been made over the past decade using animal models of ischemic stroke and several clinical trials have also been initiated, when considering the clinical application of the hNSCs-based therapy for ischemic stroke, it becomes apparent that current experimental development is still far from being clinically viable. Several issues that will determine the future success of hNSCs-based therapy remain to be solved.

It should be emphasized that some degree of spontaneous recovery occurs in virtually all patients surviving stroke but varies from modest improvement to almost complete restoration. This reflects the extremely complex nature of post-stroke recovery, which involves plastic changes in surviving neurons and neurons on the contralateral side, redistribution of brain representation, release of growth factors and anti-inflammatory factors from immune cells, synaptogenesis and changes in synaptic strength and changes in dendritic arborization and spines, as well as generation of new neurons and glial and endothelial cells from endogenous stem cells (Moskowitz et al. 2010). Stem cell-based treatments, which affect these processes and the environment of the ischemic tissue and penumbra area, could potentially improve functional outcome after stroke.

Availability of graft material has always been considered a significant issue. However, with the development of adult cell reprogramming techniques (iPSC, see above), this issue could be overcome in the near future. More research is needed in establishing safe and reliable protocols for generating cell lines from iPSCs.

Major safety issue for hNSCs-based therapy that persists is the high percentage of proliferative activity within the grafts. However, the new developments in hNSCs culturing techniques and cell programming towards specific neuronal phenotypes will likely reduce the risk of tumorigenesis after transplantation. On the downside, such approach will inevitably create the requirement of narrowly specialized hNSC lines. Because the brain is so heterogeneous in neuronal composition in different brain structures, cell lines would need to be selected depending on the location of ischemic injury. Additionally, the more differentiated or committed the graft material is, the higher the risk of reducing the graft survival.

Another clinically relevant issue is the timely determination whether a patient needs therapeutic intervention or the impaired function could be recovered spontaneously in time. As preclinical research shows, generally, early transplantation is more beneficial in regard to both graft survival and functional recovery. However, it would be difficult to argue for hNSC therapy initiation in specific cases, without having reliable and highly predictive protocols to determine the patient prognosis and therefore the need for therapy.

11.7 Highlight Summary Table of Key Factors for Successful Application of Stem Cell Therapy for Stroke (Table 11.2)

Table 11.2 Key factors for successful application of stem cell therapy for stroke

Route of transplantation	Concept
Intraparenchymal	Injecting cell suspension directly into the brain parenchyma near the injury site. The main advantage of this route of transplantation approach is the ability to precisely control the graft placement. The major drawback of using this delivery route is its invasiveness
Intraventricular	Injected in the ventricular system. Injected NSCs reach the stroke-lesioned site, but are less efficient for the recovery
Intravascular	The main advantage of intravascular delivery is the ability to introduce greater number of cells through the least invasive route. Potential accumulation of grafted cell in internal organs is a significant risk factor. The number of cells reaching the lesioned brain is also concern
Host-graft interaction	Microenvironment of the host tissue can influence the fate of grafted NSCs. Detrimental effect of immune reaction on the grafted NSC may significantly reduce the efficacy of transplantation.
Graft-induced functional recovery	
Trophic effects	Through the release of various growth and other factors, graft can reduce injury, dampen the inflammation, and promote neuronal regrowth and plasticity in the host tissue
Neuronal replacement	Neuronal differentiation of the graft into mature neurons and functional integration into the host neuronal circuits. The direct evidence of such integration in host neuronal circuitry and contribution to post-stroke recovery through restoration of damaged neuronal network is still lacking

References

Alunni A, Bally-Cuif L (2016) A comparative view of regenerative neurogenesis in vertebrates. Development 143:741–753

Anderson L, Burnstein RM, He X et al (2007) Gene expression changes in long term expanded human neural progenitor cells passaged by chopping lead to loss of neurogenic potential in vivo. Exp Neurol 204:512–524

Arvidsson A, Collin T, Kirik D et al (2002) Neuronal replacement from endogenous precursors in the adult brain after stroke. Nat Med 8:963–970

Braeuninger S, Kleinschnitz C (2009) Rodent models of focal cerebral ischemia: procedural pitfalls and translational problems. Exp Transl Stroke Med 1:8

Buhnemann C, Scholz A, Bernreuther C et al (2006) Neuronal differentiation of transplanted embryonic stem cell-derived precursors in stroke lesions of adult rats. Brain 129:3238–3248

Butti E, Bacigaluppi M, Rossi S et al (2012) Subventricular zone neural progenitors protect striatal neurons from glutamatergic excitotoxicity. Brain 135:3320–3335

Carpenter MK, Inokuma MS, Denham J et al (2001) Enrichment of neurons and neural precursors from human embryonic stem cells. Exp Neurol 172:383–397

Chang DJ, Lee N, Park IH et al (2013) Therapeutic potential of human induced pluripotent stem cells in experimental stroke. Cell Transplant 22:1427–1440

Chau MJ, Deveau TC, Song M et al (2014) iPSC transplantation increases regeneration and functional recovery after ischemic stroke in neonatal rats. Stem Cells 32:3075–3087

Chen SJ, Chang CM, Tsai SK et al (2010) Functional improvement of focal cerebral ischemia injury by subdural transplantation of induced pluripotent stem cells with fibrin glue. Stem Cells Dev 19:1757–1767

Chu K, Kim M, Chae SH et al (2004) Distribution and in situ proliferation patterns of intravenously injected immortalized human neural stem-like cells in rats with focal cerebral ischemia. Neurosci Res 50:459–465

Chu K, Kim M, Jeong SW et al (2003) Human neural stem cells can migrate, differentiate, and integrate after intravenous transplantation in adult rats with transient forebrain ischemia. Neurosci Lett 343:129–133

Chung Y, Klimanskaya I, Becker S et al (2006) Embryonic and extraembryonic stem cell lines derived from single mouse blastomeres. Nature 439:216–219

Daadi MM, Davis AS, Arac A et al (2010) Human neural stem cell grafts modify microglial response and enhance axonal sprouting in neonatal hypoxic-ischemic brain injury. Stroke 41: 516–523

Daadi MM, Li Z, Arac A et al (2009) Molecular and magnetic resonance imaging of human embryonic stem cell-derived neural stem cell grafts in ischemic rat brain. Mol Ther 17: 1282–1291

Daadi MM, Maag AL, Steinberg GK (2008) Adherent self-renewable human embryonic stem cell-derived neural stem cell line: functional engraftment in experimental stroke model. PLoS One 3: e1644

Darsalia V, Allison SJ, Cusulin C et al (2011) Cell number and timing of transplantation determine survival of human neural stem cell grafts in stroke-damaged rat brain. J Cereb Blood Flow Metab 31:235–242

Darsalia V, Kallur T, Kokaia Z (2007) Survival, migration and neuronal differentiation of human fetal striatal and cortical neural stem cells grafted in stroke-damaged rat striatum. Eur J Neurosci 26:605–614

Dhara SK, Hasneen K, Machacek DW et al (2008) Human neural progenitor cells derived from embryonic stem cells in feeder-free cultures. Differentiation 76:454–464

Doeppner TR, Kaltwasser B, Teli MK et al (2015) Post-stroke transplantation of adult subventricular zone derived neural progenitor cells—a comprehensive analysis of cell delivery routes and their underlying mechanisms. Exp Neurol 273:45–56

Eckert A, Huang L, Gonzalez R et al (2015) Bystander effect fuels human induced pluripotent stem cell-derived neural stem cells to quickly attenuate early stage neurological deficits after stroke. Stem Cells Transl Med 4:841–851

Erdo F, Buhrle C, Blunk J et al (2003) Host-dependent tumorigenesis of embryonic stem cell transplantation in experimental stroke. J Cereb Blood Flow Metab 23:780–785

Gerrard L, Rodgers L, Cui W (2005) Differentiation of human embryonic stem cells to neural lineages in adherent culture by blocking bone morphogenetic protein signaling. Stem Cells 23:1234–1241

Guruswamy R, Elali A (2017) Complex roles of microglial cells in ischemic stroke pathobiology: new insights and future directions. Int J Mol Sci 18(3):E496

Hara K, Matsukawa N, Yasuhara T et al (2007) Transplantation of post-mitotic human neuroteratocarcinoma-overexpressing Nurr1 cells provides therapeutic benefits in experimental stroke: in vitro evidence of expedited neuronal differentiation and GDNF secretion. J Neurosci Res 85:1240–1251

Hayashi J, Takagi Y, Fukuda H et al (2006) Primate embryonic stem cell-derived neuronal progenitors transplanted into ischemic brain. J Cereb Blood Flow Metab 26:906–914

Hicks AU, Lappalainen RS, Narkilahti S et al (2009) Transplantation of human embryonic stem cell-derived neural precursor cells and enriched environment after cortical stroke in rats: cell survival and functional recovery. Eur J Neurosci 29:562–574

Jensen MB, Yan H, Krishnaney-Davison R et al (2013) Survival and differentiation of transplanted neural stem cells derived from human induced pluripotent stem cells in a rat stroke model. J Stroke Cerebrovasc Dis 22:304–308

Jeong SW, Chu K, Jung KH et al (2003) Human neural stem cell transplantation promotes functional recovery in rats with experimental intracerebral hemorrhage. Stroke 34:2258–2263

Jiang M, Lv L, Ji H et al (2011) Induction of pluripotent stem cells transplantation therapy for ischemic stroke. Mol Cell Biochem 354:67–75

Jiang P, Chen C, Wang R et al (2013) hESC-derived Olig2+ progenitors generate a subtype of astroglia with protective effects against ischaemic brain injury. Nat Commun 4:2196

Jin K, Sun Y, Xie L et al (2005) Comparison of ischemia-directed migration of neural precursor cells after intrastriatal, intraventricular, or intravenous transplantation in the rat. Neurobiol Dis 18:366–374

Jin K, Wang X, Xie L et al (2010) Transgenic ablation of doublecortin-expressing cells suppresses adult neurogenesis and worsens stroke outcome in mice. Proc Natl Acad Sci USA 107:7993–7998

Jin K, Wang X, Xie L et al (2006) Evidence for stroke-induced neurogenesis in the human brain. Proc Natl Acad Sci USA 103:13198–13202

Joo KM, Kang BG, Yeon JY et al (2013) Experimental and clinical factors influencing long-term stable in vitro expansion of multipotent neural cells from human adult temporal lobes. Exp Neurol 240:168–177

Kalladka D, Sinden J, Pollock K et al (2016) Human neural stem cells in patients with chronic ischaemic stroke (PISCES): a phase 1, first-in-man study. Lancet 388:787–796

Kallur T, Darsalia V, Lindvall O et al (2006) Human fetal cortical and striatal neural stem cells generate region-specific neurons in vitro and differentiate extensively to neurons after intrastriatal transplantation in neonatal rats. J Neurosci Res 84:1630–1644

Kawai H, Yamashita T, Ohta Y et al (2010) Tridermal tumorigenesis of induced pluripotent stem cells transplanted in ischemic brain. J Cereb Blood Flow Metab 30:1487–1493

Kempermann G (2015) Adult neurogenesis: an evolutionary perspective. Cold Spring Harb Perspect Biol 8:a018986

Kim DY, Park SH, Lee SU et al (2007) Effect of human embryonic stem cell-derived neuronal precursor cell transplantation into the cerebral infarct model of rat with exercise. Neurosci Res 58:164–175

Kim JY, Park J, Chang JY et al (2016) Inflammation after ischemic stroke: the role of leukocytes and glial cells. Exp Neurobiol 25:241–251

Koch P, Opitz T, Steinbeck JA et al (2009) A rosette-type, self-renewing human ES cell-derived neural stem cell with potential for in vitro instruction and synaptic integration. Proc Natl Acad Sci USA 106:3225–3230

Kondziolka D, Steinberg GK, Wechsler L et al (2005) Neurotransplantation for patients with subcortical motor stroke: a phase 2 randomized trial. J Neurosurg 103:38–45
Kuhn HG, Eisch AJ, Spalding K et al (2016) Detection and phenotypic characterization of adult neurogenesis. Cold Spring Harb Perspect Biol 8:a025981
Lam J, Lowry WE, Carmichael ST et al (2014) Delivery of iPS-NPCs to the stroke cavity within a hyaluronic acid matrix promotes the differentiation of transplanted cells. Adv Funct Mater 24: 7053–7062
Lappalainen RS, Narkilahti S, Huhtala T et al (2008) The SPECT imaging shows the accumulation of neural progenitor cells into internal organs after systemic administration in middle cerebral artery occlusion rats. Neurosci Lett 440:246–250
Lee ST, Chu K, Jung KH et al (2008) Anti-inflammatory mechanism of intravascular neural stem cell transplantation in haemorrhagic stroke. Brain 131:616–629
Li J, Zhu H, Liu Y et al (2010) Human mesenchymal stem cell transplantation protects against cerebral ischemic injury and upregulates interleukin-10 expression in *Macacafascicularis*. Brain Res 1334:65–72
Liu SP, Fu RH, Wu DC et al (2014) Mouse-induced pluripotent stem cells generated under hypoxic conditions in the absence of viral infection and oncogenic factors and used for ischemic stroke therapy. Stem Cells Dev 23:421–433
Lundberg C, Winkler C, Whittemore SR et al (1996) Conditionally immortalized neural progenitor cells grafted to the striatum exhibit site-specific neuronal differentiation and establish connections with the host globus pallidus. Neurobiol Dis 3:33–50
Ma S, Zhong D, Chen H et al (2013) The immunomodulatory effect of bone marrow stromal cells (BMSCs) on interleukin (IL)-23/IL-17-mediated ischemic stroke in mice. J Neuroimmunol 257:28–35
Majid A, He YY, Gidday JM et al (2000) Differences in vulnerability to permanent focal cerebral ischemia among 3 common mouse strains. Stroke 31:2707–2714
Minger SL, Ekonomou A, Carta EM et al (2007) Endogenous neurogenesis in the human brain following cerebral infarction. Regen Med 2:69–74
Mohamad O, Drury-Stewart D, Song M et al (2013) Vector-free and transgene-free human iPS cells differentiate into functional neurons and enhance functional recovery after ischemic stroke in mice. PLoS One 8:e64160
Moskowitz MA, Lo EH, Iadecola C (2010) The science of stroke: mechanisms in search of treatments. Neuron 67:181–198
Murrell W, Palmero E, Bianco J et al (2013) Expansion of multipotent stem cells from the adult human brain. PLoS One 8:e71334
Nakayama D, Matsuyama T, Ishibashi-Ueda H et al (2010) Injury-induced neural stem/progenitor cells in post-stroke human cerebral cortex. Eur J Neurosci 31:90–98
Nelson PT, Kondziolka D, Wechsler L et al (2002) Clonal human (hNT) neuron grafts for stroke therapy: neuropathology in a patient 27 months after implantation. Am J Pathol 160:1201–1206
Oki K, Tatarishvili J, Wood J et al (2012) Human-induced pluripotent stem cells form functional neurons and improve recovery after grafting in stroke-damaged brain. Stem Cells 30:1120–1133
Onifer SM, Whittemore SR, Holets VR (1993) Variable morphological differentiation of a raphe-derived neuronal cell line following transplantation into the adult rat CNS. Exp Neurol 122: 130–142
Parent JM, Vexler ZS, Gong C et al (2002) Rat forebrain neurogenesis and striatal neuron replacement after focal stroke. Ann Neurol 52:802–813
Pera MF, Andrade J, Houssami S et al (2004) Regulation of human embryonic stem cell differentiation by BMP-2 and its antagonist noggin. J Cell Sci 117:1269–1280
Polentes J, Jendelova P, Cailleret M et al (2012) Human induced pluripotent stem cells improve stroke outcome and reduce secondary degeneration in the recipient brain. Cell Transplant 21: 2587–2602
Pomper MG, Hammond H, Yu X et al (2009) Serial imaging of human embryonic stem-cell engraftment and teratoma formation in live mouse models. Cell Res 19:370–379

Preynat-Seauve O, De Rham C, Tirefort D et al (2009) Neural progenitors derived from human embryonic stem cells are targeted by allogeneic T and natural killer cells. J Cell Mol Med 13:3556–3569

Qin J, Ma X, Qi H et al (2015) Transplantation of induced pluripotent stem cells alleviates cerebral inflammation and neural damage in hemorrhagic stroke. PLoS One 10:e0129881

Ryu S, Lee SH, Kim SU et al (2016) Human neural stem cells promote proliferation of endogenous neural stem cells and enhance angiogenesis in ischemic rat brain. Neural Regen Res 11:298–304

Saporta S, Borlongan CV, Sanberg PR (1999) Neural transplantation of human neuroteratocarcinoma (hNT) neurons into ischemic rats: a quantitative dose-response analysis of cell survival and behavioral recovery. Neuroscience 91:519–525

Sauter A, Rudin M (1995) Strain-dependent drug effects in rat middle cerebral artery occlusion model of stroke. J Pharmacol Exp Ther 274:1008–1013

Schuldiner M, Eiges R, Eden A et al (2001) Induced neuronal differentiation of human embryonic stem cells. Brain Res 913:201–205

Schwarting S, Litwak S, Hao W et al (2008) Hematopoietic stem cells reduce postischemic inflammation and ameliorate ischemic brain injury. Stroke 39:2867–2875

Seminatore C, Polentes J, Ellman D et al (2010) The postischemic environment differentially impacts teratoma or tumor formation after transplantation of human embryonic stem cell-derived neural progenitors. Stroke 41:153–159

Shichinohe H, Kuroda S, Yano S et al (2006) Improved expression of gamma-aminobutyric acid receptor in mice with cerebral infarct and transplanted bone marrow stromal cells: an autoradiographic and histologic analysis. J Nucl Med 47:486–491

Shihabuddin LS, Holets VR, Whittemore SR (1996) Selective hippocampal lesions differentially affect the phenotypic fate of transplanted neuronal precursor cells. Exp Neurol 139:61–72

Skardelly M, Gaber K, Burdack S et al (2011) Long-term benefit of human fetal neuronal progenitor cell transplantation in a clinically adapted model after traumatic brain injury. J Neurotrauma 28:401–414

Song M, Kim YJ, Kim YH et al (2010) Using a neodymium magnet to target delivery of ferumoxide-labeled human neural stem cells in a rat model of focal cerebral ischemia. Hum Gene Ther 21:603–610

Steinbeck JA, Koch P, Derouiche A et al (2012) Human embryonic stem cell-derived neurons establish region-specific, long-range projections in the adult brain. Cell Mol Life Sci 69:461–470

Stilley CS, Ryan CM, Kondziolka D et al (2004) Changes in cognitive function after neuronal cell transplantation for basal ganglia stroke. Neurology 63:1320–1322

Tailor J, Kittappa R, Leto K et al (2013) Stem cells expanded from the human embryonic hindbrain stably retain regional specification and high neurogenic potency. J Neurosci 33:12407–12422

Takahashi K, Yamanaka S (2006) Induction of pluripotent stem cells from mouse embryonic and adult fibroblast cultures by defined factors. Cell 126:663–676

Tatarishvili J, Oki K, Monni E et al (2014) Human induced pluripotent stem cells improve recovery in stroke-injured aged rats. Restor Neurol Neurosci 32:547–558

Thomson JA, Itskovitz-Eldor J, Shapiro SS et al (1998) Embryonic stem cell lines derived from human blastocysts. Science 282:1145–1147

Tonchev AB, Yamashima T, Zhao L et al (2003) Differential proliferative response in the post-ischemic hippocampus, temporal cortex, and olfactory bulb of young adult macaque monkeys. Glia 42:209–224

Tornero D, Tsupykov O, Granmo M et al (2017) Synaptic inputs from stroke-injured brain to grafted human stem cell-derived neurons activated by sensory stimuli. Brain 140:692–706

Tornero D, Wattananit S, Gronning Madsen M et al (2013) Human induced pluripotent stem cell-derived cortical neurons integrate in stroke-injured cortex and improve functional recovery. Brain 136:3561–3577

Wang X, Mao X, Xie L et al (2012) Conditional depletion of neurogenesis inhibits long-term recovery after experimental stroke in mice. PLoS One 7:e38932

Yuan T, Liao W, Feng NH et al (2013) Human induced pluripotent stem cell-derived neural stem cells survive, migrate, differentiate, and improve neurologic function in a rat model of middle cerebral artery occlusion. Stem Cell Res Ther 4:73

Chapter 12
Modeling Complex Neurological Diseases with Stem Cells: A Study of Bipolar Disorder

Cameron D. Pernia, Neal H. Nathan, Brian T. D. Tobe, Alicia M. Winquist, Richard L. Sidman, Yoshio Goshima, and Evan Y. Snyder

Abstract The pathogenesis of bipolar disorder (BPD) is unknown. Using human-induced pluripotent stem cells (hiPSCs) to unravel pathological mechanisms in polygenic diseases is challenging, with few successful studies to date. However, hiPSCs from BPD patients responsive to lithium have offered unique opportunities to discern lithium's mechanism of action and hence gain insight into BPD pathology. By profiling the proteomics of BPD–hiPSC-derived neurons, we found that lithium alters the phosphorylation state of collapsin response mediator protein-2 (CRMP2). The "set point" for the ratio of pCRMP2:CRMP2 is elevated uniquely in hiPSC-derived neurons from lithium responsive (Li-R) BPD patients, but not other psychiatric and neurological disorders. Utilizing neurons differentiated from human patient stem cells as an in vitro platform, we were able to elucidate the mechanism driving

C. D. Pernia · N. H. Nathan · A. M. Winquist
Sanford Burnham Prebys Medical Discovery Institute, La Jolla, CA, USA

Sanford Consortium for Regenerative Medicine, La Jolla, CA, USA

B. T. D. Tobe
Sanford Burnham Prebys Medical Discovery Institute, La Jolla, CA, USA

Sanford Consortium for Regenerative Medicine, La Jolla, CA, USA

Department of Psychiatry, Veterans Administration Medical Center, La Jolla, CA, USA

R. L. Sidman
Department of Neurology, Beth Israel-Deaconess Medical Center, Boston, MA, USA

Y. Goshima
Department of Molecular Pharmacology & Neurobiology, Yokohama City University Graduate School of Medicine, Yokohama, Japan

E. Y. Snyder (✉)
Sanford Burnham Prebys Medical Discovery Institute, La Jolla, CA, USA

Sanford Consortium for Regenerative Medicine, La Jolla, CA, USA

Department of Pediatrics, University of California, San Diego, La Jolla, CA, USA
e-mail: esnyder@sbpdiscovery.org

L. Buzanska (ed.), *Human Neural Stem Cells*, Results and Problems in Cell Differentiation 66, https://doi.org/10.1007/978-3-319-93485-3_12

the pathogenesis and pathophysiology of lithium-responsive BPD, heretofore unknown. Importantly, the findings in culture were validated in human postmortem material as well as in animal models of BPD behavior. These data suggest that the "lithium response pathway" in BPD governs CRMP2's phosphorylation, which regulates cytoskeletal organization, particularly in dendritic spines, leading to modulated neural networks that may underlie Li-R BPD pathogenesis. This chapter reviews the methodology of leveraging a functional agent, lithium, to identify unknown pathophysiological pathways with hiPSCs and how to translate this disease modeling approach to other neurological disorders.

12.1 Introduction

Biomedical research and scientific knowledge of the brain has long lagged behind our understanding of the rest of human biology. This has caused the advances for treating neurological diseases to be behind those of other clinical fields such as oncology. The discrepancies between the limited clinical therapies available for diseases of the brain and the rest of medicine center around the century's long obstacle of living human central nervous system (CNS) tissue being unattainable for study across many patients and control conditions. Thus, for a long time, animal models were the only tractable biological source for scientists to investigate the brain. Although many therapeutic breakthroughs have been made based on animal models, there are clear limitations to the level of clinical insight that animal models can provide. The human brain is more structurally complex than that of any model organism, and much of the phenomena inherent to human neuropathology are either not present or observable in lower order organisms (Hofman 1985, 1988). The gap between biomedical research of the brain and other organs was always going to remain until the technology was developed to study the human CNS ethically under conditions where cells could be systematically modified and rigorously scrutinized.

The advent of induced pluripotent stem cells (iPSCs) has provided a great tool for neuroscience research (Takahashi et al. 2007). With the ability to reprogram very accessible somatic cells (e.g., skin, blood, hair follicles) into pluripotent cells (iPSCs) that can give rise to otherwise inaccessible cells (e.g., cerebral neurons and glia), large studies from many prospectively chosen living patients and their living affected and unaffected relatives, as well as numerous control patients, could be completed and biological complexity could be revealed (Bennett Jr et al. 2016; Byers et al. 2011; Cao et al. 2016; Chin et al. 2016; Codazzi et al. 2016). Now the opportunity exists for scientists to have an inexhaustible reservoir of theoretically all cell types to interrogate for epistemic and therapeutic gain. Many neurological diseases have since been modeled or studied with iPSCs, with some studies helping in the identification of novel therapeutics (Fig. 12.1) (Desbordes and Studer 2013; Egawa et al. 2012; Kaufmann et al. 2015; Yahata et al. 2011).

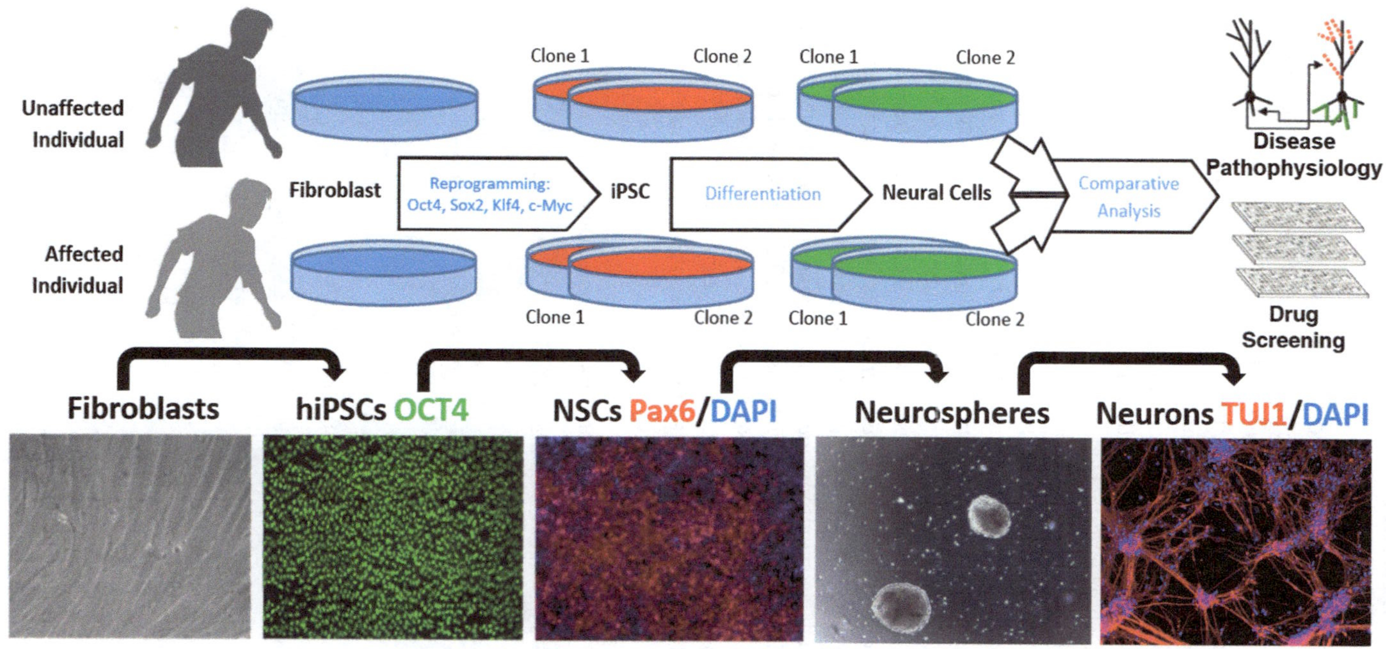

Fig. 12.1 Example workflow for stem cell-based study of a genetic neurological disease wherein human somatic cells (fibroblasts) are reprogrammed into hiPSCs with Yamanaka Transcription Factors and then sequentially differentiated into neural stem cells and then neurons to enable disease modeling and therapeutic insight

12.2 Bipolar Disorder and Lithium

Initially, iPSCs demonstrated the most promise with monogenic diseases, maladies that are caused by a single genetic mutation or alteration that is heritable across generations (Chung et al. 2016; Muotri et al. 2010). While early studies showed great potential for the use of iPSCs in translational medicine, only a small portion of neurological conditions are monogenic in nature. For iPSC technology to be a truly significant advance for biomedical research, iPSCs would have to be capable of modeling complex polygenic disorders for which the molecular pathology of the disease is unknown or multifaceted.

Recently, published work by Tobe and colleagues provides a great example of the power of iPSCs to illuminate an enigmatic neurological disorder (Tobe et al. 2017). Utilizing an iPSC approach to generate neural cell cultures, Tobe and company were able to identify, for the first time, the molecular pathological mechanism of bipolar disorder (BPD). Bipolar disorder is one of the most common psychiatric disorders, affecting 2.6% of adults in the Western world. The hallmark behavior associated with bipolar disorder is oscillation between depressive and manic moods or episodes (American Psychiatric Association 2013). Mania can be especially dangerous for bipolar individuals as symptoms include such cognitive impairments as delusional thoughts, impulsivity, and even psychosis (Fig. 12.2). Bipolar disorder is also the most lethal of all the psychiatric disorders, due to increased suicide rates; a bipolar disorder diagnosis lowers life expectancy by 9 years. The World Health Organization estimates that bipolar disorder is the world's 7th leading cause of disability and

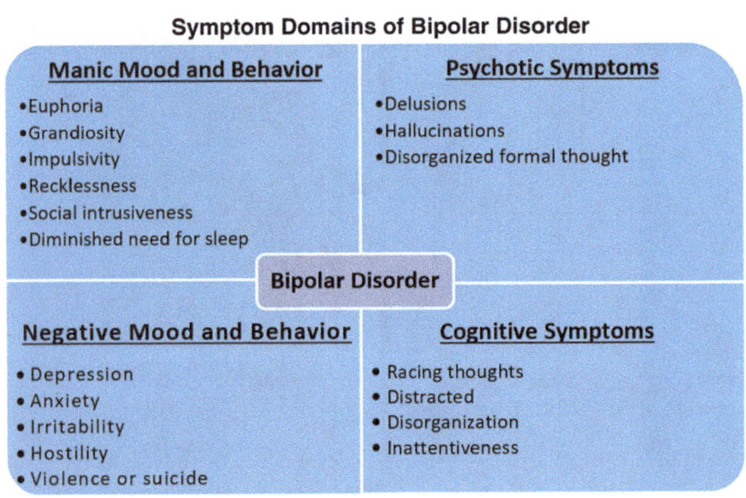

Fig. 12.2 Summary of the diverse symptoms with which bipolar disorder patients may present. Bipolar disorder is so named because of the switching between a "positive pole" of manic behavior and the "negative pole" of apathy or depression. The possibility of suicide also makes BPD a very serious condition (American Psychiatric Association 2013)

loss of productivity ("The World Health Report 1995—bridging the gaps" 1995). According to the Diagnostic and Statistics Manual created by the American Psychiatric Association, there are four major diagnostic categories of bipolar disorder, which range in severity of symptoms (American Psychiatric Association 2013). Bipolar Disorder 1 is considered the most severe form of the disease, as affected individuals experience multiple manic episodes in their lifetimes.

Currently, the most effective mood stabilizer used to treat bipolar disorder mania is a lithium salt. However, only approximately 30–40% of affected individuals respond therapeutically to lithium; however, those who do respond experience stabilization in their mood (Gershon et al. 2009). This clinical separation in response to lithium causes many psychiatrists to distinguish lithium responsive individuals (Li-R BPD) and lithium non-responsive individuals (Li-NR BPD) as having separate diseases despite presenting with similar behavioral phenotypes. Although lithium is efficacious, its safety index is very narrow, making it also a potentially dangerous—even lethal—drug. Lithium must be prescribed long term at high dosages, and chronic lithium usage may lead to many undesirable side effects, such as renal failure, endocrine disorders, and obesity. Thus, there is a significant unmet need for a therapy that is both safe for lifelong use and as effective as lithium in stabilizing manic behavior.

Interestingly, lithium is the third smallest element in the universe and can go anywhere in the body when orally administered. The reality that one of the simplest structures in the universe has profound restorative effects on the most complicated machine in existence, the human brain——and its most complex and treasured product, cognition—has piqued the interest of neuroscientists. For over a century, lithium has been utilized as a mood stabilizer, but only until recently has its putative mechanism of action for treating bipolar disorder been revealed. This is not to say that researchers have not been able to identify how lithium is biologically active; the issue was that lithium impacts a myriad of biochemical processes (in multiple organ systems, even in invertebrates), which made it exceptionally challenging to determine which effect was therapeutically relevant to human BPD (Fig. 12.3). Biological

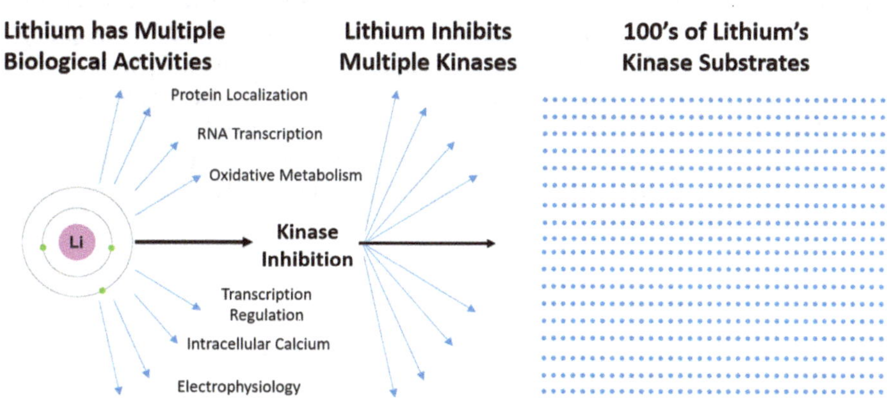

Fig. 12.3 The myriad of biological processes and downstream substrates influenced by lithium. The manifold processes impacted by lithium highlight the need to discover a drug that more selectively affects biological processes relevant to bipolar disorder without off-target influences

processes and substrates with which lithium is associated include RNA regulation, protein localization, kinase signaling cascades, membrane capacitance, enzyme activity, and chromatin structure (Brown and Tracy 2013; Klein and Melton 1996). Thus, finding lithium's mechanism of action for treating bipolar disorder had been like finding a needle in a haystack.

12.3 Modeling Bipolar Disorder with Stem Cells

Bipolar disorder has a high level of heritability. When surveyed in monozygotic discordant twin studies, bipolar disorder had a 70–80% inheritance rate, uncommonly high for a polygenic disorder (Althoff et al. 2005; Smoller and Finn 2003). Nevertheless, genome-wide association studies (GWAS) over a decade have been unsuccessful in determining a single gene responsible for the disorder. The Tobe team, therefore decided to look at a higher level of gene function—at the gene products and their regulation, i.e., proteomics. In our study, skin fibroblasts or leukocytes were isolated from unaffected and bipolar disorder individuals and were then reprogrammed with episomes containing the Yamanaka factors to convert the cells into iPSCs. After generating iPSC lines from the different genetic backgrounds, the iPSCs were differentiated into mature cortical interneuron cultures. The generated neurons were MAP2, CUX1, and VGLUT positive when immunocytochemistry was performed (Figs. 12.4 and 12.5).

With iPSC-derived human neural cultures containing the genetic basis of bipolar disorder, we were the first research group to be able to study the lithium response pathway in bipolar disorder-affected neurons in vitro. To do so, a two-dimensional gel electrophoresis (2-D DIGE) was performed, allowing the proteomic profiles of lithium treated and untreated bipolar disorder neural cultures to be juxtaposed (Fig. 12.5). Valuably, the 2-D DIGE assay is sensitive enough to identify not only total protein level differences between groups, but also differences in levels of a protein's post-translational modifications (i.e., phosphorylation) between groups. Follow-up of the 2-D DIGE with mass spectrometry allowed us to identify, for the first time, which proteins and which of their post-translational modifications (PTMs) are influenced by the lithium response pathway in bipolar disorder cortical cultures. The hypothesis being, the same proteins we identify as impacted in the lithium response pathway should be some of the same proteins involved in the molecular pathology of bipolar disorder.

12.3.1 Collapsin Response Mediator Protein-2

To analyze the list of proteins identified in the 2-D DIGE, a bioinformatics evaluation was performed to determine which protein is most central to the lithium

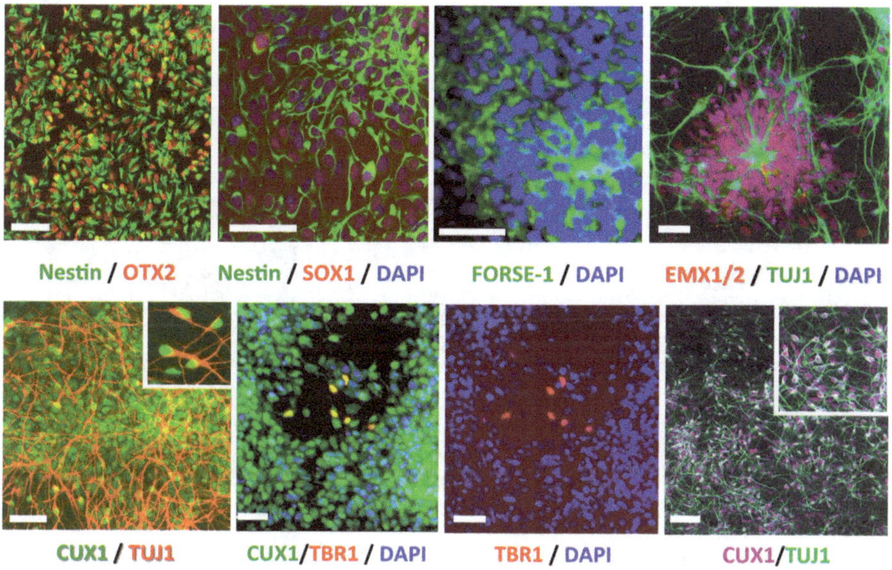

Fig. 12.4 Immunofluorescent images of the stepwise generation of hiPSC BPD-derived neurons with validation markers. Although initially devised for human embryonic stem cells (hESCs), the neural induction protocol is equally applicable to hiPSCs with identical yields. Nearly all cells first express the early neural markers OTX2, SOX1, and Nestin in monolayer. Subsequent expression of Forse1, EMX1/2, and CUX1 supports their dorsal forebrain cortical (upper layers) neuronal identity, particularly when co-expressed with the developing neuron marker Tuj1. Only a few cells are TBR1+, a marker of lower cortical layer neurons. All cells in a given field are identified by a DAPI nuclear stain (blue) (Tobe et al. 2017) (photomicrographs by Ilyas Singec)

response pathway in bipolar disorder neurons. The non-biased approach determined collapsin response mediator protein-2 (CRMP2) as the keystone protein (Fig. 12.6). CRMP2 was identified by Yoshio Goshima in 1995 as integral to growth cone collapse and axonal guidance (Goshima 1997; Goshima et al. 1995). CRMP2 has been associated with such key neurodevelopmental functions as neurite outgrowth and retraction, microtubule dynamics, actin assembly, calcium channel regulation, neurotransmitter release, and kinesin-dependent transport in neurons (Khanna et al. 2012). CRMP2 is predominantly found in the CNS and has multiple isoforms (72 kDa and 62 kDa), with the higher weight isoform only found in neurons. Moreover, CRMP2 is regulated post-translationally via kinases, phosphatases, and proteases (Cole et al. 2006; Uchida et al. 2005; Wilson et al. 2012; Zhu et al. 2010).

CRMP2's canonical biochemistry was discovered by Goshima in 2005, and its activity is determined by its ability to bind to tubulin. When CRMP2 is able to bind to tubulin, CRMP2 stabilizes tubulin's GTPase activity, allowing tubulin polymerization and subsequent axonal outgrowth (Uchida et al. 2005; Yamashita et al. 2012). This ability of CRMP2 to physically bind to tubulin is determined by the phosphorylation status of CRMP2. When CRMP2 is phosphorylated, CRMP2 is unable to bind to tubulin due to steric hindrance from the phosphate group, which

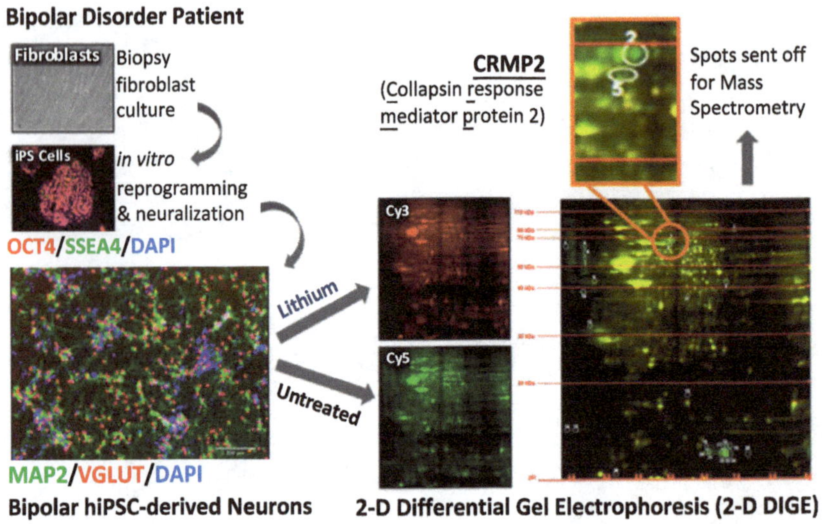

Fig. 12.5 2-Dimensional gel electrophoresis (2-D DIGE) to compare proteomic profiles of lithium treated and untreated bipolar disorder neural cultures derived from patient hiPSCs. A 2-D DIGE enables enhanced proteomic resolution as this method initially separates proteins by isoelectric point, which is determined for each protein by its constitutive amino acids, and then separates proteins orthogonally via sodium-dodecyl sulfate polyacrylamide gel electrophoresis (SDS-PAGE) to segregate proteins by molecular weight. This two-dimensional or two-parametric separation provides significant sensitivity. To detect the isolated proteins, before the 2-D DIGE, protein lysates are mixed with and labeled by cyanine dyes (CyDyes) such as Cy3 and Cy5 as depicted in red and green and an overlapped yellow. Protein spots of interest were then selected and extracted for mass spectrometry (Tobe et al. 2017)

makes tubulin filaments become unstable and unable to polymerize, leading to growth cone collapse and arrested axonal outgrowth. Once CRMP2 is unphosphorylated at key residues, it is active and free to bind to tubulin. CRMP2 is susceptible to external cellular cues such as Semaphorin-3A, which is a known growth cone collapse ligand. Semaphorin-3A binds to plasma membrane receptors such as neuroplexin, which initiates a kinase signaling cascade that converges on glycogen synthase kinase 3 beta (GSK3B) and cyclin-dependent kinase 5 (CDK5). An important aspect of CRMP2's biochemistry is that for it to be inactivated it has to be serially phosphorylated first by CDK5 at the priming residue serine-522, which only then allows GSK3B to truly inactivate CRMP2 via phosphorylation at residue threonine-514 (CRMP2-T514). GSK3B cannot inactivate CRMP2-T514 unless CRMP2 has been primed by CDK5 at serine-522 (Fig. 12.7).

Western blot analysis showed that the baseline ratio of phosphorylated CRMP2-T514 (pCRMP2) to total CRMP2 levels was significantly higher in iPSC-derived neurons from Li-R bipolar disorder cell lines compared to that of any other genetic background, including the Li-NR bipolar disorder cell line. To delineate the lithium response pathway to and from CRMP2, lithium was applied to neurons derived from

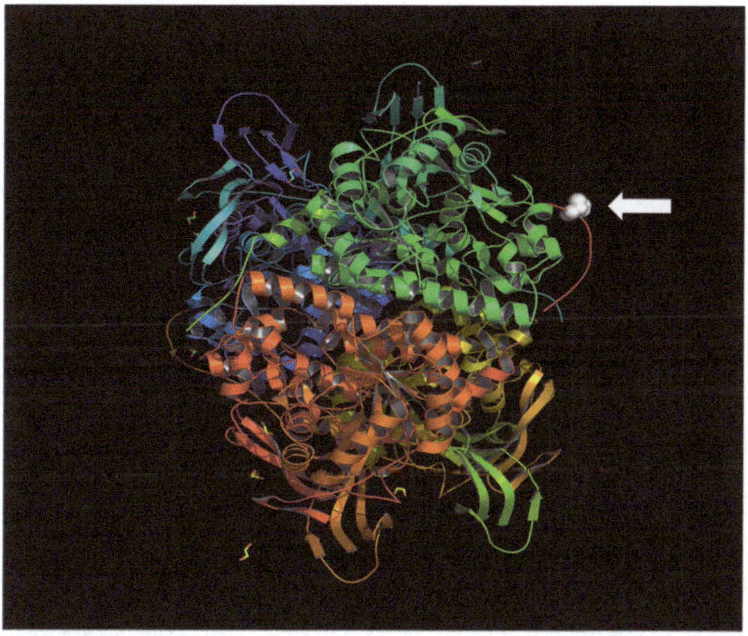

Fig. 12.6 Computed crystallization structure of CRMP2 from amino acid 13 to amino acid 516. The amino acid component which we demonstrate is important to CRMP2's, and lithium's, mechanism of action is Threonine-514, which is labeled white (Zheng et al. 2018)

Li-R iPSCs and compared with iPSC-derived neurons from Li-NR BPD and unaffected patients, as well as contrasted with other psychiatric and nonpsychiatric neurological conditions. Across all cell lines generated from the 19 different patients utilized in this study, lithium lowered the ratio of pCRMP2 to total CRMP2 levels while never perturbing total CRMP2 protein levels (Fig. 12.8a). Other, more specific, GSK3B inhibitors also lowered the ratio of pCRMP2:CRMP2, but not other psychotropic drugs such as haloperidol, risperidone, and valproic acid. Of note, clinically applicable lithium treatment lowered the level of pCRMP2:CRMP2 in Li-R bipolar disorder neurons compared to that of healthy neurons in basal conditions (Fig. 12.8b). These data suggested that the set point for the ratio of pCRMP2: CRMP2 may be abnormally and uniquely high in Li-R bipolar disorder individuals and, at least with respect to iPSC-based analysis, a molecular hallmark of the disease.

An interesting anecdote about the experimental design undertaken in this study was the choice of performing a 2-D DIGE, which appears apropos in hindsight. When analyzing total CRMP2 levels across the various iPSC lines treated with and without lithium, there is no difference in total CRMP2 levels. Therefore, if a different screening assay had been performed other than the 2-D DIGE, specifically a different test not sensitive to PTMs, CRMP2 never would have been identified as a key constituent in the lithium response pathway in bipolar disorder neurons. This reasoning could explain why our research team was the first to identify CRMP2's association with bipolar disorder pathophysiology.

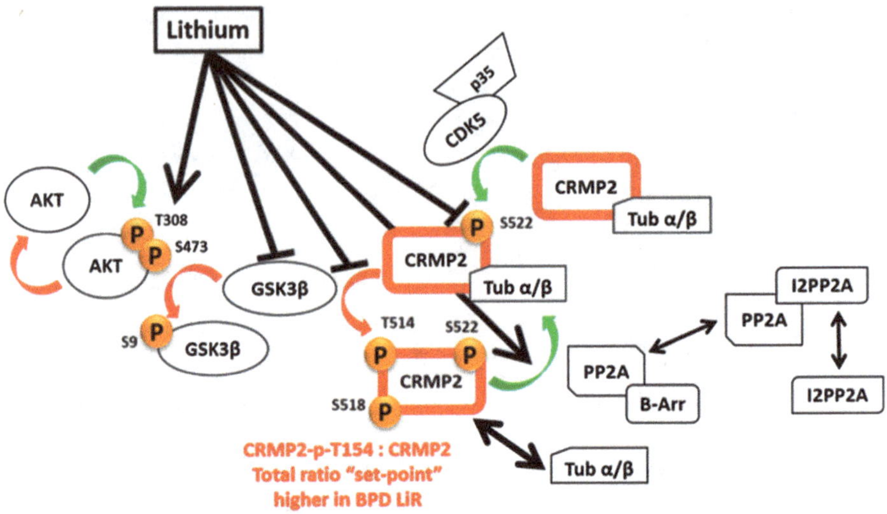

Fig. 12.7 Proposed model of the lithium response pathway in BPD, regulating CRMP2's phosphorylation state and, hence, its association with cytoskeletal elements. The proposed action of lithium in this context, as mediated by its presumptive direct and indirect targets, is to promote activation (non-phosphorylated) of CRMP2 at T514 to reduce the CRMP2-p-T514:CRMP2 ratio, the set point which we propose is excessively high in lithium responsive BPD. Lithium exerts this effect directly or indirectly through GSK3B, blocking phosphorylation of S522 on CRMP2, or by increasing phosphatase (PP2A) action (Tobe et al. 2017)

12.3.2 CRMP2 and Neuronal Morphology

While determining the location of the different phospho-forms of CRMP2 in neurons, we discovered that, while both active and inactive forms of CRMP2 were present throughout all parts of the neuron only the active non-phosphorylated form of CRMP2 was present in the neurite structures referred to as dendritic spines. Spines are the structures on dendrites where synapses between neurons are formed, making them essential for neurons to accomplish their signaling function (Rochefort and Konnerth 2012). Our hypothesis moving forward was that CRMP2 activity is involved in spine function via cytoskeletal regulation, as it is with axonal outgrowth. To test this, spine density was evaluated in the neurons of mice with their CRMP2 gene experimentally knocked-out (KO), and compared to controls; the neurons in mice lacking CRMP2 had decreased spine density. One of our collaborators, Glenn Konopaske, studied human bipolar disorder brains and found that they have increased pCRMP2 levels and, hence, abnormal ratios, which is analogous to what we saw in our in vitro model of bipolar disorder (Konopaske et al. 2014). Konopaske also found structural aberrations, specifically that bipolar disorder individuals have decreased dendritic spine density in their cortical neurons. However, in BPD patients on lithium, the ratios and the spines were normalized. These observations, taken together, led us to speculate about the fundamental cellular and molecular

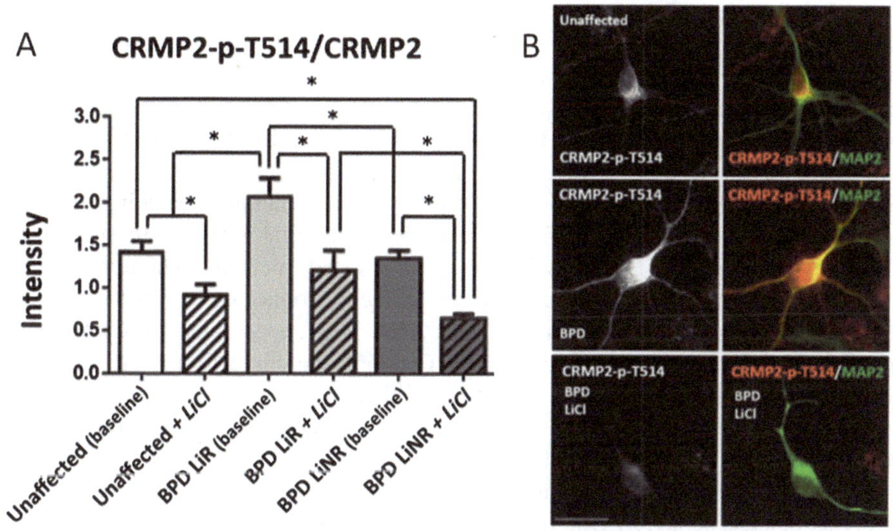

Fig. 12.8 (**a**) Graphical and statistical comparison of the ratio of pCRMP2-514 to CRMP2 between cell and treatment groups. These figures demonstrate that lithium treatment significantly lowers the ratio of pCRMP2-514 to CRMP2. Nineteen different verified patient samples (from varying healthy and diseased backgrounds), with one to three iPSC lines generated per patient, were utilized in this study. (**b**) Representative immunofluorescent staining of pCRMP2-514 (red) and the mature neuronal marker MAP2 (green) in hiPSC-derived unaffected neurons, and lithium responsive BPD patient neurons with or without lithium treatment (Tobe et al. 2017) (photomicrographs by Stephen Haggarty)

pathophysiological defect in BPD patients. Synapses are important for communication between neurons, and a decrease in the dendritic spine density across the brain or in a specific part of the brain could lead to aberrations in overall network signaling. Therefore, bipolar disorder and other neurological diseases may not be rooted in cell loss or individual cell dysfunction per se, but rather in the loss of interneuronal networking, functionality, and circuitry.

12.3.3 CRMP2 Controls Bipolar Disorder Behavior

All the evidence up to this point only *correlated* the association of CRMP2 with bipolar disorder and lithium's mechanism of action. However, we aimed to determine if there is a causative relationship. Therefore, the next question was whether CRMP2 is required for lithium-mediated behavioral changes in an accepted lithium-responsive bipolar disorder animal model. One such model is the methamphetamine-induced manic-hyper locomotion model in mice. What makes this behavioral assay the gold standard for analyzing bipolar disorder behavior is that mice pretreated with lithium before methamphetamine exposure do not enter a manic state. Therefore, we employed a transgenic mouse whose CRMP2 was incapable of being phosphorylated at precisely the motifs we had postulated were lithium's sites of action relevant to BPD. If we were correct, this mouse (called the CRMP2 "knock-in [KI]" mouse because the serine residue at 522 (Ser522) of CRMP2 was replaced with Ala) should behave as if it were on chronic lithium in this behavioral test; in other words, we were able to test whether CRMP2 activation (i.e., *preventing* CRMP2 inactivation via S522 phosphorylation, which it appeared methamphetamine was doing) recapitulated lithium treatment in this bipolar disorder behavioral assay (Yamashita et al. 2012). Astoundingly, the CRMP2-KI mice resisted methamphetamine-induced mania, suggesting that CRMP2's activity is *causally* connected to bipolar disorder behavior. Just as lithium decreases pCRMP2 levels, it appears that the CRMP2-KI mice (with their constitutively over-active CRMP2) are clinically representative of individuals who are receiving lithium, as lithium treatment leads to increased activation of CRMP2, or stated another way, decreased over-inactivation of CRMP2, hence restoring normal inactive:active CRMP2 ratios (Fig. 12.9). To the best of our knowledge, this is the first genetic intervention ever to rescue BPD-associated behavior. One of the negative effects with this mutation is related to neuronal structure. A neuron is polarized with a dendritic field and an axonal field, and it is important for neurons to not have their dendrites overlap with their axons to prevent self-synapses. However, these CRMP2-KI neurons have a partial loss of their dendritic field, or their polarity, enabling some overlap between a given neuron's own dendrites and axons. Furthermore, cortical neurons in CRMP2-KI mice show an increase in the number of primary dendrites, which may alter neuronal and network signaling profiles. Hence, a drug that regulates CRMP2 activity in a conditional or dose-dependent manner would be ideal, rather than a genetic intervention.

Changes in CRMP2 are known to alter other aspects of neurons that are critical for their signaling. The brains of adult mice with the *Crmp2* gene removed (CRMP2-KO) were characterized by a fivefold increase in the bifurcation of dendrites, creating increased dendritic branching points at the expense of main trunk dendrites (Nakamura et al. 2016). Moreover, the loss of dendritic spine density was also prominent as CRMP2-KO mice had a 25% reduction in spines (Fig. 12.10). Inactive pCRMP2-T514, which dissociates from tubulin, is not detectable in dendritic spines while unphosphorylated CRMP2 is, suggesting that when CRMP2 becomes phosphorylated (i.e., inactivated), it exits or is excluded from the spines. Lithium, as noted above, decreases the quotient of phosphorylated inactive CRMP2 to unphosphorylated active CRMP2 in vitro and in vivo (hence increasing the amount of unphosphorylated active CRMP2), inducing a 60% increase in dendritic spine area and a 36% increase in spine density. This lithium response is lost in CRMP2-KO neurons, indicating that CRMP2 is a necessary component of lithium's influence on spine formation. Importantly, independent studies that concurrently generated *Crmp2* knockout (CRMP2-KO) mice also found molecular, cellular, structural, and behavioral deficits, many of which are reminiscent of neural features and symptoms associated with psychiatric disorders.

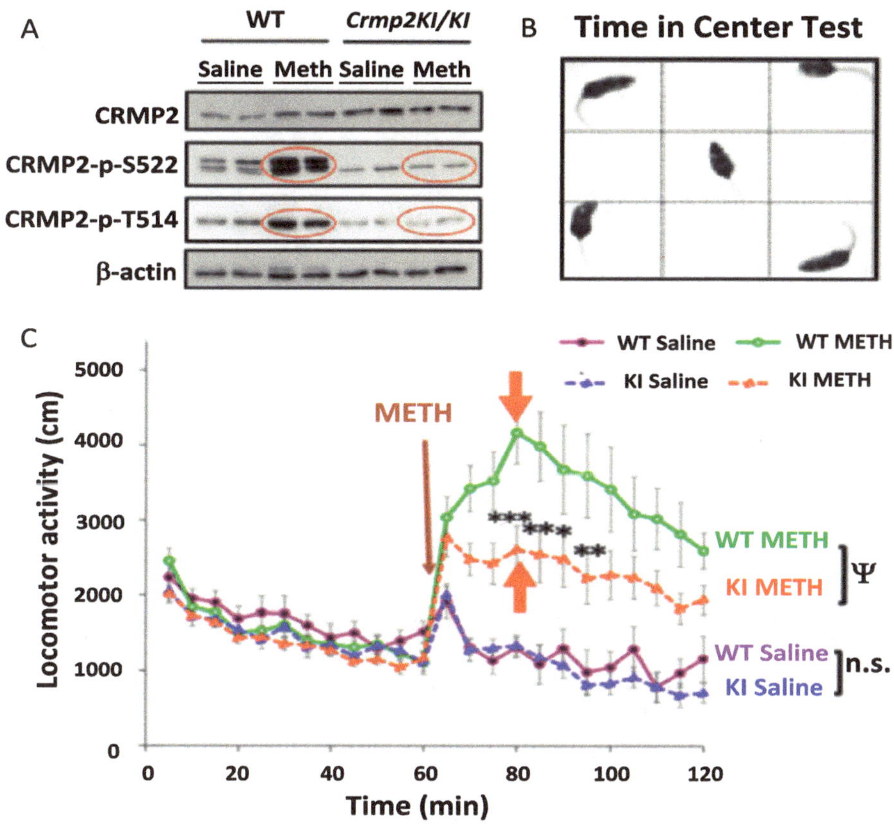

Fig. 12.9 (**a**) Western blot confirming that methamphetamine treatment increases (within 60 min) phosphorylation of CRMP2 at S522 and T514 in WT but not CRMP2-KI mouse brains. (**b**) Open-field test for quantifying non-manic behavior (time spent exploring the unprotected center) vs. manic behavior (little time in the center, more time "manically" circling the periphery). (**c**) Meth-treated CRMP2-KI mice ("KI", red dots) display less locomotor activity, in terms of distance traveled over time, compared with Meth-treated WT littermates. These comparisons of CRMP2-KI and WT mouse locomotor behavior and western blot results of the actual amount of phosphorylated CRMP2 after Meth administration confirm that the mutation preventing CRMP2 phosphorylation (emulating lithium's posited site of action) decreases BPD-like behaviors (Tobe et al. 2017)

12.4 Discussion

While it is exciting to demonstrate CRMP2's causative role in bipolar disorder, the biochemical underpinnings of how aberrant CRMP2 activity leads to macroscopic behavioral phenotypes still need to be ascertained. Previous findings looking into CRMP2's functions within neurons that are relevant to bipolar disorder pathology provided additional clues to answer this question. Work by Rajesh Khanna and others in 2009 looked into how CRMP2 interacts with calcium channels, signaling,

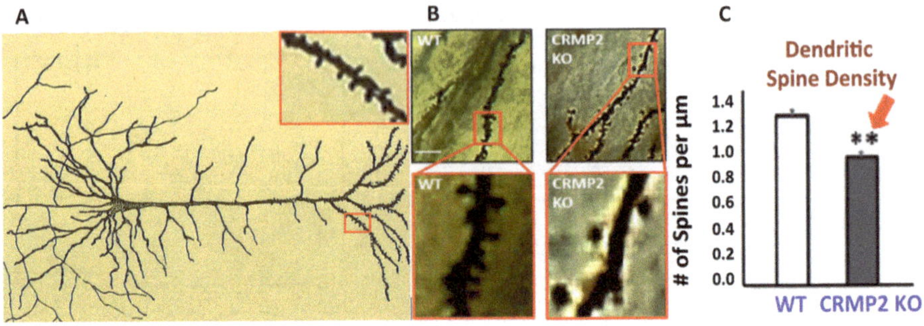

Fig. 12.10 Neuron morphology and CRMP2's impact on spines. (**a**) Cajal's drawing of a silver-stained cortical neuron; box in red shows synaptic spines on a neurite (Ramón y Cajal 1995). (**b**) Representative silver-stained mouse neurons, demonstrating CRMP2-KO neurons have fewer spines compared to WT. (**c**) Quantification of spine density; CRMP2-KO neurons have decreased spine density compared to that of control neurons (Tobe et al. 2017)

and kinetics within neurons, which is a relationship that we also evaluated in our study (Fig. 12.11) (Brittain et al. 2009, 2011). Their early work showed that CRMP2 and the calcium channel CaV2.2 physically associate with one another and that overexpressing CRMP2 in neurons causes changes in electrophysiology and neurotransmitter release (Brittain et al. 2012; Wilson et al. 2011). Interestingly, the changes in the electrophysiology and neurotransmitter signaling due to CRMP2 overexpression in these neurons could be negated by treatment with the specific CaV2.2 inhibitor conotoxin. Thus, when they measured glutamate release in these neurons after stimulation, neurons that had increased CRMP2 expression released more glutamate, but if those specific neurons were treated with conotoxin, synaptic glutamate levels came down to levels similar to those of the sham control. And so, what we think we have discovered within our study is that, although there are many proteins and activities that are impacted in the lithium response pathway in BPD neurons, CRMP2 is one of the keystone proteins that is important for leading to BPD-associated behavior and that it may be a top druggable target. However, what is interesting about this notion is that other specific GSK3B inhibitors have been brought to market with limited success for other diseases. Therefore, there is a chance that sites downstream of CRMP2 might be even more viable targets. Although we have demonstrated CRMP2's importance with regard to BPD, calcium channels, and overall neuronal structure, it is still insufficiently understood what happens downstream of modulating CRMP2 activity.

Nevertheless, these results provide the basis for tractable biomarker assay development utilizing CRMP2 to pCRMP2-514 ratios in reprogrammed patient-derived cells from convenient somatic cell sources such as skin, peripheral blood, urine, and hair, as a diagnostic aid for predicting drug responsiveness in BPD-modeling neurons in vitro. A qualitative, not just quantitative, distinction between Li-R and Li-NR BPD based on an abnormally high set point for an otherwise physiologic post-translational modification of a cytoskeletal regulator (uniquely in Li-R BPD) invites speculation that Li-NR BPD is actually a separate disease or, more likely,

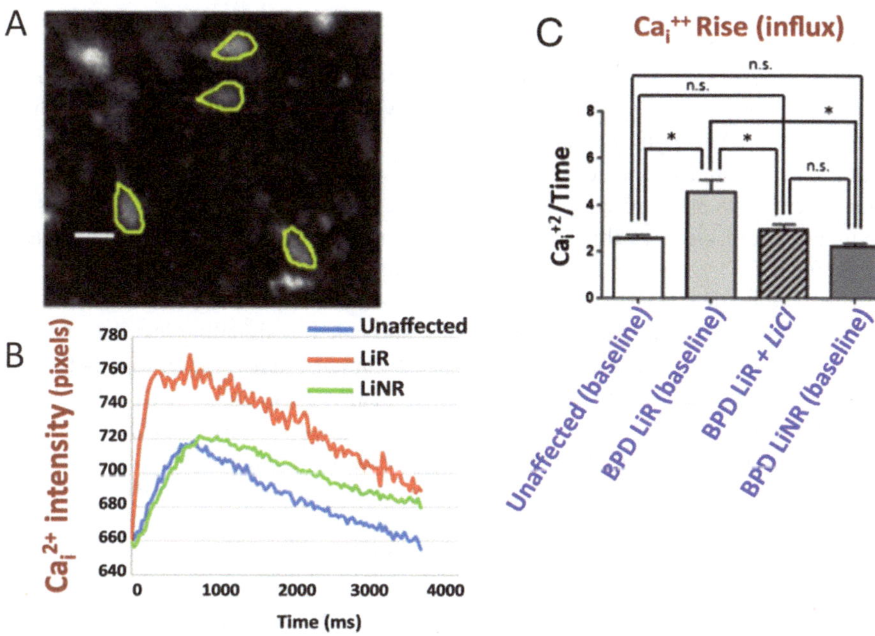

Fig. 12.11 (**a**) Image of active hiPSC-derived neurons undergoing calcium activity; yellow circles are of somas with high intracellular levels of calcium. (**b**) Representative calcium event kinetics between unaffected, Li-R BPD-, and Li-NR BPD-derived neurons. (**c**) Quantification of calcium kinetics; data show that Li-R BPD neurons uniquely have aberrant rates of calcium influx, which is normalized with lithium treatment (Tobe et al. 2017)

a group of diseases that "pheno-copy" BPD but are pathophysiologically unrelated to the lithium response pathway.

This leads to an overall larger theory we have regarding some neurological diseases in general and not just BPD. For a very long time, the field has looked at or studied neurological disorders from the perspective of an accumulation of loss or dysfuntion of individual neurons that simply needed to be replaced. However, our work with BPD has made us revise our view of what neural regenerative medicine should be trying to achieve; most neurological diseases may not be rooted in cell loss per se but rather in the loss of inter-neuronal networking. For example, in BPD there is no neuronal loss or even significant neuronal dysfunction intrinsically—the neurons we generate from BPD hiPSCs look basically normal; however, small aberrations in CRMP2 activity lead them to interconnect with each other in a dysfunctional manner. When these abnormalities occur over millions of neurons in either a specific brain region or across the brain, BPD symptomatology results. Perhaps other neurological disorders are really "networkopathies." Thus, the goal of neural regenerative medicine should not be restored neurogenesis or single neuron replacement but rather protecting or restoring neural networks and circuitry. Addressing problems in network connectivity, signaling, and dynamics may represent a more promising way to envision approaches to neurotherapeutics, particularly complex neurological disorders such as bipolar disorder.

This review highlights the ability of induced pluripotent stem cells to recapitulate neuropsychiatric disease and enable rigorous, well-controlled molecular and cellular interrogation. Through the use of easily acquired patient-derived somatic cells (e.g., skin fibroblasts and leukocytes), which were then reprogrammed to provide pluripotent cells and subsequently differentiated into neurons, we gained access to cellular pathophysiology that would otherwise have remained opaque. We were able to perform 2-D DIGE and bioinformatics analyses to identify the pathologically relevant molecule CRMP2, and characterize protein function ranging from its impact in single neurons at the electrophysiological and calcium signaling level all the way to the level of behaving animals and diseased human brains. Taken together, comprehensive mechanistic insight into bipolar disorder—a polygenic, not solely a monogenic, disease—resulted. Hence, we believe that this work helps to demonstrate that stem cell biology may be useful for modeling the molecular underpinnings of an even broader range of diseases than has heretofore been tried. Of course, once one can model the molecular pathways that underly a disorder, then the same models may serve as assays for discovering drugs that can modify those pathways in a therapeutic manner. For this use of stem cells in "regenerative medicine", it is not the stem cell per se that goes into the patient, but rather the drug discovered by the stem cell that goes in. We hope that stem cell-based studies like ours reinforce the power and promise of "disease in a dish" modeling empowered by thoughtful and rigorous stem cell biology.

References

Althoff RR, Faraone SV, Rettew DC, Morley CP, Hudziak JJ (2005) Family, twin, adoption, and molecular genetic studies of juvenile bipolar disorder. Bipolar Disord 7(6):598–609. https://doi.org/10.1111/j.1399-5618.2005.00268.x

American Psychiatric Association (2013) Bipolar and related disorders. In: Diagnostic and statistical manual of mental disorders. American Psychiatric Association, Washington, DC

Bennett JP Jr, O'Brien LC, Brohawn DG (2016) Pharmacological properties of microneurotrophin drugs developed for treatment of amyotrophic lateral sclerosis. Biochem Pharmacol 117:68–77. https://doi.org/10.1016/j.bcp.2016.08.001

Brittain JM, Piekarz AD, Wang Y, Kondo T, Cummins TR, Khanna R (2009) An atypical role for collapsin response mediator protein 2 (CRMP-2) in neurotransmitter release via interaction with presynaptic voltage-gated calcium channels. J Biol Chem 284(45):31375–31390. https://doi.org/10.1074/jbc.M109.009951

Brittain JM, Duarte DB, Wilson SM, Zhu W, Ballard C, Johnson PL et al (2011) Suppression of inflammatory and neuropathic pain by uncoupling CRMP-2 from the presynaptic Ca(2)(+) channel complex. Nat Med 17(7):822–829. https://doi.org/10.1038/nm.2345

Brittain JM, Wang Y, Eruvwetere O, Khanna R (2012) Cdk5-mediated phosphorylation of CRMP-2 enhances its interaction with CaV2.2. FEBS Lett 586(21):3813–3818. https://doi.org/10.1016/j.febslet.2012.09.022

Brown KM, Tracy DK (2013) Lithium: the pharmacodynamic actions of the amazing ion. Ther Adv Psychopharmacol 3(3):163–176. https://doi.org/10.1177/2045125312471963

Byers B, Cord B, Nguyen HN, Schule B, Fenno L, Lee PC et al (2011) SNCA triplication Parkinson's patient's iPSC-derived DA neurons accumulate alpha-synuclein and are susceptible to oxidative stress. PLoS One 6(11):e26159. https://doi.org/10.1371/journal.pone.0026159
Cao L, McDonnell A, Nitzsche A, Alexandrou A, Saintot PP, Loucif AJ et al (2016) Pharmacological reversal of a pain phenotype in iPSC-derived sensory neurons and patients with inherited erythromelalgia. Sci Transl Med 8(335):335ra356. https://doi.org/10.1126/scitranslmed.aad7653
Chin EW, Marcy G, Yoon SI, Ma D, Rosales FJ, Augustine GJ, Goh EL (2016) Choline ameliorates disease phenotypes in human iPSC models of rett syndrome. Neuromol Med 18(3):364–377. https://doi.org/10.1007/s12017-016-8421-y
Chung SY, Kishinevsky S, Mazzulli JR, Graziotto J, Mrejeru A, Mosharov EV et al (2016) Parkin and PINK1 patient iPSC-derived midbrain dopamine neurons exhibit mitochondrial dysfunction and alpha-synuclein accumulation. Stem Cell Rep 7(4):664–677. https://doi.org/10.1016/j.stemcr.2016.08.012
Codazzi F, Hu A, Rai M, Salerno Scarzella F, Mangiameli E, Pelizzoni I et al (2016) Friedreich ataxia induced pluripotent stem cell-derived neurons show a cellular phenotype that is corrected by a benzamide HDAC inhibitor. Hum Mol Genet 25(22):4847–4855. https://doi.org/10.1093/hmg/ddw308
Cole AR, Causeret F, Yadirgi G, Hastie CJ, McLauchlan H, McManus EJ et al (2006) Distinct priming kinases contribute to differential regulation of collapsin response mediator proteins by glycogen synthase kinase-3 in vivo. J Biol Chem 281(24):16591–16598. https://doi.org/10.1074/jbc.M513344200
Desbordes SC, Studer L (2013) Adapting human pluripotent stem cells to high-throughput and high-content screening. Nat Protoc 8(1):111–130. https://doi.org/10.1038/nprot.2012.139
Egawa N, Kitaoka S, Tsukita K, Naitoh M, Takahashi K, Yamamoto T et al (2012) Drug screening for ALS using patient-specific induced pluripotent stem cells. Sci Transl Med 4(145):145ra104. https://doi.org/10.1126/scitranslmed.3004052
Gershon S, Chengappa KN, Malhi GS (2009) Lithium specificity in bipolar illness: a classic agent for the classic disorder. Bipolar Disord 11(Suppl 2):34–44. https://doi.org/10.1111/j.1399-5618.2009.00709.x
Goshima Y (1997) Regulation of growth cone motility by the repulsive factor collapsin/semaphorin: a role of collapsin response mediator protein (CRMP). Seikagaku 69(12):1389–1392 http://www.ncbi.nlm.nih.gov/pubmed/9483985
Goshima Y, Nakamura F, Strittmatter P, Strittmatter SM (1995) Collapsin-induced growth cone collapse mediated by an intracellular protein related to UNC-33. Nature 376(6540):509–514. https://doi.org/10.1038/376509a0
Hofman MA (1985) Size and shape of the cerebral cortex in mammals. I. The cortical surface. Brain Behav Evol 27(1):28–40 http://www.ncbi.nlm.nih.gov/pubmed/3836731
Hofman MA (1988) Size and shape of the cerebral cortex in mammals. II. The cortical volume. Brain Behav Evol 32(1):17–26 http://www.ncbi.nlm.nih.gov/pubmed/3056571
Kaufmann M, Schuffenhauer A, Fruh I, Klein J, Thiemeyer A, Rigo P et al (2015) High-throughput screening using iPSC-derived neuronal progenitors to identify compounds counteracting epigenetic gene silencing in fragile X syndrome. J Biomol Screen 20(9):1101–1111. https://doi.org/10.1177/1087057115588287
Khanna R, Wilson SM, Brittain JM, Weimer J, Sultana R, Butterfield A, Hensley K (2012) Opening Pandora's jar: a primer on the putative roles of CRMP2 in a panoply of neurodegenerative, sensory and motor neuron, and central disorders. Future Neurol 7(6):749–771. https://doi.org/10.2217/FNL.12.68
Klein PS, Melton DA (1996) A molecular mechanism for the effect of lithium on development. Proc Natl Acad Sci USA 93(16):8455–8459. http://www.ncbi.nlm.nih.gov/pubmed/8710892, http://www.pnas.org/content/93/16/8455.full.pdf

Konopaske GT, Lange N, Coyle JT, Benes FM (2014) Prefrontal cortical dendritic spine pathology in schizophrenia and bipolar disorder. JAMA Psychiat 71(12):1323–1331. https://doi.org/10.1001/jamapsychiatry.2014.1582

Muotri AR, Marchetto MC, Coufal NG, Oefner R, Yeo G, Nakashima K, Gage FH (2010) L1 retrotransposition in neurons is modulated by MeCP2. Nature 468(7322):443–446. https://doi.org/10.1038/nature09544

Nakamura H, Yamashita N, Kimura A, Kimura Y, Hirano H, Makihara H et al (2016) Comprehensive behavioral study and proteomic analyses of CRMP2-deficient mice. Genes Cells 21 (10):1059–1079. https://doi.org/10.1111/gtc.12403

Ramón y Cajal S (1995) Histology of the nervous system of man and vertebrates. Oxford University Press, New York

Rochefort NL, Konnerth A (2012) Dendritic spines: from structure to in vivo function. EMBO Rep 13(8):699–708. https://doi.org/10.1038/embor.2012.102

Smoller JW, Finn CT (2003) Family, twin, and adoption studies of bipolar disorder. Am J Med Genet C Semin Med Genet 123C(1):48–58. https://doi.org/10.1002/ajmg.c.20013

Takahashi K, Tanabe K, Ohnuki M, Narita M, Ichisaka T, Tomoda K, Yamanaka S (2007) Induction of pluripotent stem cells from adult human fibroblasts by defined factors. Cell 131 (5):861–872. https://doi.org/10.1016/j.cell.2007.11.019

Tobe BTD, Crain AM, Winquist AM, Calabrese B, Makihara H, Zhao WN et al (2017) Probing the lithium-response pathway in hiPSCs implicates the phosphoregulatory set-point for a cytoskeletal modulator in bipolar pathogenesis. Proc Natl Acad Sci USA 114(22):E4462–E4471. https://doi.org/10.1073/pnas.1700111114

Uchida Y, Ohshima T, Sasaki Y, Suzuki H, Yanai S, Yamashita N et al (2005) Semaphorin3A signalling is mediated via sequential Cdk5 and GSK3beta phosphorylation of CRMP2: implication of common phosphorylating mechanism underlying axon guidance and Alzheimer's disease. Genes Cells 10(2):165–179. https://doi.org/10.1111/j.1365-2443.2005.00827.x

WHO (1995) The World Health Report 1995—bridging the gaps. World Health Forum 16 (4):377–385 http://www.ncbi.nlm.nih.gov/pubmed/8534341

Wilson SM, Brittain JM, Piekarz AD, Ballard CJ, Ripsch MS, Cummins TR et al (2011) Further insights into the antinociceptive potential of a peptide disrupting the N-type calcium channel-CRMP-2 signaling complex. Channels (Austin) 5(5):449–456. https://doi.org/10.4161/chan.5.5.17363

Wilson SM, Xiong W, Wang Y, Ping X, Head JD, Brittain JM et al (2012) Prevention of posttraumatic axon sprouting by blocking collapsin response mediator protein 2-mediated neurite outgrowth and tubulin polymerization. Neuroscience 210:451–466. https://doi.org/10.1016/j.neuroscience.2012.02.038

Yahata N, Asai M, Kitaoka S, Takahashi K, Asaka I, Hioki H et al (2011) Anti-Abeta drug screening platform using human iPS cell-derived neurons for the treatment of Alzheimer's disease. PLoS One 6(9):e25788. https://doi.org/10.1371/journal.pone.0025788

Yamashita N, Ohshima T, Nakamura F, Kolattukudy P, Honnorat J, Mikoshiba K, Goshima Y (2012) Phosphorylation of CRMP2 (collapsin response mediator protein 2) is involved in proper dendritic field organization. J Neurosci 32(4):1360–1365. https://doi.org/10.1523/JNEUROSCI.5563-11.2012

Zheng Y, Sethi R, Mangala L, Taylor C, Goldsmith J, Wang M et al (2018) Tuning microtuble dynamics to enhance cancer therapy by modulating FER-mediated CRMP2 phosphorylation. Nat Commun 9(1):476. https://doi.org/10.1038/s41467-017-02811-7

Zhu LQ, Zheng HY, Peng CX, Liu D, Li HL, Wang Q, Wang JZ (2010) Protein phosphatase 2A facilitates axonogenesis by dephosphorylating CRMP2. J Neurosci 30(10):3839–3848. https://doi.org/10.1523/JNEUROSCI.5174-09.2010

Chapter 13
Neural Stem Cell Dysfunction in Human Brain Disorders

Ewa Liszewska and Jacek Jaworski

Abstract Neural stem cells (NSCs) give rise to the entire nervous system. Animal models suggest that defects in NSC proliferation and differentiation contribute to several brain disorders (e.g., microcephaly, macrocephaly, autism, schizophrenia, and Huntington's disease). However, animal models of such diseases do not fully recapitulate all disease-related phenotypes because of substantial differences in brain development between rodents and humans. Therefore, additional human-based evidence is required to understand the mechanisms that are involved in the development of neurological diseases that result from human NSC (hNSC) dysfunction. Human-induced pluripotent stem cells provide a new model to investigate the contribution of hNSCs to various neurological pathologies. In this chapter, we review the role of hNSCs in both neurodevelopment- and neurodegeneration-related human brain pathologies, with an emphasis on recent evidence that has been obtained using embryonic stem cell- or induced pluripotent stem cell-derived hNSCs and progenitors.

13.1 Introduction

Neural stem cells (NSCs) are multipotent cells that give rise to the entire nervous system during development and contribute to physiological neuron renewal in specific brain areas. The maintenance and expansion of the NSC pool by self-renewal and the timing and mechanism by which NSCs become committed to differentiation are tightly regulated and critical for proper development of the nervous system. Defects in NSC proliferation and differentiation have been shown to be responsible for brain pathologies in animal models of various neurodevelopmental disorders. The dysfunction of NSCs has been observed in other brain diseases that are typically linked to improper neuronal transmission (e.g., autism, epilepsy, schizophrenia, and bipolar disorder) and neuronal death

E. Liszewska (✉) · J. Jaworski (✉)
International Institute of Molecular and Cell Biology, Warsaw, Poland
e-mail: eliszewska@iimcb.gov.pl; jaworski@iimcb.gov.pl

L. Buzanska (ed.), *Human Neural Stem Cells*, Results and Problems in Cell Differentiation 66, https://doi.org/10.1007/978-3-319-93485-3_13

(i.e., neurodegenerative disorders). Animal models of disease provide key insights into the pathogenesis of neurological disorders, but they do not fully recapitulate all phenotypes that are observed in humans because of substantial differences in brain development between rodents and humans. Therefore, additional human-based evidence is required to understand the mechanisms that are involved in the development of neurological diseases that result from human neural stem cell (hNSC) dysfunction. Recent advances in stem cell technologies (e.g., optimized protocols for cell reprogramming and differentiation) have provided new tools to investigate the contribution of hNSCs to various neurological pathologies. Combined with genome engineering and high-throughput methods to analyze gene and protein expression globally, patient-derived cellular models have begun to reveal the ways in which hNSCs are affected, not only in the course of neurodevelopmental diseases but also in other nervous system pathologies. In this chapter, we review the role of hNSCs in human brain pathologies, with an emphasis on recent evidence that has been obtained using embryonic stem cell (ESC)- and induced pluripotent stem cell (iPSC)-derived hNSCs.

13.2 Neurogenesis and Neural Stem Cell-Related Disorders

The mature nervous system is composed of several cell types (e.g., neurons and glia) that originate from NSCs. The transition from multipotent and proliferative NSCs to fully differentiated and functional neurons is called neurogenesis. The majority of neurons are generated during early embryonic development and early postnatal stages. In the human brain, as in the brains of other mammals, a few neurogenic niches remain active in adulthood and produce neurons throughout life, although less efficiently (Bergmann et al. 2015; Lim and Alvarez-Buylla 2016; Gonçalves et al. 2016). During embryonic development, the first neurogenic niche is formed after closure of the neural tube (Florio and Huttner 2014). The single layer of neuroepithelial cells that surround the lumen of the neural tube, known as the neuroepithelium, represents the first NSCs. These cells first undergo so-called symmetrical proliferative divisions that lead to transformation of the neural tube into multilayered tissue. As neurogenesis progresses, neuroepithelial cells begin to divide asymmetrically and generate radial glia. Radial glia reside in a layer that lines the ventricle, called the ventricular zone (VZ), and are able to undergo proliferative or differentiating division. During differentiating/asymmetric division, radial glia give rise to one radial glia cell and one neuron or one intermediate progenitor. Intermediate progenitors migrate to a layer basal to the VZ, called the subventricular zone (SVZ), and proliferate to increase the number of neurons (Florio and Huttner 2014). As brain development progresses, the production of new neurons by these stem cell niches declines, but in some brain areas the SVZ retains its neurogenic competence postnatally. In adult rodents, neurogenesis occurs in the SVZ, and newly born γ-aminobutyric acid (GABA)ergic neurons populate the olfactory bulbs (Lim and Alvarez-Buylla 2016). However, the neurogenic potential of the SVZ in humans remains controversial (Bergmann et al. 2015). To date, there is no definitive evidence of the migration of new neurons from the human SVZ to olfactory bulbs, which are atrophied relative to the olfactory bulbs in rodents and other mammals that rely more heavily on olfaction. Some studies suggest that by 2 years of age, no new neurons are born in the SVZ (Sanai et al. 2011; Bergmann et al. 2012). However, recent studies suggest that such neurons exist but populate the striatum (i.e., an area adjacent to the SVZ) instead of the olfactory bulbs (Ernst et al. 2014). In addition to the VZ and SVZ, two additional neurogenic niches generate neurons during central nervous system (CNS) development: the subgranular zone (SGZ) of the dentate gyrus (DG) in the hippocampus and external granule cell layer of the cerebellum, which generate DG and cerebellar granule neurons, respectively. Dentate gyrus stem cells originate from the VZ and migrate toward the hilar region. Interestingly, the SGZ of the DG generates neurons that integrate into an existing circuit throughout life not only in rodents but also in humans (Bergmann et al. 2015). Although the role of this adult neurogenesis is still debated, rodent studies suggest that these adult-born neurons are needed for brain plasticity (Gonçalves et al. 2016).

Neurodevelopmental disorders often have a clearly established genetic cause, and brain tumors are the main CNS pathologies that are linked to NSCs and neuroprogenitor cell (NPC) abnormalities (Swartling et al. 2015; Prajumwongs et al. 2016). One example of the former group is primary microcephaly (MCPH), which is either an autosomal recessive or X chromosome-linked disease. The characteristic features of MCPH include a smaller head size (mostly attributable to a reduction of the cerebral cortex) and deficits in intellectual, language, and motor skills. Magnetic resonance imaging scans of MCPH patients revealed a smaller brain volume but normal brain organization, suggesting that MCPH is not a disorder of neuronal migration or organization but rather a disorder of neuronal number. Primary microcephaly is caused by different genes that encode proteins that are expressed in cortical NSCs/NPCs and are involved in regulating the cell cycle and the function of the centrosome or mitotic spindle orientation (e.g., *MCPH1*, *ASPM*, *CDK5RAP2*, *CENPJ*, and *STIL*). Cellular and animal model studies demonstrated that MCHP is related to the defective proliferation of NSCs/NPCs and disturbances in the balance or premature transition from symmetric to asymmetric neuronal progenitor cell division, resulting in a reduction of the progenitor pool, a decrease in the number of neurons, and reduced cell survival (Faheem et al. 2015). Another group of neurodevelopmental disorders comprises pathologies that are associated with the upregulation of mechanistic/mammalian target of rapamycin (mTOR) kinase signaling, including tuberous sclerosis complex (TSC), neurofibromatosis type 1 (NF1), and phosphatase and tensin homolog deleted on chromosome ten (*PTEN*) mutation-related syndromes (Switon et al. 2017). Based on animal models and human-derived brain samples, these disorders were shown to be linked to NSC/NPC dysfunction (Switon et al. 2017). For example, the formation of brain tumors and macrocephaly in tuberous sclerosis was suggested to be caused by the improper proliferation and differentiation of NSCs (Switon et al. 2017). The observation that autism spectrum disorder (ASD) is common to several neurodevelopmental disorders (e.g., TSC, Fragile X syndrome [FXS]) prompted the search for connections between ASD and hNSC pathologies in addition to analysis of disturbances of synapse formation and neural network function.

The link between aberrant neurogenesis and neurodegenerative disorders, such as Alzheimer's disease (AD), Parkinson's disease (PD), and Huntington's disease (HD), is not often considered. Convincing evidence has shown that the human HD brain exhibits greater NPC proliferation that is proportional to the severity of the gene defect that is responsible for the disease and proportional to the severity of the pathology of the disease (Curtis et al. 2003). The data for PD are less consistent, but fewer proliferating cells in the SVZ were observed in postmortem human brain samples (Höglinger et al. 2004). Mice that expressed a mutation of leucine-rich repeat kinase 2 (*Lrrk2*; G2019S mutation), which is encoded by a PD susceptibility gene, had fewer proliferating cells in the SGZ and SVZ (Winner et al. 2011). Mice that overexpressed A30P α-synuclein, another PD susceptibility gene, did not present significant differences in the number of neurons that were generated in the SVZ, but the integration and survival of postnatally born dopaminergic neurons were affected (Neuner et al. 2014). Analyses of hNSCs and NPCs in mouse models of AD and patients' brain specimens yielded contradictory results, suggesting either increases or decreases in proliferation that depended on the animal model and stage of the disease (Liu and Song 2016; Tincer et al. 2016; Hollands et al. 2016). Additional disease models are needed to understand the primary effects of the disease on hNSCs and NPCs and their contribution to disease progression. Promising new models have emerged. In the following sections, we discuss these technical advancements in more detail.

13.3 Human Neural Stem Cells In Vitro: Patient in a Dish

For many years, animal models and postmortem human samples were the only source of accessible material to investigate the role of NSCs in human CNS diseases. Although both of these provided important information, they also have limitations. Disease models that consist of knockout animals often do not fully recapitulate all phenotypes that are observed in humans because of substantial differences in rodent and human neurogenesis. These problems could be potentially overcome by using human cell models in studies of neurological disorders. However, obtaining human brain tissue samples is limited by inaccessibility, difficulties in obtaining material, or poor status of the tissue samples. Thus, postmortem studies of neurological disorders are mainly conducted. This creates a problem for understanding disease etiology and progression because postmortem samples give only a "snapshot" of mainly the end stage of the disease that does not inform about the underlying mechanism of pathology. Pathological changes that are observed in these samples could be secondary or mask the primary causes of the disease. Therefore, knowledge of neuropathological abnormalities and their progression during the course of a human disease is limited, and the development of new human models that are based on defined cell populations that are affected by the disease (e.g., hNSCs and hNPCs) is important. Until recently, in vitro-cultured hNSCs that are obtained either directly from the patient nervous system or as a result of the differentiation of hESCs were used to overcome the aforementioned obstacles. However, two recent discoveries substantially changed this situation. Yamanaka and colleagues showed that differentiated somatic cells can be reprogrammed to pluripotency, and several efficient protocols for the production and differentiation of iPSCs were developed (Takahashi and Yamanaka 2006; Takahashi et al. 2007; Kulcenty et al. 2015). This allowed researchers to obtain human disease-specific iPSCs from which hNSCs and hNPCs could be derived. It also provided the opportunity to analyze the early stages of disease progression in cells with genomes that are prone or can lead to disorder development and investigate the molecular mechanisms that could be the origin of the disease. New tools for genome engineering (e.g., Clustered Regularly Interspaced Short Palindromic Repeats [CRISPR]-CRISPR-associated protein 9 [Cas9] technology) were developed that allowed the introduction of "disease-causing mutations" to already available hNSC lines. In Sect. 13.3.1, we describe iPSCs and their differentiation to hNSCs. Descriptions of CRISPR-Cas9 technology can be found in several extensive recent reviews (Hsu et al. 2014; Muffat et al. 2016; Komor et al. 2017).

13.3.1 Induced Pluripotent Stem Cells as a Source of Human Neural Stem Cells

Takahashi and Yamanaka were pioneers of the reprogramming method. They demonstrated that the combined ectopic expression of four transcription factors [octamer-binding transcription factor 4 (Oct-4; also known as POU5F1), sex-determining region Y box-2 (Sox-2), Kruppel-like factor 4 (Klf-4), and proto-oncogene c-Myc (c-Myc)] was sufficient to reprogram mouse fibroblasts back to the pluripotent state. The newly obtained cells were called iPSCs (Takahashi and Yamanaka 2006; Fig. 13.1). Shortly thereafter, this approach was repeated using human fibroblasts, resulting in the generation of human iPSCs (hiPSCs). Human iPSCs resemble human ESCs (hESCs) in many aspects, including morphology, proliferation, pluripotency markers, gene expression profiles, and the ability to differentiate into three germ layers (Takahashi et al. 2007). Soon after, iPSCs were obtained from somatic cells of patients who suffered from a variety of neurological disorders, including both neurodevelopmental and neurodegenerative disorders (e.g., Rett syndrome, FXS, HD, and PD; Park et al. 2008; Han et al. 2011; Dolmetsch and Geschwind 2011; Marchetto et al. 2011). Numerous studies subsequently established many protocols for the differentiation of hiPSCs into hNSCs/hNPCs with variable efficiency and quality of the obtained cells. The traditional protocol required the generation of embryoid bodies (EBs) by culturing iPSC colonies under non-adherent conditions in iPSC culture medium without basic fibroblast growth factor (bFGF) (Fig. 13.1). In some protocols, to increase induction of the neural linage, retinoic acid was added for a couple of the last days of EB formation, or the

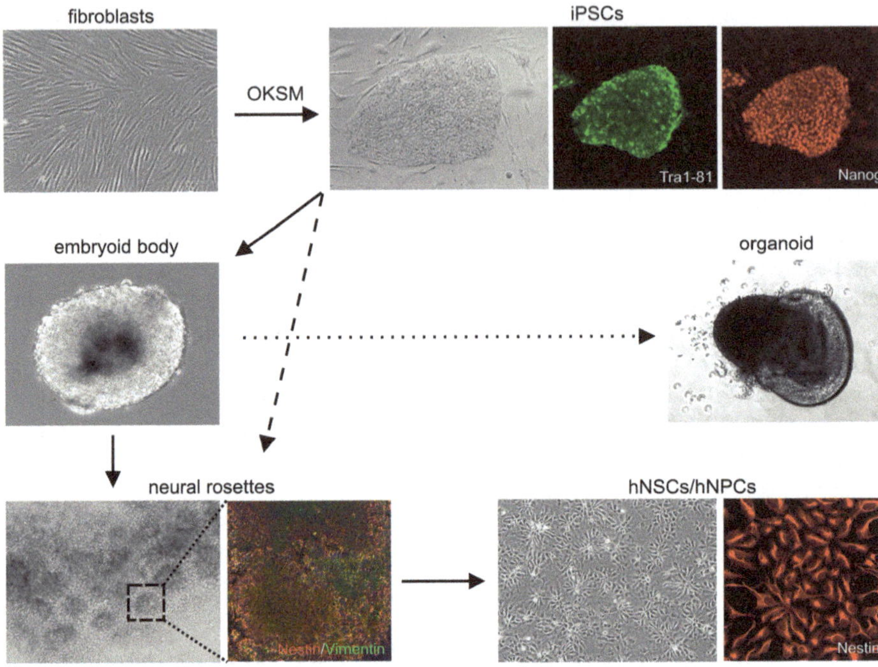

Fig. 13.1 Application of reprogramming technology for hNSC/hNPC production. Differentiated cells (e.g., fibroblasts) can be reprogrammed to induced pluripotent stem cells (iPSCs), which express pluripotency markers such as Tra1-81 and Nanog. iPSCs can be next differentiated via embryoid bodies or directly (e.g., by dual SMAD inhibition protocol; dashed line arrow) to neural rosettes resembling neuroepithelium expressing nestin and vimentin, markers of NSCs. From neural rosettes NSCs lines can be established. Using spinning bioreactors embryoid bodies can be also used to produce 3D brain organoids (dotted line arrow). *See text for more details*

EBs were transferred to serum-free medium that was supplemented with bFGF and epidermal growth factor (EGF; Yuan et al. 2013). The EBs were then plated on matrigel- or laminin/poly-L-ornithine-coated dishes and cultured to form neural rosettes that were reminiscent of early neural tube organization (Fig. 13.1). The composition of the medium that was used for rosette formation varied between protocols (e.g., serum-free medium supplemented with bFGF/EGF or retinoic acid (Salimi et al. 2014). The rosettes were then manually collected and dissociated. The NSCs that were obtained could be further cultured as neurospheres or as a monolayer in serum-free medium supplemented with bFGF and EGF; some protocols also supplemented the medium with brain-derived neurotrophic factor (BDNF). Further steps depended on the particular experimental needs (i.e., required population of neurons). The simplest approach for NSC differentiation requires the withdrawal of bFGF and EGF from the culture medium. In a few weeks, iPSC-derived NSCs could be differentiated into a mixture of different classes of neurons and astrocytes (Fig. 13.1). The enrichment of a particular neuronal type could be obtained by additional treatment with growth factors, small molecules, or inhibitors of certain signaling pathways during the differentiation process or cell fractioning based on live staining for protein markers (Mertens et al. 2016). A widely used method for the derivation of NSCs/NPCs from hiPSCs is currently a dual Sma and Mad Related proteins (SMAD) inhibition protocol that bypasses the EB step (Fig. 13.1). This protocol is based on the synergistic action of two inhibitors of SMAD signaling, for example, Noggin (antagonist of *bone morphogenetic protein* [BMP] signaling) and SB431542 (inhibitor of transforming growth factor β [TGFβ] receptor), which efficiently and rapidly induce neural conversion in an adherent monolayer of hiPSCs (Chambers et al. 2009). The SMAD inhibitors were also successfully used to modify the conventional protocol of NSC derivation from hiPSCs, in which they were added during the steps of EB and rosette formation.

13.3.2 Brain Organoids

hESC- and hiPSC-derived NSCs, NPCs, and neurons have already been shown to be useful for modeling both human neurogenesis and disease (Han et al. 2011; Dolmetsch and Geschwind 2011; Marchetto et al. 2011). However, two-dimensional (2D) culture models lack important information on cell positioning within tissue, morphogen gradients, and complex interactions between different cell types in the nervous system. This is a serious limitation to modeling the dysfunction of such a complex environment as the brain. These limitations can be overcome by using organoids. Organoids are miniature three-dimensional (3D) cell cultures that are derived from pluripotent cells, in which the cytoarchitecture of particular organs is at least partially recreated, based on the ability of ESCs/iPSCs to self-organize and differentiate. In 2013, the first cerebral organoids were obtained from iPSC-derived EBs that were cultured in spinning bioreactors (Lancaster et al. 2013). The first organoids were heterogeneous and contained several brain microregions per organoid. Analyses of protein marker expression revealed that these microregions resembled such structures as the dorsal cortex, hippocampus, choroid plexus, and ventral telencephalon (Lancaster et al. 2013). The initial protocol had relatively low efficiency and was quite expensive, but subsequent improvements to the protocol overcame some of these drawbacks (Hartley and Brennand 2016). Importantly, organoids cytoarchitectonically resemble the developing brain. Analyses of single-cell gene expression and epigenetic markers confirmed that the genetic program that was activated in cerebral organoids was very similar to the fetal cortex (Camp et al. 2015; Luo et al. 2016). Therefore, organoids or other 3D types of cultures are receiving more attention in modeling neurodevelopmental brain disorders, such as genetic or ZIKA virus-induced microcephaly (Lancaster et al. 2013; Garcez et al. 2016; Qian et al. 2016), lissencephaly (Bershteyn et al. 2017), non-syndromic ASD (Mariani et al. 2015), and neurodegenerative disorders (e.g., AD; Raja et al. 2016) (Table 13.1).

Table 13.1 Use of organoids to model neuropathology

Neuropathology/ manipulation	3D culture type	NSC/NPCs phenotype	References
Microcephaly (*CDK5RAP2* mutation)	iPSC-derived cerebral organoids	• Premature differentiation to neurons	Lancaster et al. (2013)
Microcephaly (ZIKA virus infection)	iPSC-derived forebrain-specific organoids	• Preferential NCS infection • Lower NCS proliferation rate • NCS cell death	Garcez et al. (2016) and Qian et al. (2016)
Miller–Dieker syndrome	iPSC-derived cerebral organoids	• Increased apoptosis of NSCs • Increased number of horizontal divisions of NSCs • Outer radial glia cytokinesis delay	Bershteyn et al. (2017)
CRISPR-Cas9 *CHD8* deletion (ASD model)	iPSC-derived cerebral organoids	• Not analyzed	Wang et al. (2017)
Idiopathic ASD	iPSC-derived telencephalic organoids	• Decreased cell cycle length in day 11 organoids • Enhanced differentiation toward GABAergic neurons	Mariani et al. (2015)
CRISPR-Cas9 *PTEN* deletion	hESC-derived cerebral organoids	• Enhanced proliferation of NSC/NPCs • Transient delay in NPCs differentiation • Organoid surface folding • Increased organoid size	Li et al. (2017b)
AD (*APP duplication*)	iPSC-derived organoids	• Not analyzed	Raja et al. (2016)
PD (*LRRK2* G2019S)	3D human ectodermal spheres	• Not analyzed	Son et al. (2017)

13.4 Neural Stem Cell-related Disorders: Lessons from In Vitro Modeling

13.4.1 Neurodevelopmental Disorders and Autism

The iPSC-based modeling of neurodevelopment appears to be an ideal tool for studying disorders of CNS development. The protocols that are used to differentiate ESCs and iPSCs into hNSCs/hNPCs and organoid cultures allow the reliable reproduction of certain stages of neurogenesis and gliogenesis that resemble those in the developing human brain. Such protocols also provide the opportunity to study disorders that cannot be easily modeled in rodent systems because of an unknown genetic cause or differences between human and rodent brain development (e.g., during cortical development). Consequently, the iPSC technology has been extensively used to model several neurodevelopmental diseases, including MCPH, FXS, Timothy syndrome, TSC, Rett syndrome, Williams–Beuren syndrome, Williams–Beuren region duplication syndrome, and several non-syndromic ASD cases (Prajumwongs et al. 2016; Wen et al. 2016; Ben-Reuven and Reiner 2016) (Table 13.2). One important feature that emerged from these studies was that gene expression was significantly dysregulated in NSCs in a majority of these disorders (Table 13.2). The affected genes are known to regulate neurodevelopment, differentiation, cell adhesion, inflammation, and oncogenesis. The dysregulation of certain effector genes/processes is common to several datasets, even if the initial cause of the disease and intermediate signaling hubs are different (e.g., the dysregulation of GABA neuron differentiation may depend on FoxoG1 or CHD8 in nonsyndromic autism; Mariani et al. 2015; Wang et al. 2017). The findings of these studies are too numerous to be described in detail herein, but several recent articles extensively reviewed the modeling of neurodevelopmental disorders in vitro (Wen et al. 2016; Young-Pearse and Morrow 2016; Ben-Reuven and Reiner 2016). In the following sections, we discuss only a few of these studies as examples of the ways in which "brain in the dish" technology helps reveal NSC/NPC pathology in neurodevelopment.

Microcephaly is characterized by a marked reduction of brain size. Several genes that cause this disease have been identified. However, mice that carry mutations of these identified genes have failed to recapitulate the severe reduction of brain size that is seen in humans, which substantially limits studies of the cellular and molecular mechanisms that lead to disease development. Lancaster et al. (2013) were the first to apply cerebral organoid technology to model this neurodevelopmental disorder. These authors showed that a heterozygous mutation of *CDK5RAP2* (i.e., a newly identified microcephaly-related gene) led to the premature differentiation of NPCs into neurons, which likely prevented the generation of a sufficient number of cells that are needed to attain a proper brain size. Li et al. (2016b) analyzed the mechanism of primary microcephaly that was caused by a mutation of citron kinase and found that the duration of cytokinesis may play a role in brain undergrowth.

Hypertrophy of the brain or specific brain regions has been observed in many neurodevelopmental disorders, which is often linked to hyperactivation of the mTOR pathway (Switon et al. 2017). A greater head circumference is associated with a higher risk for ASD. Li et al. (2017b) generated cerebral organoids with mutated *PTEN*. The loss of *PTEN* increases activity of the mTOR signaling pathway and may cause a variety of disorders, including Cowden syndrome, Lhermitte–Duclos disease, Bannayan–Riley–Ruvalcaba syndrome, and various cancers (Endersby and Baker 2008; Pilarski et al. 2013). *PTEN* mutations were also described in approximately 1% of human ASD patients. The majority of patients with *PTEN* mutations exhibit macrocephaly, often with polymicrogyria. Macrocephaly can be modeled in rodents (Switon et al. 2017), but polymicrogyria cannot. Brain organoids that are derived from PTEN-deficient cells present a larger size and form gyri-like structures through extensive organoid surface expansion that is caused by an extended time of NPC proliferation (Li et al. 2017b). Similar effects of Akt hyperactivation in cerebral organoids were observed (Li et al. 2017b).

Table 13.2 Use of hESC/iPSC-NSC/NPCs to model neurodevelopmental disorders

Disease	Starting cell type	NSC/NPCs phenotype	References
ASD	hiPSC	• Decreased cell cycle length	Mariani et al. (2015)
ASD (*CHD8* knockdown)	hiPSC	• Transcriptional program alteration	Sugathan et al. (2014)
ASD (*TRPC6* loss)	hiPSC	• Calcium signaling alteration • Transcriptional program alteration	Griesi-Oliveira et al. (2015)
ASD (Timothy syndrome)	hiPSC	• Changed calcium signaling • Calcium-dependent transcriptional • Decreased lower cortical layer neuron generation	Paşca et al. (2011)
MCPH (*CIT*; c317g>T; c376A>C; c689A>T)	hiPSC	• Abnormal cytokinesis with delayed mitosis • Multipolar spindles • Increased apoptosis	Li et al. (2016b)
Down Syndrome	hESCs	• Increased apoptosis • Downregulation of forebrain developmental genes	Halevy et al. (2016)
Williams–Beuren syndrome (7q11.23 CNV - deletion)	hiPSC	• Transcriptional program alteration	Adamo et al. (2015)
Williams–Beuren region duplication syndrome (7q11.23 CNV - duplication)	hiPSC	• Transcriptional program alteration	Adamo et al. (2015)
Fragile X Syndrome	hESCs	• Gradual silencing of *FMR1* • Delayed differentiated to neurons • Inappropriate balance between Sox2/Sox9 levels	Telias et al. (2013, 2015)
Fragile X Syndrome	hiPSC	• Inappropriate neuronal differentiation	Sheridan et al. (2011)
Rett syndrome	hiPSC	• Increased rate of L1 retrotransposition	Muotri et al. (2010)
Bipolar disorder	hiPSC	• Transcriptional program alteration suggesting changes in hNCS differentiation to different neuronal subtypes	Chen et al. (2014)
Bipolar disorder	hiPSC	• Transcriptional program alteration	Madison et al. (2015)
Schizophrenia (idiopathic)	hiPSC	• Decreased migration • Accelerated neural differentiation, • Increased canonical Wnt signaling • Increased abundance of translational machinery	Brennand et al. (2015), Topol et al. (2015a, b, 2016)

(continued)

Table 13.2 (continued)

Disease	Starting cell type	NSC/NPCs phenotype	References
		• Increased expression of miRNA-9 • Elevated oxidative stress • Mitochondrial damage • Transcriptome alterations	
Schizophrenia (idiopathic)	hiPSC	• High variability in stress responses to environmental/extra-cellular stressors on a single cell level	Hashimoto-Torii et al. (2014)
Schizophrenia (4-bp deletion in *DISC1*)	hiPSC	• Increased expression of miR-219 • Reduced proliferation	Murai et al. (2016)
Schizophrenia (15q11.2 CNV, loss of *CYFIP1*)	hiPSC	• Adherens junctions disruption • Loss of apical polarity	Yoon et al. (2014)
Schizophrenia (heterozygous deletion of *CNTNAP2;* loss of exons 14-15)	hiPSC	• Decreased migration • Changes in *CNTNAP2* isoform expression	Lee et al. (2015)
TSC (*TSC2*$^{-/-}$; deletion by zinc finger nuclease)	hESC	• Increased ectodermal rosettes area • Decreased number of differentiated neurons • Increased differentiation toward astroglial lineage	Costa et al. (2016)
TSC (*TSC2*$^{-/-}$; deletion by zinc finger nuclease)	hESC	• Reduced neuronal maturation • Increased astrogliosis • Transcriptional and translational program alterations	Grabole et al. (2016)
TSC (*TSC2*$^{+/-}$; c.1444-2A>C)	hiPSC	• High proliferation rate of NSC	Li et al. (2017a)
Paroxysmal kinesigenic dyskinesia (*PRTT*$^{+/-}$; c.487C>T; c.573dupT)	hiPSC	• Impairment of neuronal differentiation by dual SMAD inhibition protocol • Transcriptional program alterations	Li et al. (2016a)

As shown in Table 13.2, several studies investigated the pathology of NSCs and NPCs in ASD. For example, Paşca et al. (2011) analyzed hiPSCs-NSCs/NPCs that were obtained from Timothy syndrome patients. Timothy syndrome is caused by a missense mutation of the L-type calcium channel Cav1.2 (Splawski et al. 2004). Although cell proliferation and migration and neuron formation from NPCs appeared to be intact in Timothy syndrome cells, changes in calcium signaling led to profound changes in calcium-dependent transcription. This in turn caused subtle changes in neuronal differentiation. Gene expression analysis at the level of single cells revealed a significant decrease in the proportion of neurons that expressed lower cortical layer markers. According to the authors, this phenomenon may be responsible for fewer interhemispheric connections via the corpus callosum that are observed in Timothy syndrome and other ASD patients. Another study investigated organoids from non-syndromic ASD patients and found subtle changes in neuronal differentiation and a greater number of GABAergic neurons (Mariani et al. 2015).

In vitro-cultured hESC- and iPSC-derived hNSCs/hNPCs provide interesting insights into FXS. Fragile X syndrome is caused by inactivation of the *FMR1* gene, which is attributable to expansion of the CGG-triplet repeat in the 5-′-untranslated region. In hESCs, derived from blastocysts identified via preimplantation genetic diagnosis as carrying *FMR1* mutation, *FMR1* gene silencing occurred only upon ESC differentiation (Telias et al. 2013, 2015). Neuronal differentiation from hESCs was shown to recapitulate the gradual loss of FMRP, an *FMR1* product (Telias et al. 2013, 2015). In contrast, hiPSCs have already an inactivated *FMR1* locus as the "parental" cells from which they were derived. In some cases, however, random unsilencing can occur which makes the iPSC-based model less predictable (Sheridan et al. 2011). Fragile X syndrome has been considered to be a disease of neuronal networks, but studies have demonstrated that the pathology begins already in the hNPC stage. The differentiation of hESCs-NPCs in FXS was shown to be delayed because of an inappropriate balance between the levels of Sox2 (differentiation inhibitor) and Sox9 (differentiation activator) proteins (Telias et al. 2015).

13.4.2 Schizophrenia

The potential contribution of hNSC dysfunction to schizophrenia, bipolar disorder, and ASD has been investigated (Table 13.2). This section focuses on recent studies of schizophrenia, which is likely the most intensively studied psychiatric disorder with regard to hNSCs/hNPCs. Changes that are observed in postmortem schizophrenia patient specimens include ventricle enlargement, brain size/weight reduction, heterotopias through cortical areas, and reductions of dendritic arbors and dendritic spines (Harrison and Weinberger 2005). The data suggest that the pathogenesis of schizophrenia begins during prenatal brain development. The knockout of schizophrenia-related genes (e.g., Disrupted in schizophrenia 1 [*DISC-1*]) in mice causes aberrant postnatal and adult neurogenesis (Duan et al. 2007; Kim et al. 2009). The lower proliferation of NSCs has also been reported in the DG in schizophrenia patients, based on the number of cells that express Ki67, a marker of proliferating cells (Reif et al. 2006). Schizophrenia is a mental disorder that is characterized by disturbances in social behavior and problems differentiating reality from hallucinations. Both environmental and genetic factors have been linked to a higher risk for this complex psychiatric disorder (Lewis and Levitt 2002; Harrison and Weinberger 2005). Schizophrenia affects up to 1% of the human populations. There is an urgent need for models that better recapitulate this human disease to further understand the mechanism of pathogenesis and for potential drug screening.

Several lines of iPSCs have been developed from sporadic schizophrenia patients and schizophrenia patients with a clear genetic cause (Table 13.2). Initial studies focused on analyzing differentiated neurons, revealing deficits in neuronal connectivity that could be "cured" by the antipsychotic drug loxapine (Brennand et al. 2011). This study was followed by several analyses of hiPSCs-NPCs that were derived from sporadic schizophrenia patients, which revealed several differences from control cells that were obtained from healthy people. These differences included a decrease in NPC migration, accelerated neural differentiation, an increase in canonical Wnt signaling, an increase in the abundance of translational machinery, higher expression of microRNAs (e.g., miRNA-9), elevated oxidative stress, and mitochondrial damage (Brennand et al. 2015; Topol et al. 2015a, b, 2016). The increase in miRNA-9 expression correlated with deficits in migration, and the downregulation of miRNA-9 effectively reversed this phenotype (Topol et al. 2016). Another interesting observation at the single-cell level was that schizophrenia-derived hiPSCs-NPCs, unlike control cells, had very high variability in response to environmental/extracellular stressors, which is consistent with the significant impact of different types of prenatal stress on the development of schizophrenia (Hashimoto-Torii et al. 2014). Notably, none of the studies cited above reported changes in the rate of proliferation of NSCs/NPCs. Such changes have been reported in schizophrenia patients (Allen et al. 2016).

In addition to NSCs/NPCs that are obtained from sporadic schizophrenia patients, cells from patients with genetically driven cases have been analyzed (Table 13.2). Yoon and coworkers suggested an interesting link between schizophrenia and the presence of cortical heterotopias in patients. Heterotopic neurons may result from either inappropriate NPC/neuron migration or premature differentiation. Schizophrenia that was related to the 15q11.2 copy number variant provided insights into the way in which such a phenotype can be induced by NSC defects (Yoon et al. 2014). An analysis of neural rosettes that were derived from patients' iPSCs revealed defects in apical polarity and the formation of adherens junctions of NSCs that stemmed from *CYFIP1* haploinsufficiency. Subsequent analyses in a mouse model revealed that the lack of CYFIP1 led to the detachment and mispositioning of radial glia, followed by the aberrant positioning of NPCs and neurons in the developing cortex. Notably, however, such changes were specific to cells that were derived from patients with the 15q11.2 copy number variant but not those whose schizophrenia was attributable to the loss of *DISC1*, further confirming the complex etiology of schizophrenia (Yoon et al. 2014). hNPCs that were derived from patients with mutated *DISC1* have been analyzed. Murai et al. (2016) reported that such iPSCs-hNPCs exhibit a significant decrease in proliferation and increase in neuronal differentiation compared with control cells. According to these authors, such changes resulted from the higher expression of miR-219. These observations appear to be corroborated by a previous study of neurogenesis deficits in *Disc-1* knockdown mice (Mao et al. 2009) but are in striking contrast to the work of Srikanth et al. (2015). In the latter study, using hNPCs with *DISC-1* that was edited to mimic the disease-related mutation, these authors observed an increase in proliferation and a decrease in differentiation. Further work needs to be done to explain this discrepancy. To date, most standard iPSC-based models have not proved significant changes in the proliferation of hNSCs in schizophrenia.

13.4.3 Neurodegenerative Diseases

Most studies of neurodegenerative diseases that have used human stem cells (embryonic or reprogrammed) focused on disease phenotypes that are observed in mature neurons in affected patients. However, some observations of hESC or hiPSC differentiation and analyses of postmortem material prompted speculations that hNSCs can be affected much earlier in neurodegenerative disorders before actual symptoms of neurodegeneration are observed (Table 13.3). The pathological functioning of hNSCs can also make some people more vulnerable to neurodegeneration or at least explain some of the disease symptoms that cannot be explained simply by neuronal cell death.

To date, most progress in revealing the role of hNSCs in neurodegenerative disorders has been made in the case of HD. Huntington's disease is an autosomal dominant genetic disorder that affects six in 100,000 people. The best known symptoms of HD are erratic, random, and uncontrollable movements. As the disease progresses, physical symptoms become more severe and include rigidity, posture abnormalities, and the loss of muscle control. Several behavioral symptoms are also observed in HD, including irritability, apathy, anxiety, depression, and obsessive/compulsive behaviors. Huntington's disease is caused by a mutation of the *Huntingtin* gene (*HTT*). The mutation encompasses the expansion of CAG repeats that encode polyglutamine tracts. The *HTT* gene in unaffected individuals has fewer than 26 such repeats. The disease manifests when the number of repeats exceeds 35, the onset and severity of which depend on the number of repeats. With more than

Table 13.3 Use of hESC/iPSC-NSC/NPCs to model neurodegenerative disorders

Neuropathology/ manipulation	starting cell type	NSC/NPCs phenotype	References
Huntington disease	hiPSC	• Lower number of MAP2-positive cells	Chae et al. (2012)
Huntington disease	hIPSC	• Vulnerable to cell death caused by BDNF withdrawal • NeuroD1 downregulation • Transcriptional program alteration	Mattis et al. (2015 and Lim et al. (2017)
Huntington disease	hESC	• Transcriptional program alteration • Mitotic spindle mispositioning	Lopes et al. (2016)
Parkinson disease (LRRK2 G2019S)	hIPSC	• Decreased proliferation at latter passages • Lost ability for differentiation at latter passages • Changes in nucleus shape	Liu et al. (2012)
SPG11	hIPSC	• Decreased proliferation • Decreased differentiation • Transcriptional program alteration	Mishra et al. (2016)

60 CAG repeats, symptoms may occur as early as 20 years of age. *HTT* encodes huntingtin protein, which has several cellular functions (e.g., intracellular transport, cell division, and transcription; (Saudou and Humbert 2016). Mutated huntingtin forms potentially toxic aggregates in cells, but cellular damage begins before that. Cellular symptoms of HD likely stem from a combination of the gain of function of mutated HTT and the partial loss of function of wild-type HTT. Additionally, mutated *HTT* RNA may contribute to the HD phenotype and disease progression (Fiszer and Krzyzosiak 2013; Martí 2016).

The first iPSCs from HD patients were obtained in 2012 and differentiated into neurons (The HD iPSC Consortium 2012; Jeon et al. 2012; Chae et al. 2012). Patient-specific neurons presented greater vulnerability to cellular stress that was caused by glutamate, prolonged culture, trophic factor withdrawal, and oxidative stress (the HD iPSC Consortium 2012; Jeon et al. 2012). However, the pathology was not restricted to diseased neurons; it also occurred in hNSCs. For example, Chae et al. (2012) used a classic differentiation protocol and found that HD-derived cells gave rise to significantly fewer MAP 2-positive cells, suggesting that HD hNSCs might have deficits in proper differentiation. Recent work that focused specifically on hNSCs that were derived from early-onset HD patients revealed additional findings (Mattis et al. 2015; Lim et al. 2017). Although HD hNSCs could be efficiently differentiated into neurons and glia, a substantial number of nestin-expressing neural progenitors were preserved compared with control hNSCs. The HD hNPCs were vulnerable to cell death that was caused by BDNF withdrawal and subsequent glutamate toxicity. This phenotype could be prevented by the specific inhibition of mutated HTT expression. Similar results were found with NPCs that were obtained from embryos in a mouse model of HD. High-throughput analyses of mRNA and protein in HD hiPSCs that differentiated into a neural lineage revealed prominent changes in the expression of genes and proteins that are related to neuronal development and neuronal function (e.g., NeuroD1 downregulation) compared with control cells (Lim et al. 2017). Comparisons with the gene expression profiles of the normally developing striatum suggested that the differentiation of HD hNSCs is indeed either delayed or abnormal. Intriguingly, the small molecule Isx-9, which is known to upregulate NeuroD1 expression, reversed the changes in gene expression in HD-hNSCs and their vulnerability to BDNF withdrawal (Lim et al. 2017). Isx-9 also reversed behavioral phenotypes in an R6/2 mouse model of HD. The work of the HD iPSC Consortium revealed that HD pathology begins much earlier than when visible neurological and physical symptoms occur, likely during neural development, and can be at least partially reversed by small-molecule compounds. Recently, Lopes et al. (2016) suggested that aberrant gene expression that is related to neurogenesis and neuronal function may contribute to abnormalities that are seen in HD NSCs. These authors reported the improper mitotic spindle orientation of hESCs-NSCs from HD patients. Thus, in HD, NSCs might be affected before neurodegeneration occurs, but unclear are which and how particular symptoms of HD are linked to this hNSC defect. One speculation is that a similar defect in NSCs appears in HD during adult neurogenesis, thus contributing to the production of new striatal neurons. If so, then patients would have a limited capacity to cope with the greater loss of these neurons, which may aggravate HD symptoms.

The proper function of adult NSCs may also be compromised in PD. Parkinson's disease is a neurodegenerative disorder that primarily affects motor function and causes shaking, rigidity, and problems with walking. The non-motor symptoms of PD include depression, anxiety, and hyposomia. More than 90% of PD cases do not have a clear genetic cause (i.e., idiopathic PD). The remaining ~10% of cases are caused by mutations of different genes, including α-synuclein, parkin, *LRRK2*, and *PTEN*-induced putative kinase 1 (*PINK1*). The proteins that are encoded by these genes have different cellular functions. The ultimate result of such gene mutations is the death of dopaminergic neurons. Such cell death is likely caused by a decrease in the clearance of cellular protein deposits, increases in oxidative stress, and mitochondrial dysfunction (Calì et al. 2011). Several studies have reported the successful generation of iPSCs from Parkinsonian patients (Soldner et al. 2009; Hargus et al. 2010; Nguyen et al. 2011; Liu et al. 2012). In most of these studies, no defects in NSC proliferation or differentiation have been reported. However, early passages of hNSCs were used in these studies. Liu et al. (2012) aged hiPSCs-NSCs that were derived from PD patients with a *LRRK2* mutation (G2019S) by serial passaging, and the disease phenotype appeared. Up to passage 14, no significant differences were found in the number of colonies that formed or neuronal differentiation between wild-type and PD hNSCs. However, after passage 14, "diseased" hNSCs lost the ability to efficiently proliferate and generate adult neurons. This loss was accompanied by changes in the shape of cell nuclei. Interestingly, similar changes in nuclear shape were found in neurogenic areas in PD patients' brains (Liu et al. 2007). Thus, in PD, adult hNSC dysfunction may progress and lead to non-motor PD symptoms.

Similar to PD, most AD cases are sporadic, the underlying cause of which is unknown. The smaller proportion of cases is caused by mutations in genes that encode amyloid precursor protein (APP) and presenilins 1 and 2. These mutations lead to the greater production of β-amyloid (Aβ42), which is neurotoxic and responsible for neurodegeneration. Neuronal loss eventually leads to progressive dementia. Postmortem brain analyses and various in vitro and in vivo models have led to two contradictory conclusions: AD is linked to either an increase or decrease in hNSC proliferation and/or differentiation (Liu and Song 2016; Tincer et al. 2016). In vitro experiments showed that the exposure of hNPCs to Aβ42 decreased their proliferation and neuronal differentiation in in vitro cultures (Haughey et al. 2002). However, in several AD patients, no deficits in the proliferation or differentiation of iPSC-derived hNPCs have been reported (Yagi et al. 2011; Israel et al. 2012; Kondo et al. 2013; Duan et al. 2014; Sproul et al. 2014). There may be several explanations for these discrepancies. For example, similar to NSCs in PD, the AD phenotype may become apparent after in vitro aging. Another possibility is that the levels of Aβ42 that are produced by hNSCs in culture may be too low to affect function. Therefore, more studies are needed to clarify whether neurogenesis and NSC/NPC function are affected in AD.

Recently, hNPC dysfunction has been reported in another neurodegenerative disorder, hereditary spastic paraplegia (SPG), which is caused by a mutation of spastic paraplegia gene 11 (*SPG11*). In addition to a neurodegenerative phenotype that is caused by the degeneration of axons in corticospinal tracts, mutated SPG11 causes cortical atrophy and thinning of the corpus callosum. This particular form of SPG may be associated with both neurodegeneration and problems in neurodevelopment. Gene expression analyses of patient hiPSCs-NPCs revealed changes in the expression of genes that are related to neurogenesis and differentiation (Mishra et al. 2016). Subsequent analyses of the proliferation and differentiation of patients' NPCs confirmed that these cells proliferated and differentiated less efficiently. Interestingly, this phenotype could be reversed by GSK3 inhibitors (Mishra et al. 2016).

13.5 Future Directions

Over the last decade, tremendous progress has been made in the generation of human-derived models of neurological disorders using iPSCs. Undoubtedly, the list of diseases that can be modeled by iPSCs will expand further. Some important points should be considered with regard to existing and future models. In many cases, cells have been obtained from insufficient cohorts of patients, which limits generalization of the findings. Moreover, insufficient details of analyses of NSC/NPC phenotypes have been reported. The results of very detailed transcriptomic studies were not tested with regard to functional significance. Finally, genetic correction has rarely been performed to demonstrate that defects in NSCs/NPCs indeed depend on a particular mutation. Further efforts should be made to address these shortcomings, and 2D cultures should be validated in 3D settings.

Acknowledgments This work was supported by National Science Centre "Sonata" grant no. 2013/11/D/NZ3/01079 to EL and "Opus" grant no. 2016/21/B/NZ3/03639 to JJ. JJ was a recipient of the Foundation for Polish Science "Mistrz" Professorial Subsidy and Fellowship.

References

Adamo A, Atashpaz S, Germain P-L et al (2015) 7q11.23 dosage-dependent dysregulation in human pluripotent stem cells affects transcriptional programs in disease-relevant lineages. Nat Genet 47:132–141. https://doi.org/10.1038/ng.3169

Allen KM, Fung SJ, Weickert CS (2016) Cell proliferation is reduced in the hippocampus in schizophrenia. Aust N Z J Psychiatry 50:473–480. https://doi.org/10.1177/0004867415589793

Ben-Reuven L, Reiner O (2016) Modeling the autistic cell: iPSCs recapitulate developmental principles of syndromic and nonsyndromic ASD. Dev Growth Differ 58:481–491. https://doi.org/10.1111/dgd.12280

Bergmann O, Liebl J, Bernard S et al (2012) The age of olfactory bulb neurons in humans. Neuron 74:634–639. https://doi.org/10.1016/j.neuron.2012.03.030

Bergmann O, Spalding KL, Frisén J (2015) Adult neurogenesis in humans. Cold Spring Harb Perspect Biol 7:a018994. https://doi.org/10.1101/cshperspect.a018994

Bershteyn M, Nowakowski TJ, Pollen AA et al (2017) Human iPSC-derived cerebral organoids model cellular features of lissencephaly and reveal prolonged mitosis of outer radial glia. Cell Stem Cell 20:435–449.e4. https://doi.org/10.1016/j.stem.2016.12.007

Brennand KJ, Simone A, Jou J et al (2011) Modelling schizophrenia using human induced pluripotent stem cells. Nature 473:221–225. https://doi.org/10.1038/nature09915

Brennand K, Savas JN, Kim Y et al (2015) Phenotypic differences in hiPSC NPCs derived from patients with schizophrenia. Mol Psychiatry 20:361–368. https://doi.org/10.1038/mp.2014.22

Calì T, Ottolini D, Brini M (2011) Mitochondria, calcium, and endoplasmic reticulum stress in Parkinson's disease. Biofactors Oxf Engl 37:228–240. https://doi.org/10.1002/biof.159

Camp JG, Badsha F, Florio M et al (2015) Human cerebral organoids recapitulate gene expression programs of fetal neocortex development. Proc Natl Acad Sci USA 112:15672–15677. https://doi.org/10.1073/pnas.1520760112

Chae J-I, Kim D-W, Lee N et al (2012) Quantitative proteomic analysis of induced pluripotent stem cells derived from a human Huntington's disease patient. Biochem J 446:359–371. https://doi.org/10.1042/BJ20111495

Chambers SM, Fasano CA, Papapetrou EP et al (2009) Highly efficient neural conversion of human ES and iPS cells by dual inhibition of SMAD signaling. Nat Biotechnol 27:275–280. https://doi.org/10.1038/nbt.1529

Chen HM, DeLong CJ, Bame M et al (2014) Transcripts involved in calcium signaling and telencephalic neuronal fate are altered in induced pluripotent stem cells from bipolar disorder patients. Transl Psychiatry 4:e375. https://doi.org/10.1038/tp.2014.12

Costa V, Aigner S, Vukcevic M et al (2016) mTORC1 inhibition corrects neurodevelopmental and synaptic alterations in a human stem cell model of tuberous sclerosis. Cell Rep 15:86–95. https://doi.org/10.1016/j.celrep.2016.02.090

Curtis MA, Penney EB, Pearson AG et al (2003) Increased cell proliferation and neurogenesis in the adult human Huntington's disease brain. Proc Natl Acad Sci USA 100:9023–9027. https://doi.org/10.1073/pnas.1532244100

Dolmetsch R, Geschwind DH (2011) The human brain in a dish: the promise of iPSC-derived neurons. Cell 145:831–834. https://doi.org/10.1016/j.cell.2011.05.034

Duan X, Chang JH, Ge S et al (2007) Disrupted-in-schizophrenia 1 regulates integration of newly generated neurons in the adult brain. Cell 130:1146–1158. https://doi.org/10.1016/j.cell.2007.07.010

Duan L, Bhattacharyya BJ, Belmadani A et al (2014) Stem cell derived basal forebrain cholinergic neurons from Alzheimer's disease patients are more susceptible to cell death. Mol Neurodegener 9:3. https://doi.org/10.1186/1750-1326-9-3

Endersby R, Baker SJ (2008) PTEN signaling in brain: neuropathology and tumorigenesis. Oncogene 27:5416–5430. https://doi.org/10.1038/onc.2008.239

Ernst A, Alkass K, Bernard S et al (2014) Neurogenesis in the striatum of the adult human brain. Cell 156:1072–1083. https://doi.org/10.1016/j.cell.2014.01.044

Faheem M, Naseer MI, Rasool M et al (2015) Molecular genetics of human primary microcephaly: an overview. BMC Med Genomics. https://doi.org/10.1186/1755-8794-8-S1-S4

Fiszer A, Krzyzosiak WJ (2013) RNA toxicity in polyglutamine disorders: concepts, models, and progress of research. J Mol Med Berl Ger 91:683–691. https://doi.org/10.1007/s00109-013-1016-2

Florio M, Huttner WB (2014) Neural progenitors, neurogenesis and the evolution of the neocortex. Development 141:2182–2194. https://doi.org/10.1242/dev.090571

Garcez PP, Loiola EC, Madeiro da Costa R et al (2016) Zika virus impairs growth in human neurospheres and brain organoids. Science 352:816–818. https://doi.org/10.1126/science.aaf6116

Gonçalves JT, Schafer ST, Gage FH (2016) Adult neurogenesis in the hippocampus: from stem cells to behavior. Cell 167:897–914. https://doi.org/10.1016/j.cell.2016.10.021

Grabole N, Zhang JD, Aigner S et al (2016) Genomic analysis of the molecular neuropathology of tuberous sclerosis using a human stem cell model. Genome Med 8:94. https://doi.org/10.1186/s13073-016-0347-3

Griesi-Oliveira K, Acab A, Gupta AR et al (2015) Modeling non-syndromic autism and the impact of TRPC6 disruption in human neurons. Mol Psychiatry 20:1350–1365. https://doi.org/10.1038/mp.2014.141

Halevy T, Biancotti J-C, Yanuka O et al (2016) Molecular characterization of down syndrome embryonic stem cells reveals a role for RUNX1 in neural differentiation. Stem Cell Rep 7:777–786. https://doi.org/10.1016/j.stemcr.2016.08.003

Han SSW, Williams LA, Eggan KC (2011) Constructing and deconstructing stem cell models of neurological disease. Neuron 70:626–644. https://doi.org/10.1016/j.neuron.2011.05.003

Hargus G, Cooper O, Deleidi M et al (2010) Differentiated Parkinson patient-derived induced pluripotent stem cells grow in the adult rodent brain and reduce motor asymmetry in Parkinsonian rats. Proc Natl Acad Sci 107:15921–15926. https://doi.org/10.1073/pnas.1010209107

Harrison PJ, Weinberger DR (2005) Schizophrenia genes, gene expression, and neuropathology: on the matter of their convergence. Mol Psychiatry 10:40–68. https://doi.org/10.1038/sj.mp.4001558

Hartley BJ, Brennand KJ (2016) Neural organoids for disease phenotyping, drug screening and developmental biology studies. Neurochem Int. https://doi.org/10.1016/j.neuint.2016.10.004

Hashimoto-Torii K, Torii M, Fujimoto M et al (2014) Roles of heat shock factor 1 in neuronal response to fetal environmental risks and its relevance to brain disorders. Neuron 82:560–572. https://doi.org/10.1016/j.neuron.2014.03.002

Haughey NJ, Nath A, Chan SL et al (2002) Disruption of neurogenesis by amyloid beta-peptide, and perturbed neural progenitor cell homeostasis, in models of Alzheimer's disease. J Neurochem 83:1509–1524

Höglinger GU, Rizk P, Muriel MP et al (2004) Dopamine depletion impairs precursor cell proliferation in Parkinson disease. Nat Neurosci 7:726–735. https://doi.org/10.1038/nn1265

Hollands C, Bartolotti N, Lazarov O (2016) Alzheimer's disease and hippocampal adult neurogenesis; exploring shared mechanisms. Front Neurosci. https://doi.org/10.3389/fnins.2016.00178

Hsu PD, Lander ES, Zhang F (2014) Development and applications of CRISPR-Cas9 for genome engineering. Cell 157:1262–1278. https://doi.org/10.1016/j.cell.2014.05.010

Israel MA, Yuan SH, Bardy C et al (2012) Probing sporadic and familial Alzheimer's disease using induced pluripotent stem cells. Nature 482:216–220. https://doi.org/10.1038/nature10821

Jeon I, Lee N, Li J-Y et al (2012) Neuronal properties, in vivo effects, and pathology of a Huntington's disease patient-derived induced pluripotent stem cells. Stem Cells 30:2054–2062. https://doi.org/10.1002/stem.1135

Kim JY, Duan X, Liu CY et al (2009) DISC1 regulates new neuron development in the adult brain via modulation of AKT-mTOR signaling through KIAA1212. Neuron 63:761–773. https://doi.org/10.1016/j.neuron.2009.08.008

Komor AC, Badran AH, Liu DR (2017) CRISPR-based technologies for the manipulation of eukaryotic genomes. Cell 168:20–36. https://doi.org/10.1016/j.cell.2016.10.044

Kondo T, Asai M, Tsukita K et al (2013) Modeling Alzheimer's disease with iPSCs reveals stress phenotypes associated with intracellular Aβ and differential drug responsiveness. Cell Stem Cell 12:487–496. https://doi.org/10.1016/j.stem.2013.01.009

Kulcenty K, Wróblewska J, Mazurek S et al (2015) Molecular mechanisms of induced pluripotency. Contemp Oncol Poznan Pol 19:A22–A29. https://doi.org/10.5114/wo.2014.47134

Lancaster MA, Renner M, Martin C-A et al (2013) Cerebral organoids model human brain development and microcephaly. Nature 501:373–379. https://doi.org/10.1038/nature12517

Lee IS, Carvalho CMB, Douvaras P et al (2015) Characterization of molecular and cellular phenotypes associated with a heterozygous CNTNAP2 deletion using patient-derived hiPSC neural cells. NPJ Schizophr 1:15019

Lewis DA, Levitt P (2002) Schizophrenia as a disorder of neurodevelopment. Annu Rev Neurosci 25:409–432. https://doi.org/10.1146/annurev.neuro.25.112701.142754

Li C, Ma Y, Zhang K et al (2016a) Aberrant transcriptional networks in step-wise neurogenesis of paroxysmal kinesigenic dyskinesia-induced pluripotent stem cells. Oncotarget. https://doi.org/10.18632/oncotarget.10680

Li H, Bielas SL, Zaki MS et al (2016b) Biallelic mutations in citron kinase link mitotic cytokinesis to human primary microcephaly. Am J Hum Genet 99:501–510. https://doi.org/10.1016/j.ajhg.2016.07.004

Li Y, Cao J, Chen M et al (2017a) Abnormal neural progenitor cells differentiated from induced pluripotent stem cells partially mimicked development of TSC2 neurological abnormalities. Stem Cell Rep. https://doi.org/10.1016/j.stemcr.2017.02.020

Li Y, Muffat J, Omer A et al (2017b) Induction of expansion and folding in human cerebral organoids. Cell Stem Cell 20:385–396.e3. https://doi.org/10.1016/j.stem.2016.11.017

Lim DA, Alvarez-Buylla A (2016) The adult ventricular-subventricular zone (V-SVZ) and olfactory bulb (OB) neurogenesis. Cold Spring Harb Perspect Biol. https://doi.org/10.1101/cshperspect.a018820

Lim RG, Salazar LL, Wilton DK et al (2017) Developmental alterations in Huntington's disease neural cells and pharmacological rescue in cells and mice. Nat Neurosci. https://doi.org/10.1038/nn.4532

Liu H, Song N (2016) Molecular mechanism of adult neurogenesis and its association with human brain diseases. J Cent Nerv Syst Dis 8:5–11. https://doi.org/10.4137/JCNSD.S32204

Liu J, Lamb D, Chou MM et al (2007) Nerve growth factor-mediated neurite outgrowth via regulation of Rab5. Mol Biol Cell 18:1375–1384

Liu G-H, Qu J, Suzuki K et al (2012) Progressive degeneration of human neural stem cells caused by pathogenic LRRK2. Nature 491:603–607. https://doi.org/10.1038/nature11557

Lopes C, Aubert S, Bourgois-Rocha F et al (2016) Dominant-negative effects of adult-onset huntingtin mutations alter the division of human embryonic stem cells-derived neural cells. PLoS One 11:e0148680. https://doi.org/10.1371/journal.pone.0148680

Luo C, Lancaster MA, Castanon R et al (2016) Cerebral organoids recapitulate epigenomic signatures of the human fetal brain. Cell Rep 17:3369–3384. https://doi.org/10.1016/j.celrep.2016.12.001

Madison JM, Zhou F, Nigam A et al (2015) Characterization of bipolar disorder patient-specific induced pluripotent stem cells from a family reveals neurodevelopmental and mRNA expression abnormalities. Mol Psychiatry 20:703–717. https://doi.org/10.1038/mp.2015.7

Mao Y, Ge X, Frank CL et al (2009) Disrupted in schizophrenia 1 regulates neuronal progenitor proliferation via modulation of GSK3beta/beta-catenin signaling. Cell 136:1017–1031. https://doi.org/10.1016/j.cell.2008.12.044

Marchetto MC, Brennand KJ, Boyer LF, Gage FH (2011) Induced pluripotent stem cells (iPSCs) and neurological disease modeling: progress and promises. Hum Mol Genet 20:R109–R115. https://doi.org/10.1093/hmg/ddr336

Mariani J, Coppola G, Zhang P et al (2015) FOXG1-dependent dysregulation of GABA/glutamate neuron differentiation in autism spectrum disorders. Cell 162:375–390. https://doi.org/10.1016/j.cell.2015.06.034

Martí E (2016) RNA toxicity induced by expanded CAG repeats in Huntington's disease. Brain Pathol Zurich Switz 26:779–786. https://doi.org/10.1111/bpa.12427

Mattis VB, Tom C, Akimov S et al (2015) HD iPSC-derived neural progenitors accumulate in culture and are susceptible to BDNF withdrawal due to glutamate toxicity. Hum Mol Genet 24:3257–3271. https://doi.org/10.1093/hmg/ddv080

Mertens J, Marchetto MC, Bardy C, Gage FH (2016) Evaluating cell reprogramming, differentiation and conversion technologies in neuroscience. Nat Rev Neurosci 17:424–437. https://doi.org/10.1038/nrn.2016.46

Mishra HK, Prots I, Havlicek S et al (2016) GSK3ß-dependent dysregulation of neurodevelopment in SPG11-patient induced pluripotent stem cell model: Neurodevelopmental defects in motor neuron disease. Ann Neurol 79:826–840. https://doi.org/10.1002/ana.24633

Muffat J, Li Y, Jaenisch R (2016) CNS disease models with human pluripotent stem cells in the CRISPR age. Curr Opin Cell Biol 43:96–103. https://doi.org/10.1016/j.ceb.2016.10.001

Muotri AR, Marchetto MCN, Coufal NG et al (2010) L1 retrotransposition in neurons is modulated by MeCP2. Nature 468:443–446. https://doi.org/10.1038/nature09544

Murai K, Sun G, Ye P et al (2016) The TLX-miR-219 cascade regulates neural stem cell proliferation in neurodevelopment and schizophrenia iPSC model. Nat Commun 7:10965. https://doi.org/10.1038/ncomms10965

Neuner J, Filser S, Michalakis S et al (2014) A30P α-Synuclein interferes with the stable integration of adult-born neurons into the olfactory network. Sci Rep 4:3931. https://doi.org/10.1038/srep03931

Nguyen HN, Byers B, Cord B et al (2011) LRRK2 mutant iPSC-derived DA neurons demonstrate increased susceptibility to oxidative stress. Cell Stem Cell 8:267–280. https://doi.org/10.1016/j.stem.2011.01.013

Park I-H, Arora N, Huo H et al (2008) Disease-specific induced pluripotent stem cells. Cell 134:877–886. https://doi.org/10.1016/j.cell.2008.07.041

Paşca SP, Portmann T, Voineagu I et al (2011) Using iPSC-derived neurons to uncover cellular phenotypes associated with Timothy syndrome. Nat Med 17:1657–1662. https://doi.org/10.1038/nm.2576

Pilarski R, Burt R, Kohlman W et al (2013) Cowden syndrome and the PTEN hamartoma tumor syndrome: systematic review and revised diagnostic criteria. JNCI J Natl Cancer Inst 105:1607–1616. https://doi.org/10.1093/jnci/djt277

Prajumwongs P, Weeranantanapan O, Jaroonwitchawan T, Noisa P (2016) Human embryonic stem cells: a model for the study of neural development and neurological diseases. Stem Cells Int 2016:2958210. https://doi.org/10.1155/2016/2958210

Qian X, Nguyen HN, Song MM et al (2016) Brain-region-specific organoids using mini-bioreactors for modeling ZIKV exposure. Cell 165:1238–1254. https://doi.org/10.1016/j.cell.2016.04.032

Raja WK, Mungenast AE, Lin Y-T et al (2016) Self-organizing 3D human neural tissue derived from induced pluripotent stem cells recapitulate Alzheimer's disease phenotypes. PLoS One 11: e0161969. https://doi.org/10.1371/journal.pone.0161969

Reif A, Fritzen S, Finger M et al (2006) Neural stem cell proliferation is decreased in schizophrenia, but not in depression. Mol Psychiatry 11:514–522. https://doi.org/10.1038/sj.mp.4001791

Salimi A, Nadri S, Ghollasi M et al (2014) Comparison of different protocols for neural differentiation of human induced pluripotent stem cells. Mol Biol Rep 41:1713–1721. https://doi.org/10.1007/s11033-014-3020-1

Sanai N, Nguyen T, Ihrie RA et al (2011) Corridors of migrating neurons in the human brain and their decline during infancy. Nature 478:382–386. https://doi.org/10.1038/nature10487

Saudou F, Humbert S (2016) The biology of huntingtin. Neuron 89:910–926. https://doi.org/10.1016/j.neuron.2016.02.003

Sheridan SD, Theriault KM, Reis SA et al (2011) Epigenetic characterization of the FMR1 gene and aberrant neurodevelopment in human induced pluripotent stem cell models of fragile X syndrome. PLoS One 6:e26203. https://doi.org/10.1371/journal.pone.0026203

Soldner F, Hockemeyer D, Beard C et al (2009) Parkinson's disease patient-derived induced pluripotent stem cells free of viral reprogramming factors. Cell 136:964–977. https://doi.org/10.1016/j.cell.2009.02.013

Son M-Y, Sim H, Son YS et al (2017) Distinctive genomic signature of neural and intestinal organoids from familial Parkinson's disease patient-derived induced pluripotent stem cells. Neuropathol Appl Neurobiol. https://doi.org/10.1111/nan.12396

Splawski I, Timothy KW, Sharpe LM et al (2004) Ca(V)1.2 calcium channel dysfunction causes a multisystem disorder including arrhythmia and autism. Cell 119:19–31. https://doi.org/10.1016/j.cell.2004.09.011

Sproul AA, Jacob S, Pre D et al (2014) Characterization and molecular profiling of PSEN1 familial Alzheimer's disease iPSC-derived neural progenitors. PLoS One 9:e84547. https://doi.org/10.1371/journal.pone.0084547

Srikanth P, Han K, Callahan DG et al (2015) Genomic DISC1 disruption in hiPSCs alters Wnt signaling and neural cell fate. Cell Rep 12:1414–1429. https://doi.org/10.1016/j.celrep.2015.07.061

Sugathan A, Biagioli M, Golzio C et al (2014) *CHD8* regulates neurodevelopmental pathways associated with autism spectrum disorder in neural progenitors. Proc Natl Acad Sci 111:E4468–E4477. https://doi.org/10.1073/pnas.1405266111

Swartling FJ, Čančer M, Frantz A et al (2015) Deregulated proliferation and differentiation in brain tumors. Cell Tissue Res 359:225–254. https://doi.org/10.1007/s00441-014-2046-y

Switon K, Kotulska K, Janusz-Kaminska A et al (2017) Molecular neurobiology of mTOR. Neuroscience 341:112–153. https://doi.org/10.1016/j.neuroscience.2016.11.017

Takahashi K, Yamanaka S (2006) Induction of pluripotent stem cells from mouse embryonic and adult fibroblast cultures by defined factors. Cell 126:663–676. https://doi.org/10.1016/j.cell.2006.07.024

Takahashi K, Tanabe K, Ohnuki M et al (2007) Induction of pluripotent stem cells from adult human fibroblasts by defined factors. Cell 131:861–872. https://doi.org/10.1016/j.cell.2007.11.019

Telias M, Segal M, Ben-Yosef D (2013) Neural differentiation of fragile X human embryonic stem cells reveals abnormal patterns of development despite successful neurogenesis. Dev Biol 374:32–45. https://doi.org/10.1016/j.ydbio.2012.11.031

Telias M, Mayshar Y, Amit A, Ben-Yosef D (2015) Molecular mechanisms regulating impaired neurogenesis of fragile X syndrome human embryonic stem cells. Stem Cells Dev 24:2353–2365. https://doi.org/10.1089/scd.2015.0220

The HD iPSC Consortium (2012) Induced pluripotent stem cells from patients with Huntington's disease show CAG-repeat-expansion-associated phenotypes. Cell Stem Cell 11:264–278. https://doi.org/10.1016/j.stem.2012.04.027

Tincer G, Mashkaryan V, Bhattarai P, Kizil C (2016) Neural stem/progenitor cells in Alzheimer's disease. Yale J Biol Med 89:23–35

Topol A, English JA, Flaherty E et al (2015a) Increased abundance of translation machinery in stem cell-derived neural progenitor cells from four schizophrenia patients. Transl Psychiatry 5:e662. https://doi.org/10.1038/tp.2015.118

Topol A, Zhu S, Tran N et al (2015b) Altered WNT signaling in human induced pluripotent stem cell neural progenitor cells derived from four schizophrenia Patients. Biol Psychiatry 78:e29–e34. https://doi.org/10.1016/j.biopsych.2014.12.028

Topol A, Zhu S, Hartley BJ et al (2016) Dysregulation of miRNA-9 in a subset of schizophrenia patient-derived neural progenitor cells. Cell Rep 15:1024–1036. https://doi.org/10.1016/j.celrep.2016.03.090

Wang P, Mokhtari R, Pedrosa E et al (2017) CRISPR/Cas9-mediated heterozygous knockout of the autism gene CHD8 and characterization of its transcriptional networks in cerebral organoids derived from iPS cells. Mol Autism. https://doi.org/10.1186/s13229-017-0124-1

Wen Z, Christian KM, Song H, Ming G (2016) Modeling psychiatric disorders with patient-derived iPSCs. Curr Opin Neurobiol 36:118–127. https://doi.org/10.1016/j.conb.2015.11.003

Winner B, Melrose HL, Zhao C et al (2011) Adult neurogenesis and neurite outgrowth are impaired in LRRK2 G2019S mice. Neurobiol Dis 41:706–716. https://doi.org/10.1016/j.nbd.2010.12.008

Yagi T, Ito D, Okada Y et al (2011) Modeling familial Alzheimer's disease with induced pluripotent stem cells. Hum Mol Genet 20:4530–4539. https://doi.org/10.1093/hmg/ddr394

Yoon K-J, Nguyen HN, Ursini G et al (2014) Modeling a genetic risk for schizophrenia in iPSCs and mice reveals neural stem cell deficits associated with adherens junctions and polarity. Cell Stem Cell 15:79–91. https://doi.org/10.1016/j.stem.2014.05.003

Young-Pearse TL, Morrow EM (2016) Modeling developmental neuropsychiatric disorders with iPSC technology: challenges and opportunities. Curr Opin Neurobiol 36:66–73. https://doi.org/10.1016/j.conb.2015.10.006

Yuan T, Liao W, Feng N-H et al (2013) Human induced pluripotent stem cell-derived neural stem cells survive, migrate, differentiate, and improve neurologic function in a rat model of middle cerebral artery occlusion. Stem Cell Res Ther 4:73. https://doi.org/10.1186/scrt224

Chapter 14
Human Fetal Neural Stem Cells for Neurodegenerative Disease Treatment

Daniela Ferrari, Maurizio Gelati, Daniela Celeste Profico, and Angelo Luigi Vescovi

Abstract Clinical trials for Parkinson's disease, which used primary brain fetal tissue, have demonstrated that neural stem cell therapy could be suitable for neurodegenerative diseases. The use of fetal tissue presents several issues that have hampered the clinical development of this approach. In addition to the ethical concerns related to the required continuous supply of fetal specimen, the necessity to use cells from multiple fetuses in a single graft greatly compounded the problem. Cell viability and composition vary in different donors, and, further, the heterogeneity in the donor cells increased the probability of immunological rejection or contamination. An ideal cell source for cell therapy is one that is renewable, thus eliminating the need for transplantation of primary fetal tissue, and that also allows for viability, sterility, cell composition, and cell maturation to be controlled, while being inherently not tumorigenic. The availability of continuous and standardized clinical grade normal human neural cells, able to combine the plasticity of fetal tissue with an extensive proliferating capacity and functional stability, would be of paramount importance for the translation of cell therapy for central nervous system (CNS) disorders into the clinic. Here we describe a well-established protocol to

---

All authors contributed equally to the work.

D. Ferrari (✉)
Department of Biotechnology and Biosciences, University of Milano-Bicocca, Milan, Italy
e-mail: daniela.ferrari@unimib.it

M. Gelati
IRCCS Casa Sollievo della Sofferenza Hospital, San Giovanni Rotondo (FG), Italy

Stem Cell Laboratory, Cell Factory and Biobank, S.Maria Hospital, Terni, Italy

D. C. Profico
IRCCS Casa Sollievo della Sofferenza Hospital, San Giovanni Rotondo (FG), Italy

A. L. Vescovi (✉)
Department of Biotechnology and Biosciences, University of Milano-Bicocca, Milan, Italy

IRCCS Casa Sollievo della Sofferenza Hospital, San Giovanni Rotondo (FG), Italy

Stem Cell Laboratory, Cell Factory and Biobank, S.Maria Hospital, Terni, Italy
e-mail: angelo.vescovi@unimib.it

L. Buzanska (ed.), *Human Neural Stem Cells*, Results and Problems in Cell Differentiation 66, https://doi.org/10.1007/978-3-319-93485-3_14

produce human neural stem cells following GMP guidelines that allows us to obtain "clinical grade" cell lines.

14.1 Introduction

When considering translation of stem cell therapy, particularly neural ones, from the laboratory bench to clinical application, certain criteria have to be applied:

1. An easily and ethically available origin of stem cells.
2. Successful transplantation of cells to the damaged region demonstrated in pre-clinical and possibly clinical pilot trials.
3. The ability of cells to survive and incorporate into host parenchyma.
4. Last but not the least, stem cells should be produced according to good manufacturing practice (GMP) protocols because they are defined as advanced therapies.

14.2 Neural Stem Cells: GMP-Grade Production and Characterization

As defined by the European Regulation 1394/2007, the production of an "advanced therapy medicinal product" (ATMP), requires compliance with pharmaceutical good manufacturing practice (GMP) guidelines.

According to Eudralex, Vol. 4, Annex 13, "*The application of GMP to the manufacture of investigational medicinal products is intended to ensure that trial subjects are not placed at risk, and that the results of clinical trials are unaffected by inadequate safety, quality or efficacy arising from unsatisfactory manufacture.*"

Achieving full compliance with GMP standard, during the production process of a neural stem cell population, can be quite tricky. GMP compliance requires a deep comprehension of both cell biology and regulatory framework, in order to translate the routine cell culture method into a fully standardized production protocol.

Here we try to define a set of useful guidelines for neural stem cell GMP production. It's important to underline that the first step to do is to define the process flow chart and draw up a set of detailed standard operating procedures (SOPs). Each macro-phase of the production process should be described in a different SOP. Every member of the staff should be trained on the appropriate procedures, and retraining should be performed periodically, to ensure that every staff member is always up to date.[1]

[1]This chapter refers only to the regulatory requirements dictated by the European Union Regulation.

14.2.1 Production

14.2.1.1 Raw Materials

Raw materials should be carefully evaluated in terms of composition, endotoxin contents, cleaning procedures, and packaging. When possible, the use of sterile disposable plastic materials should be taken into account, such as culture flasks, petri dishes, or bioreactors. Although this approach seems to be expensive, this will substantially reduce cross contamination risks; also the use of any non-disposable material or device needs to be validated in terms of cleaning and decontamination procedures. One of the most critical material for the NSC production is the culture medium; usually this is a DMEM/F12 medium, with growth factors (EGF, bFGF), BSA, and hormones (Gelati et al. 2013; Gritti et al. 2000; Vescovi et al. 1999a, b). The production of in-house-made culture media can be time-consuming due to the large number of reagents needed for the medium preparation. On the other hand, there are only few companies producing culture medium components in compliance with GMP requirements, and usually these components are declared only for R&D or IVD use.

Because every production protocol is different, the choice of materials, devices, and culture medium should be made on a risk-based approach, evaluating also the laboratory layout, material transfer procedures, number of batch for the year, and the historical data available from the production unit or the R&D department.

14.2.1.2 Human Neural Biopsy Supply

NSC can be isolated from fetal neural tissue. Usually the preferred source is tissue from aborted fetuses because of the possibility to operate on a time-scheduled program or a specimen obtained by an official tissue biobank.

Obviously the use of tissue retrieved from aborted fetuses can elicit ethical concerns, causing difficulties in the clinical application.

In the last years, our group set up a suitable protocol for NSC isolation from tissue specimens deriving exclusively from spontaneous abortion or in utero death (Mazzini et al. 2015), which can exclude any ethical concerns on the tissue origin (Gelati et al. 2013).

Regardless of the source, tissue samples are required to be freshly isolated. After the retrieval, tissue needs to be delivered to the designed cell factory, as soon as possible, in an appropriate culture medium with antibiotics and antifungal solution. Long delivery time requires also to put the tissue in a portable refrigerator, which allows to monitor the sample temperature for the entire transport.

It is strongly recommended to validate the transport procedure, especially in case of long-term transport; this will guarantee the suitability of the procedure. Special care should be care in the definition of the transport medium composition, to avoid any possible toxic effect related to the antibiotics/antifungal on the cells.

14.2.1.3 Production Method

Once delivered to the cell factory, tissues need to be immediately processed, in order to preserve cell vitality. Every tissue/cell manipulation should be conducted under a pharmaceutical grade A laminar flow cabinet, in a pharmaceutical grade B room, or in an isolator.

Tissue should be washed to remove the transport medium and the residual of antibiotics and/or antifungal, then transferred to a petri dish, and dissected. Depending on the fetus gestational age and the tissue preservation, it may be possible to identify and cut apart specific anatomical structure rich in neural stem cells, such as the sub-ventricular zone (SVZ). The selection of this anatomical structure will allow to have a bigger batch size and the most efficient production.

Once the tissue is dissected and disaggregated, the obtained cell suspension should be counted and plated at the appropriate concentration (usually 10^4 cells/cm^2) and incubated in controlled atmosphere (CO_2 5%, O_2 5%, T 37.0 °C, RH > 95%) for 5–10 days. Cultures should be carefully monitored, especially in the very first days, so as to verify the presence of any contaminants and to evaluate the cell growth. Depending on the anatomical region of origin and the gestational age of the fetus, cells will show a different growth rate, forming neurospheres in a period comprised between 5 and 10 days. Regardless of the growth rate, cells should be disaggregated and plated when neurospheres reach a diameter of about 100–120 μm. After at least six passages, the neurospheres' shape and diameter will be uniform enough to be cryopreserved; it's recommended to use a validated controlled-rate freezer in order to monitor the freezing process and to transfer cells in nitrogen vapor right after the freezing, for long-term storage. Once the patient has been recruited for the treatment according to an authorized clinical trial, cells need to be thawed, washed, and plated for 24–48 h. After that, additional 5–10 passages are recommended in order to allow cells to restart their physiological cell cycle and to expand to the required amount.

Due to the short lifespan of the cells, the drug should be formulated right before the administration to the patient. Cells should be washed from the culture medium, then resuspended in the appropriate buffer at the required concentration, and immediately delivered to the operatory room.

14.2.1.4 Process Scale-Up

The above described GMP production method for hNSC is a manual expansion process, through operator manipulation of traditional culture vessel. This approach led us to conduct a phase I clinical trial on amyotrophic lateral sclerosis patients (EudraCT number 2009-014484-39), successfully concluded in mid-2015.

The transition to phase II or III clinical trials requires larger number of cells for larger number of patients, which involves the use of automated culture systems.

Culturing stem cells as suspension of free-floating aggregates could theoretically facilitates large-scale production of cells in closed bioreactors; still this could elicit several problems such as:

1. How to provide automatic neurosphere disaggregation without enzymes? The use of enzymes is the easiest way to achieve disaggregation directly into the bioreactor, by simply injecting the enzyme into the culture medium. In our experience, the use of enzymes such as trypsin or collagenase for neurosphere disaggregation should be avoided in order to prevent NSC to experience chemical stress. Also, the addition of chemical/enzymatic molecules to the culture medium requires additional preclinical in vivo and in vitro tests for safety and toxicology. The classic mechanical disaggregation leads no consequences in terms of chemical stress to the cells and has been already successfully validated and used in clinical trials; unfortunately the use of an automated production system doesn't allow the classic manual disaggregation, so a compromise has to be found, in favor of one solution or the other, depending mostly on the availability of both preclinical in vitro and in vivo data and economic resources.
2. Can stirrer bioreactor provide adequate oxygen exchange in culture? Culturing cells in large medium volumes can negatively affect the oxygen distribution into the culture. Usually the combination of gas insufflation and mechanical agitation of the vessel are used to improve gas distribution. Still, usually the automated system commercially available takes account only for carbon dioxide rather than oxygen distribution. The use of any kind of automated system should be carefully validated in terms of oxygen distribution to the entire vessel.
3. How to verify neurosphere morphology and diameter during the culture? The use of classic vessel in incubator allows the operators to check for cell growth at any time, simply using a microscope and not affecting the culture. The use of automated system prevents the operator for regular visual checks, unless they take a sample of the culture. Of course this could negatively affect the product in terms of both potential microbiological contamination and product waste.

14.2.2 Characterization

Every batch of NSC should be tested to ensure correspondence to the prescribed quality requirements. Starting from the process flow chart, it is important to identify the most critical point of the process in which process controls (IPCs) should be carried out, to monitor the ongoing culture.

IPCs should be conducted both on cells and culture media. The nature and the frequency of the tests should be defined on a risk-based approach, based on the final drug composition and the cells' physiological behavior. As described in Mazzini et al. (2015), in a recently completed clinical trial on ALS patients treated with NSCs, our process control strategy comprises:

- Microbiological tests conducted according to the European Pharmacopeia, to verify the presence of microbial contaminants and endotoxin level during the culture
- Cellular tests, including vitality evaluation and cell growth monitoring by the construction of a growth curve

At the last step of the production protocol, cells need to be tested and defined by the batch release control plan studied in order to define the drug safety, identity, and potency:

- Compendial tests, performed according to the European Pharmacopeia (or equivalent) requirements, i.e., sterility test, mycoplasma test, and LAL test
- Not compendial tests:
  - Self-renewal
  - Clonal efficiency
  - Differentiation test
  - Karyotype
  - Growth factor dependence

According to the European Regulation, it is allowed to deliver the drug substance to the patient before the conclusion of the release tests. If so, it is advisable to run a more detailed IPC strategy, to ensure the safety of the patient.

14.3 Preclinical Tests: Safety, Pharmacology, and Efficacy

Preclinical studies also including the use of animal models represent a key step within the evaluation of any drug (Daley et al. 2016; Frey-Vasconcells et al. 2012). Pharmacopeia has defined specific experimental paradigms to test toxicity as well as pharmacokinetic and pharmacodynamic properties of traditional drug compounds. These assays need to be partially adjusted, in order to be applied to cellular product such as hNSCs. In this case, the investigational medicinal product (IMP) is a biological living system, and the interaction with the host tissues and bodily clearance can be, only partially, assimilated to classical inorganic compounds.

Toxicology of hNSCs should be addressed so as to evaluate potential inflammatory reaction, immune rejection of host tissue, and carcinogenicity risk (in this case referable to the possibility that injected hNSCs might transform into tumorigenic cells).

Biodistribution properties of hNSCs depend on cell delivery route and tissue-related variables such as inflammatory state. One of the main properties of stem cells is their capacity to respond to secreted signals that regulate their ability to maintain both tissue homeostasis—through the modulation of inherent stem cell properties (self-renewal, asymmetric, or symmetric cell division)—and retention/homing mechanisms. For this reason, in addition to more classical distribution dynamics, these signals might be important elements to be taken into account when evaluating pharmacokinetic aspects of NSCs transplantation. This ability of NSCs might indeed influence redistribution of transplanted cells to niche and peripheral locations.

Pharmacodynamic activity of transplanted NSCs comprises a broad array of mechanisms, which partly depend on the inflammatory status of the nervous tissue and on the specific mechanisms at work in each given pathology. These mechanisms appear to emerge from the complex interplay between the transplanted cells and the damaged host tissue and can be tentatively subdivided into two main sequential key modes of action:

1. Integration of the cells into the host nervous system and their differentiation and maturation and long-term survival, with the potential to regenerate the tissue by directly replacing damaged/degenerated cells.
2. "Bystander" effect: local or systemic release of a number of molecules as cytokines and growth and neurotrophic factors by the transplanted/integrated cells. This can elicit immunomodulatory and neurotrophic actions, which can prevent or even revert cell death and the ensuing degeneration phenomena, thus antagonizing the key pathological events of the disease.

The *efficacy* of hNSC treatment needs to be evaluated in specific animal models for each disease. Notwithstanding this consideration, given the notion that neurodegenerative diseases share common features (such as neuroinflammation), it is possible that therapeutic mechanisms discovered for a specific disease might be translatable to others.

In the following paragraphs, we will outline the most important elements to consider in designing preclinical safety screening of hNSCs to be used in clinical setting.

14.3.1 Toxicology

14.3.1.1 Tumorigenic Risk Evaluation

One of the issues related to the use of stem cells for clinical purposes is their proliferative potential and the risk that the cells might cause proliferative lesions. These concerns have been fostered by reports of stem cell-derived tumor formations in patients, in the context of the so-called stem cell tourisms (Amariglio et al. 2009; Berkowitz et al. 2016). In regard to this issue and in order to produce safe cellular product to be applied in clinical trials, it is of utmost importance the quality of the expansion/production process (as discussed earlier in the present chapter) and the definition of proper preclinical safety assays to monitor the functional and genetic stability of each cell line (Bailey 2012; Daley et al. 2016; Frey-Vasconcells et al. 2012). Poorly characterized cells or preparations consisting of a mixture of donors/

stem cell types expose the patients to the risk of serious adverse events (Amariglio et al. 2009; Berkowitz et al. 2016) and have the additional drawback to undermine the development of safe clinical studies with stem cells. In contrast, the results obtained from the currently ongoing or concluded phase I and II clinical trials (see Tables 14.1 and 14.2) are so far reassuring on the safety of properly characterized hNSCs (for details see Sect. 14.4).

Given this premise, the potential risk of cell transformation as a consequence of extensive in vitro manipulation needs to be carefully evaluated. Unlike other somatic stem cells (Miura et al. 2006) or neural progenitors derived from pluripotent stem cells (Carpenter et al. 2009; Iida et al. 2017), NSCs have shown to be more resilient to transformation and endowed with an inherent genetic and functional stability (Foroni et al. 2007; Mazzini et al. 2015) although others have reported a moderate degree of chromosomal instability (Diaferia et al. 2011; Vukicevic et al. 2010). As from our experience, murine adult NSCs can be expanded by using the neurosphere assay technique over 100 passages without undergoing transformation and do not give rise to tumors in vivo even when immortalized through delivery of *Myc* and *Ras* oncogene (Foroni et al. 2007). Preclinical testing of human fetal hNSCs applied in phase I trial for ALS patients (Mazzini et al. 2015) has reinforced this notion. Indeed, hNSC lines used in the study were evaluated by transplantation into the striatum of athymic nude mice in long-term experiments, i.e., 6 months after transplant, a time window which is sufficient for 50 cancer stem cells derived from human glioblastoma (hGBM) to develop in significant masses. Brain histopathological observations never showed signs of tumoral transformation or aberrant growth, and histological evaluation of peripheral organs (liver, kidney, spleen, lung, heart, and peritoneal lymph nodes) did not reveal any signs of pathogenic event related to transplanted cells (Mazzini et al. 2015).

A platform of tests to address the safety of NSC lines in the context of clinical application should include:

1. Functional characterization

It is important to monitor the growth rate, verifying that no abrupt changes in the proliferation ability of the cells occur along the amplification process and confirm the maintenance of growth factor dependence for expansion.

Although all hNSC lines must be endowed with the ability of self-renewal (Gelati et al. 2013; Vescovi et al. 1999a, b), each of them display variable growth rate; in relation to this, an additional precaution could be established, for example, by defining a threshold of the growth rate above which a cell line cannot be accepted for clinical purposes. The threshold could be determined by comparison of the growth rate of cancer stem cells derived from human glioblastoma (hGBM). In any case, a fast growth is not per se indicative of tumorigenic transformation; as already mentioned, the important functional properties to be evaluated are the stability of both growth profile and growth factor dependence.

Table 14.1 Completed clinical trials using hNSC for the cure of neurodegenerative diseases

Disease	NSC source	Sponsor, PI place-period	Trial ID, phase and brief title	Transplant and follow-up	Notes
Neuronal ceroid lipofuscinosis (NCL)	HuCNS-SC® Fetal brain Single donor	StemCells, Inc. Robert Steiner, MD United States 2006/2009	NCT00337636—phase I Study of HuCNS-SC cells in patients with infantile or late infantile neuronal ceroid lipofuscinosis (NCL)	Six patients Intracerebral Single dose Multiple injections (bilaterally, lateral ventricle plus three subcortical sites) Two groups: 5×10^8 and 1×10^9 cells/patient Follow-up: 13 months	Cells: Uchida N et al., PNAS, (2000) Trial: Selden NR et al., J Neurosurg Pediatr, (2013)
Pelizaeus-Merzbacher disease (PMD)	HuCNS-SC® Fetal CNS Single donor	StemCells, Inc. Stephen Huhn, MD United States 2009–2012	NCT01005004—phase I Study of human central nervous system (CNS) stem cell transplantation in Pelizaeus-Merzbacher disease (PMD) subjects	Four patients Intracerebral Single dose Multiple injections (bilaterally, two sites in the corona radiata of the frontal white matter) 3×10^8 cells/patient Follow-up: 12 months	Cells: Uchida N et al., PNAS, (2000) Trial: Gupta et al., Sci. Transl. Med, (2012)
Cerebral palsy	Human neural progenitor Fetal forebrain	n.p.	n.p.	45 patients + 49 controls Intracerebral Single dose Single injection (unilaterally, lateral ventricle) 8–10 × 10^6 cells/patient	Not enrolled in European/US databases Luan Z et al., Cell Transpl, (2012)
Stroke	Human Fetal CNS	Yonsei University College of Medicine, Severance Hospital Park KI Korea 2006–n.p.	KCT0001429—phase I/II Clinical trial of human neural stem cell transplantation in patients with severe perinatal hypoxic-ischemic brain injury	Patients: n.p. Intracranial Single dose Multiple injections 2.5×10^7 cells/patient	

(continued)

Table 14.1 (continued)

Disease	NSC source	Sponsor, PI place-period	Trial ID, phase and brief title	Transplant and follow-up	Notes
ALS	Fetal brain Single donor	Not for profit Azienda Ospedaliera Santa Maria Angelo Vescovi Italy 2011–2015	NCT01640067—EudraCT 2009-014484-39—phase I Human neural stem cell transplantation in amyotrophic lateral sclerosis (ALS) (hNSCALS)	18 patients Intraspinal cord Single dose Multiple injections depending on group (unilateral lumbar or bilateral lumbar or unilateral cervical or bilateral cervical) 2.5×10^6–4.5×10^6 cells/patient Immunosuppression: 6 months Follow-up: 12 months	Mazzini L, et al., J Trans Med, (2015)
SCI	HuCNS-SC® Fetal CNS Single donor	StemCells, Inc. Stephen Huhn, MD Canada, Switzerland 2011–2015	NCT01321333—phase I/II Study of human central nervous system stem cells (HuCNS-SC) in patients with thoracic spinal cord injury	12 patients Intramedullary spinal cord Single dose Multiple injections 2×10^7 cells/patient Follow-up: 18 months	
SCI	HuCNS-SC	StemCells, Inc. Stephen Huhn, MD Switzerland 2012–2017	NCT01725880 Long-term follow-up of transplanted human central nervous system stem cells (HuCNS-SC) in spinal cord trauma subjects	12 patients No transplant, only follow-up (see notes)	Long-term follow-up of the trial NCT01321333 Terminated based on business decision unrelated to any safety concerns
SCI	HuCNS-SC® Fetal CNS derived Single donor	StemCells, Inc. Stephen Huhn, MD USA 2014–2017	NCT02163876—phase II Study of human central nervous system (CNS) stem cell transplantation in cervical spinal cord injury	50 patients Intraspinal cord Cervical region n.p.	Terminated based on business decision unrelated to any safety concerns

AMD	HuCNS-SC® Fetal CNS derived Single donor	StemCells, Inc. Stephen Huhn, MD United States 2012–2015	NCT01632527—phase I/II Study of human central nervous system stem cells (HuCNS-SC) in age-related macular degeneration (AMD)	Fifteen patients Subretinal transplantation, single dose 200,000 or 1 million cells/patient Follow-up: 18 months	

Source: clinical trials.gov; International clinical trial registry platform (WHO), EU clinical trial register EUdraCT, Giusto et al. (2014)
SCI spinal cord injury, *n.p.* not provided

Table 14.2 Active studies using hNSC for the cure of neurodegenerative diseases

Disease	NSC source	Sponsor, PI place-period (start/primary compl./ end)	Trial ID, status, phase, and brief title	Transplant	Notes
Stroke	CTX0E03, c-myc immortalized Fetal brain Single donor	ReNeuron Limited Keith Muir, MD United Kingdom 2010–2015–2023	NCT01151124—active, not recruiting—phase I Pilot investigation of stem cells in stroke (PISCES)	12 patients Intracerebral Single dose Single injection (Putamen) Four groups from 2 to 20 million/patients	Cells: Pollock et al., Exp Neurol (2006)
Stroke	CTX0E03 DP, c-myc immortalized Fetal, brain Single donor	ReNeuron Limited, Keith W Muir United Kingdom 2014–2015–2017	NCT02117635—recruiting—phase II Pilot investigation of stem cells in stroke phase II efficacy (PISCES-II)	Forty-one patients Intracerebral Single dose Single injection (Striatum) 20 million/patient	Cells: Pollock et al., Exp Neurol (2006)
Stroke	Human Fetal CNS	Yonsei University College of Medicine, Severance Hospital Park KI Korea 2006–n.p.	KCT0001429—phase I/II Clinical trial of human neural stem cell transplantation in patients with severe perinatal hypoxic-ischemic brain injury	Patients: n.p. Intracranial Single dose Multiple injections 2.5×10^7 cells/patient	
ALS	Fetal (8 weeks) Spinal cord derived Single donor	Neuralstem, Inc. Eva Feldman MD, PhD USA 2009–2015	NCT01348451—phase I—active not recruiting Human spinal cord derived neural stem cell transplantation for the treatment of amyotrophic lateral sclerosis (ALS)	18 patients (~35–67 years old) Intraspinal cord Single or two doses (subsequent surgical procedures) Multiple injections depending on group (unilateral lumbar or bilateral lumbar or unilateral cervical or bilateral lumbar + unilateral cervical)	Cells: Guo X, et al., —Glass JD Cerebrum, (2010), J Tissue Eng Regen Med, (2010) Trial: Riley J., et al., Neurosurgery, (2011) and (2012) Glass JD, et al., Stem Cells, (2012)

				500,000–3 × 10^6 cells/patient Immunosuppression: lifelong (discontinued in 5 patients)	Feldman EL, et al., Ann Neurol, (2014) Tadesse T, et al., Ann Clin Transl Neurol, (2014)
ALS	Fetal (8weeks) Spinal cord-derived Single donor	Neuralstem Inc. Eva Feldman MD, PhD USA Dec 2015–Nov 2016–unknown	NCT01730716—phase II—unknown—dose escalation and safety study of human spinal cord-derived neural stem cell transplantation for the treatment of amyotrophic lateral sclerosis	15 patients Intraspinal cord Up to five doses Multiple injections (bilateral cervical or bilateral cervical and lumbar) From 2 × 10^6 cells to 16 × 10^6 cells/patient	Glass JD, et al., Stem Cells, (2012) Riley J, et al., Neurosurger, (2012) Glass JD, et al., Neurology, (2016)
SCI	Human fetal CNS	Ministry of Health and Welfare, Park KI 2006–2015	KCT0000879—phase I/II—recruiting-clinical trial of human neural stem cell transplantation in patients with traumatic spinal cord injury	Intraspinal cord Single dose 3.5 × 10^7 cells/patient 60 patients	
Peripheral arterial disease (Fontaine stage II–IV)	CTX0E03 c-myc immortalized Fetal, brain Allogeneic Single donor	ReNeuron Limited, Jill JF Belch, MBChB United Kingdom	NCT01916369—phase I—active recruiting Safety trial of CTX cells in patients with lower limb ischemia	Nine patients Intramuscular (gastrocnemius muscle) Single dose Ten injections (20, 50, or 80 million cells)	
SCI	Spinal cord derived	Neural stem Inc. Joseph Ciacci USA 2014–Dec 2022	NCT01772810—recruiting—phase I Safety study of human spinal cord-derived neural stem cell transplantation for the treatment of SCI	Eight patients T2–T12 or C5–C7 transplanted	

(continued)

Table 14.2 (continued)

Disease	NSC source	Sponsor, PI place-period (start/primary compl./end)	Trial ID, status, phase, and brief title	Transplant	Notes
PD	Human neural stem cells	Soochow University Jing Chen China April 2017–Nov 2018	NCT03128450—recruiting on invitation—phase II/III A study to evaluate the safety and efficacy of human neural stem cells for Parkinson's disease patient (hNSCPD)	12 patients with moderate PD	
Stroke	Human Fetal CNS	Yonsei University College of Medicine, Severance Hospital Park KI Korea 2006–n.p.	KCT0001429—phase I/II Clinical trial of Human neural stem cell transplantation in patients with severe perinatal hypoxic-ischemic brain injury	Patients: n.p. Intracranial Single dose Multiple injections 2.5×10^7 cells/patient	

2. Genomic stability

Genomic stability should be monitored by evaluating karyotype stability over time. Additional studies, made more recently available, such as whole-genome array technique, might further help in analyzing more subtle DNA aberrations.

3. Transplantation into immune-deficient models

Immune-deficient models represent one of the elective experimental paradigms for the analysis of putative tumorigenicity of hNSC in vivo. In fact it has been successfully used to demonstrate the tumor-initiating ability of neural cancer stem cells from human glioblastomas (hGBMs) (Fogh et al. 1977; Galli et al. 2004; Shultz et al. 2007). Transplantation of very low numbers of cancer stem cells from hGBM (50 cells/animal) into the brain of these mice generates aggressive tumors in 4 months, demonstrating the sensitivity of this model in revealing tumorigenic potential of even such a small population of cells (Mazzini et al. 2015). The use of the rat model could allow to transplant higher numbers of cells, hence permitting to verify additional safety parameters such as acute toxicity (Garitaonandia et al. 2016).

4. Immune rejection

The brain has long been considered an immunologically privileged site for transplantation owing to both the presumed absence of functional lymphatic vessels able to deliver antigen from CNS tissue to secondary lymphoid organs and the presence of a blood-brain barrier (BBB) able to block infiltration of blood-borne immune cells. These notions have been lately revised, and several immune surveillance mechanisms at work in healthy and diseased CNS have been described (rev. in Hoornaert et al. 2017), including the discovery of functional lymphatic vessels connecting the meninges to deep cervical lymph nodes, as well as the increase of BBB permeability in response to bacterial lipopolysaccharide and/or pro-inflammatory cytokines such as tumor necrosis factor (TNF)-a, interferon (IFN)-g, and interleukin (IL)-1b (Hickey et al. 1991; Louveau et al. 2015). In addition activation of innate immune cells can be elicited by cell transplantation, involving recruitments of neutrophils and macrophages/microglia cells, and finally astroglial scarring as proposed for mesenchymal cell graft and NSCs (for rev. see Hoornaert et al. 2017).

In spite of this growing knowledge of immune responses arising into the CNS in response to cell grafting, results obtained in many years of hNSCs transplantations in animal models are encouraging regarding the possibility of overcoming mechanisms of cells rejection. hNSCs can successfully engraft and differentiate into the CNS of immune-competent animal models with rejection controlled by traditional immunosuppressive drugs. Importantly it has been shown that even using a transient immunosuppression protocol (cyclosporine administration for only 2 weeks after transplant), hNSCs can long-term survive into the brain and spinal cord of rat models of amyotrophic lateral sclerosis (unpublished data) and global ischemia (Rota Nodari et al. 2010). In these studies, a consistent percentage of the transplanted cells survived into the brain ($\sim$20% at 1 and 4 months), suggesting a quite stable profile of integration and survival over a long period even under mild immunosuppressive therapy. In addition immunomodulatory abilities of NSCs (rev. in Drago et al. 2013; Giusto et al. 2014) can further contribute to transplant survival.

Only the results derived from clinical trials could clarify whether hNSC transplant will be rejected by the CNS immune response (see Sect. 14.4).

14.3.2 Biodistribution and Route of Administration

hNSCs for neurodegenerative diseases can be delivered essentially by three different routes (by intravenous, intraparenchymal, or intra-cerebroventricular administration) and have been attempted in several animal models.

Intraparenchymal injection in the nervous central system, particularly into a precise anatomic site is a tricky surgical procedure. An example is the hNSC injection into the spinal cord ventral horn of ALS patients (Mazzini et al. 2015); in this case a stereotactic platform will be fixed to the spine through percutaneous posts. After laminectomy and dural opening, a "floating retracting cannula" design should be used to place a needle precisely into the ventral horn, using a rigid conformation. After placement, the cannula should be retracted and converted to its flexible form to allow it to float jointly liable with the spinal cord/central nervous system (Riley et al. 2014). Preclinical results have shown that by using this way of delivery, NSCs migrate and spread within the nervous tissue, in some cases along preferred routes such as the corpus callosum for the brain (Rota Nodari et al. 2010) or driven by specific disease-related factors released by damaged tissue (Ferrari et al. 2012). Interestingly, the distance of migration seems to be a property inherent to the specific NSC line.

Intra-cerebroventricular injection is preferred in diseases with multifocal degeneration of the brain (i.e., multiple sclerosis); in this case a stereotactic frame (Lunsford 2009) will be used. This device allows for the identification of the target and the trajectory, with the added possibility to run the trajectory through areas of interest, beside the final target in the ventricle.

Intravenous injection of cells is apparently a safe and easy route of cell delivery; NSCs systemically injected were able to reach the demyelinating areas of the CNS in animal models of multiple sclerosis such as EAE animals (Pluchino et al. 2003, 2009a), crossing the inflamed blood-brain barrier, presumably under the guidance of integrin-driven system, such as CD44 and VLA-4 (expressed on NSCs surface), and eliciting therapeutic actions. In these cases, NSCs reach several peripheral organs (heart, liver, lungs, kidney, gut, and lymph nodes) with complete clearance observed between 20 and 90 days in all the abovementioned organs except in the gut and kidney, where occasionally few scattered cells were found at 90 days posttransplant (Pluchino et al. 2009a). Into the CNS and CNS-draining lymph nodes, the NSCs remained longer, either because of the delivery route (aimed to these organs) or presumably because they were retained by local cues as part of their possible therapeutic action (Pluchino et al. 2009a, b). NSC retention into lymph nodes seems to be mediated through cell-to-cell interactions, as evidenced by anatomical complexes formed between injected NSCs and lymph node cells, such as polarized microvilli, cytoplasmic extensions, or elongated intercellular junctions (Pluchino et al. 2009b). Notwithstanding preclinical studies, this way of administration is still not convincing for clinical applications; in fact it seems that the main effect of intravenous injection is systemic, since only a small fraction of cells enter the CNS. It is important to evidence that in case of intravenous delivery, the vast majority of injected cells (~90%) finds their final location in organs other than the CNS, even when the latter is not shielded by the blood/tissue barrier (Pendharkar et al. 2010; Pluchino et al. 2003, 2009a); in addition, transplanted cells are lost because of physical stress, inflammation, hypoxia, anoikis, or immunogenic rejection. Also there are no convincing data to suggest that intravenously injected NPC migrate efficiently further from the perivascular space into white matter tracts to attain their trophic and regenerative properties.

Finally, systemic administration can lead cells to become entrapped in the lung or microvasculature, causing dangerous side effects, such as the pulmonary emboli reported following intravenous administration of adipose tissue-derived stem cells (Heslop et al. 2015).

In conclusion, even if preclinical data suggest that systemic administration could be attempted to achieve widespread distribution of NSCs for multifocal diseases, the strategy so far applied to clinical trials is to choose CNS delivery, both for safety reasons and as it allows to maximize the number of cells that reaches the damaged areas of the nervous tissue.

14.3.3 Efficacy

There is now a plenitude of preclinical evidences that shows how transplantation of NSCs can carry out antagonizing effects on both inflammation and neurodegeneration, on a long-term time frame. Albeit a comprehensive description of this vast topic goes behind the scope of the present chapter, we will briefly mention some of the most significant mechanisms underpinning the possible amelioration of these diseases upon stem cell transplantation, as evidenced in animal models of various neurodegenerative diseases. NSCs can home into the damaged areas of the CNS even when delivered systemically, both in the cerebrospinal fluid and into the blood circulation, also crossing the damaged blood-brain barrier (Ben-Hur et al. 2003; Pluchino et al. 2003, 2009a), integrate as macroglia and neurons (Parr et al. 2007, 2008; Rota Nodari et al. 2010), carry out local remyelination directly by differentiating into oligodendrocytes (Ferrari et al. 2012; Karimi-Abdolrezaee et al. 2006; Neri et al. 2010; Parr et al. 2008; Pluchino et al. 2003), and, by stimulating the host's own local oligodendroglial precursors (Einstein et al. 2009), dampen local inflammatory/immune response and astroglial scarring (Bacigaluppi et al. 2009; Capone et al. 2007; Daadi et al. 2010; Ferrari et al. 2012; Lee et al. 2008; Pluchino et al. 2003, 2009a; Rota Nodari et al. 2010).

14.4 Clinical Protocols

Encouraged by the promising preclinical data obtained in experimental models of CNS diseases, in 2006, the use of hNSCs has reached the clinical application. Since then, these cells have been applied in a variety of fatal and not-fatal incurable neurodegenerative diseases allowing to make preliminary but encouraging considerations on the safety and feasibility of this approach. Current ongoing or completed studies involving the use of hNSCs have been summarized in Tables 14.1 and 14.2, highlighting relevant parameters of the procedure, such as delivery route, transplant sites, and cell dosage.

So far, more than 100 patients have been treated with hNSCs, and no major negative complications have been derived from the surgery or donor cells (considerations based on "completed" or "active not recruiting" studies).

14.4.1 hNSC Lines Approved for Clinical Trials

hNSC lines currently applied to registered clinical studies are produced by four centers. The cell lines are generated according to processes compliant with GMP in serum-free media and have been characterized in vitro and in vivo to certify safety, differentiation abilities, and therapeutic potential.

1. hNSC lines from Azienda Ospedaliera Santa Maria. These cells (Gelati et al. 2013) are derived from donated fetal brain derived from spontaneous miscarriages. Each line is derived from a single donor; cells are expanded as neurospheres in a serum-free medium containing EGF and FGF-2 and have been used in the phase I clinical trial I EudraCT 2009-014484-39 (also registered in www.clinicaltrials.gov as NCT01640067).
2. HuCNS-SC lines (StemCells Inc.) are derived from a single-donated fetal brain enzymatically digested, FACS-sorted to retrieve CD133+ CD24-negative/low population, and subsequently expanded as neurospheres in the presence of FGF-2, EGF, and LIF (Carpenter et al. 1999; Uchida et al. 2000). Used in phase I trials: NCT00337636, NCT01005004, NCT 01321333, and NCT01632527. Two phase II trials (NCT01725880 and NCT02163876) have been terminated in advance due to economic difficulties of the company, importantly not for reason related to safety concerns [News in Nature Biotechnology (2016) Vol. 34, Nr. 7, 677–678]..
3. CTX0E03 are produced by ReNeuron. The line has been derived from fetal brain and subsequently genetically modified with a conditional immortalizing gene, c-mycER™. This transgene generates a fusion protein that stimulates cell proliferation in the presence of a synthetic drug 4-hydroxytamoxifen (4-OHT). The cell line is clonal and expands rapidly in adhesion in the presence of EGF and FGF-2, but the cells undergo growth arrest and differentiate into neurons and astrocytes after the removal of growth factors (Pollock et al. 2006). Used in phase I trial NCT01151124 and phase II NCT02117635.

4. NSC from fetal spinal cord produced by Neuralstem, Inc. are grown as adherent monolayer in a medium containing FGF-2 as single growth factor (Guo et al. 2010). Used in phase I trials NCT01348451 and NCT01772810 and phase II NCT01730716.

In this section we comment on clinical studies whose results have been published and that could highlight important safety parameters of clinical use of hNSCs.

14.4.2 Clinical Studies for ALS

ALS is a devastating motor neuron disease for which there are no efficacious therapies. Based on supportive, preclinical proof-of-principle data, up to 21 somatic stem cell-based pilot trials have been developed (Atassi et al. 2016). Three of those studies have used hNSC lines: NCT01640067 with cells produced from Azienda Ospedaliera Santa Maria Terni (Gelati et al. 2013; Mazzini et al. 2015; Vescovi et al. 1999a, b) and NCT01348451 and NCT01730716 with cells produced by Neuralstem Inc. (Feldman et al. 2014; Glass et al. 2012, 2016; Guo et al. 2010). In these studies, cells have been delivered intraparenchymal into the spinal cord of patients, according to a protocol specifically designed to minimize the trauma derived from injection (Riley et al. 2011, 2012, 2014) (see Tables 14.1 and 14.2 for details on the design). The major outcomes that can be derived from these studies are related to the safety of the approach: surgery was uncomplicated in most patients, and few side effects were reported, with the most common negative event from surgery being transient pain in the site of treatment. In the phase II trial aimed at escalating cell dosage (Glass et al. 2016), two patients, receiving the highest dosage of cells, incurred life-altering complications such as pain, sensory loss, and paraparesis. The causes are debatable and surgical trauma and inflammatory reaction remaining possibilities. The other four patients receiving the same treatment did not experience the same adverse events. Immunosuppressive treatment was administered for 6 months in one trial (Mazzini et al. 2015) and had to be discontinued in some patients of the US trial. In spite of this, cells were well tolerated, and no immunological signs of rejection were evidenced; on the contrary, postmortem analysis showed the integrity and survival of grafted cells up to 2.5 years posttransplantation (Tadesse et al. 2014). Most importantly, in spite of the above described concerns related to tumor formation derived from transplanted cells (Amariglio et al. 2009; Berkowitz et al. 2016), there were no indications of aberrant cell growth, thus suggesting that correct production procedures and thorough preclinical testing are extremely relevant to minimize the risk of cell overgrowth (Mazzini et al. 2015). Interestingly, all the studies confirmed no acceleration in the course of the disease; on the contrary some of the patients showed a transient improvement of motor functions and ALS-FRS scale (Feldman et al. 2014; Glass et al. 2016; Mazzini et al. 2015). Although the data are encouraging, it is not possible to infer efficacy data from such small pilot trials; this type of conclusions should be addressed more properly in larger phase II/III trial (see Atassi et al. 2016).

14.4.3 Clinical Studies for Pediatric Disease

In three completed trials (Gupta et al. 2012; Luan et al. 2012; Selden et al. 2013), hNSC have been delivered into brain regions of pediatric patients suffering from neuronal ceroid lipofuscinosis (NCL), Pelizaeus-Merzbacher disease (PMD), and cerebral palsy.

Both intraventricular and parenchymal (brain) delivery strategies have been used. These studies are particularly relevant because they demonstrate the safety of intraventricular administration of hNSCs. Indeed no adverse alterations were detected at doses up to 200 millions of cells/ventricle and 1 billion cells for patient.

Posttransplant MRI and CT analyses have revealed no major tissue damage derived from the procedure; in few cases, limited hemorrhagic event occurred in the early postoperative period but were resolved without clinical consequences. The surgery procedure and most probably the immunosuppression therapy induced mild to moderate systemic adverse event all manageable with standard supportive care or adjustments in immunosuppressive agent dosage. No opportunistic infections were reported. Most serious or even fatal events were related to the natural course of the underling neurodegenerative pathology (seizure and respiratory failure) and occurred at least after a year from transplantation. No neurological adverse events were observed as precipitous and focal deterioration of neurological functions or unexpected worsening of the pathological profile of the inherent neurodegeneration process.

In follow-up windows of over a year (in one case up to 3.5 years), no signs of generation of cellular atypia, neoplasia, ischemia, or inflammation were evident in the area of transplant as from histological observations, MRI, or MR spectroscopy. Histology revealed NSC migration possibly following fiber tract or blood vessels as preferred pathways of migration. In one case the cells were not detected (3–5 years), although demonstrating that even in case of cell rejection and death, the host parenchyma does not react with massive inflammation or develops other harmful reactions.

The definition of end points to evaluate the success or failure of each trial is a difficult challenge, due to the complexity of these diseases and our limited knowledge of the pathological mechanisms, which unable to follow in details the disease progression and the possible favorable impact of cell transplant on it. Notwithstanding these considerations, the data derived from these patients also suggest a mild improvement of neurological parameters both for cerebral palsy and MPD. Interestingly in three of the four MPD patients, a modest gain in neurological and neuropsychological profiles was accompanied by indications of a remyelination process ongoing in the area of the transplant, as assessed by MRI and DTI.

Overall these hNSCs used in human studies showed that no major complication is derived from transplant procedure, immunosuppression therapy, or delivered cells.

Many other studies are currently ongoing (see Table 14.2) demonstrating an increasing interest in the field of advanced therapies particularly related to somatic cell transplant using neural stem cells.

References

Amariglio N, Hirshberg A, Scheithauer BW et al (2009) Donor-derived brain tumor following neural stem cell transplantation in an ataxia telangiectasia patient. PLoS Med 6(2):e1000029. https://doi.org/10.1371/journal.pmed.1000029

Atassi N, Beghi E, Blanquer M et al (2016) Intraspinal stem cell transplantation for amyotrophic lateral sclerosis: ready for efficacy trials? Cytotherapy 18(12):1471–1475

Bacigaluppi M, Pluchino S, Peruzzotti-Jametti L et al (2009) Delayed post-ischaemic neuroprotection following systemic neural stem cell transplantation involves multiple mechanisms. Brain 132(Pt 8):2239–2251

Bailey AM (2012) Balancing tissue and tumor formation in regenerative medicine. Sci Transl Med 4(147):147fs128

Ben-Hur T, Einstein O, Mizrachi-Kol R et al (2003) Transplanted multipotential neural precursor cells migrate into the inflamed white matter in response to experimental autoimmune encephalomyelitis. Glia 41(1):73–80

Berkowitz AL, Miller MB, Mir SA et al (2016) Glioproliferative lesion of the spinal cord as a complication of "stem-cell tourism". N Engl J Med 375(2):196–198

Capone C, Frigerio S, Fumagalli S et al (2007) Neurosphere-derived cells exert a neuroprotective action by changing the ischemic microenvironment. PLoS One 2(4):e373. https://doi.org/10.1371/journal.pone.0000373

Carpenter MK, Cui X, Hu ZY et al (1999) In vitro expansion of a multipotent population of human neural progenitor cells. Exp Neurol 158(2):265–278

Carpenter MK, Frey-Vasconcells J, Rao MS (2009) Developing safe therapies from human pluripotent stem cells. Nat Biotechnol 27(7):606–613

Daadi MM, Davis AS, Arac A et al (2010) Human neural stem cell grafts modify microglial response and enhance axonal sprouting in neonatal hypoxic-ischemic brain injury. Stroke 41 (3):516–523

Daley GQ, Hyun I, Apperley JF et al (2016) Setting global standards for stem cell research and clinical translation: the 2016 ISSCR guidelines. Stem Cell Rep 6(6):787–797

Diaferia GR, Conti L, Redaelli S et al (2011) Systematic chromosomal analysis of cultured mouse neural stem cell lines. Stem Cells Dev 20(8):1411–1423

Drago D, Cossetti C, Iraci N et al (2013) The stem cell secretome and its role in brain repair. Biochimie 95(12):2271–2285

Einstein O, Friedman-Levi Y, Grigoriadis N (2009) Transplanted neural precursors enhance host brain-derived myelin regeneration. J Neurosci 29(50):15694–15702

Feldman EL, Boulis NM, Hur J et al (2014) Intraspinal neural stem cell transplantation in amyotrophic lateral sclerosis: phase 1 trial outcomes. Ann Neurol 75(3):363–373

Ferrari D, Zalfa C, Nodari LR et al (2012) Differential pathotropism of non-immortalized and immortalized human neural stem cell lines in a focal demyelination model. Cell Mol Life Sci 69 (7):1193–1210

Fogh J, Fogh JM, Orfeo T (1977) One hundred and twenty-seven cultured human tumor cell lines producing tumors in nude mice. J Natl Cancer Inst 59(1):221–226

Foroni C, Galli R, Cipelletti B et al (2007) Resilience to transformation and inherent genetic and functional stability of adult neural stem cells ex vivo. Cancer Res 67(8):3725–3733

Frey-Vasconcells J, Whittlesey KJ, Baum E (2012) Translation of stem cell research: points to consider in designing preclinical animal studies. Stem Cells Transl Med 1(5):353–358

Galli R, Binda E, Orfanelli U et al (2004) Isolation and characterization of tumorigenic, stem-like neural precursors from human glioblastoma. Cancer Res 64(19):7011–7021

Garitaonandia I, Gonzalez R, Christiansen-Weber T et al (2016) Neural stem cell tumorigenicity and biodistribution assessment for phase i clinical trial in Parkinson's disease. Sci Rep 6:34478. https://doi.org/10.1038/srep34478

Gelati M, Profico D, Projetti-Pensi M et al (2013) Culturing and expansion of "clinical grade" precursors cells from the fetal human central nervous system. Methods Mol Biol 1059:65–77

Giusto E, Donegà M, Cossetti C et al (2014) Neuro-immune interactions of neural stem cell transplants: from animal disease models to human trials. Exp Neurol 260:19–32
Glass JD (2010) The promise and the reality of stem-cell therapies for neurodegenerative diseases. Cerebrum 2010:24
Glass JD, Boulis NM, Johe K et al (2012) Lumbar intraspinal injection of neural stem cells in patients with amyotrophic lateral sclerosis: results of a phase I trial in 12 patients. Stem Cells 30 (6):144–1151
Glass JD, Hertzberg VS, Boulis NM et al (2016) Transplantation of spinal cord-derived neural stem cells for ALS: analysis of phase 1 and 2 trials. Neurology 87(4):392–400
Gritti A, Galli R, Vescovi A (2000) Culture of stem cells of the CNS. In: Fedoroff S et al (eds) Protocols for neural cell culture. Humana, Totowa, pp 173–198
Guo X, Johe K, Molnar P et al (2010) Characterization of a human fetal spinal cord stem cell line, NSI-566RSC, and its induction to functional motoneurons. J Tissue Eng Regen Med 4 (3):181–193
Gupta N, Henry RG, Strober J et al (2012) Neural stem cell engraftment and myelination in the human brain. Sci Transl Med 4(155):155ra137
Heslop JA, Hammond TG, Santeramo I et al (2015) Concise review: workshop review: understanding and assessing the risks of stem cell-based therapies. Stem Cells Transl Med 4 (4):389–400
Hickey WF, Hsu BL, Kimura H (1991) T-lymphocyte entry into the central nervous system. J Neurosci Res 28(2):254–260
Hoornaert CJ, Le Blon D, Quarta A et al (2017) Concise review: innate and adaptive immune recognition of allogeneic and xenogeneic cell transplants in the central nervous system. Stem Cells Transl Med 6(5):1434–1441
Iida T, Iwanami A, Sanosaka T et al (2017) Whole-genome DNA methylation analyses revealed epigenetic instability in tumorigenic human iPS cell-derived neural stem/progenitor cells. Stem Cells 35(5):1316–1327
Karimi-Abdolrezaee S, Eftekharpour E, Wang J et al (2006) Delayed transplantation of adult neural precursor cells promotes remyelination and functional neurological recovery after spinal cord injury. J Neurosci 26(13):3377–3389
Lee ST, Chu K, Jung KH et al (2008) Anti-inflammatory mechanism of intravascular neural stem cell transplantation in haemorrhagic stroke. Brain 131(Pt 3):616–629
Louveau A, Smirnov I, Keyes TJ et al (2015) Structural and functional features of central nervous system lymphatic vessels. Nature 523(7560):337–341
Luan Z, Liu W, Qu S et al (2012) Effects of neural progenitor cell transplantation in children with severe cerebral palsy. Cell Transplant 21(Suppl 1):S91–S98
Lunsford LD (2009) Stereotactic radiosurgery with the cyberknife for pituitary adenomas. J Korean Neurosurg Soc 45(6):405. https://doi.org/10.3340/jkns.2009.45.6.405
Mazzini L, Gelati M, Profico DC et al (2015) Human neural stem cell transplantation in ALS: initial results from a phase I trial. J Transl Med 1:13–17
Miura M, Miura Y, Padilla-Nash HM et al (2006) Accumulated chromosomal instability in murine bone marrow mesenchymal stem cells leads to malignant transformation. Stem Cells 24 (4):1095–1103
Neri M, Maderna C, Ferrari D et al (2010) Robust generation of oligodendrocyte progenitors from human neural stem cells and engraftment in experimental demyelination models in mice. PLoS One 5(4):e10145. https://doi.org/10.1371/journal.pone.0010145
Parr AM, Kulbatski I, Tator CH (2007) Transplantation of adult rat spinal cord stem/progenitor cells for spinal cord injury. J Neurotrauma 24(5):835–845
Parr AM, Kulbatski I, Zahir T et al (2008) Transplanted adult spinal cord-derived neural stem/progenitor cells promote early functional recovery after rat spinal cord injury. Neuroscience 155 (3):760–770
Pendharkar AV, Chua JY, Andres RH et al (2010) Biodistribution of neural stem cells after intravascular therapy for hypoxic-ischemia. Stroke 41(9):2064–2070

Pluchino S, Quattrini A, Brambilla E et al (2003) Injection of adult neurospheres induces recovery in a chronic model of multiple sclerosis. Nature 422(6933):688–694
Pluchino S, Gritti A, Blezer E et al (2009a) Human neural stem cells ameliorate autoimmune encephalomyelitis in non-human primates. Ann Neurol 66(3):343–354
Pluchino S, Zanotti L, Brambilla E et al (2009b) Immune regulatory neural stem/precursor cells protect from central nervous system autoimmunity by restraining dendritic cell function. PLoS One 4(6):e5959. https://doi.org/10.1371/journal.pone.0005959
Pollock K, Stroemer P, Patel S et al (2006) A conditionally immortal clonal stem cell line from human cortical neuroepithelium for the treatment of ischemic stroke. Exp Neurol 199 (1):143–155
Riley JP, Raore B, Taub JS et al (2011) Platform and cannula design improvements for spinal cord therapeutics delivery. Neurosurgery 69(2 Suppl Operative):ons147–154; discussion ons155
Riley J, Federici T, Polak M et al (2012) Intraspinal stem cell transplantation in amyotrophic lateral sclerosis: a phase I safety trial, technical note, and lumbar safety outcomes. Neurosurgery 71 (2):405–416
Riley J, Glass J, Feldman EL et al (2014) Intraspinal stem cell transplantation in amyotrophic lateral sclerosis: a phase I trial, cervical microinjection, and final surgical safety outcomes. Neurosurgery 74(1):77–87
Rota Nodari L, Ferrari D, Giani F et al (2010) Long-term survival of human neural stem cells in the ischemic rat brain upon transient immunosuppression. PLoS One 5(11):e14035. https://doi.org/10.1371/journal.pone.0014035
Selden NR, Al-Uzri A, Huhn SL et al (2013) Central nervous system stem cell transplantation for children with neuronal ceroid lipofuscinosis. J Neurosurg Pediatr 11(6):643–652
Shultz LD, Ishikawa F, Greiner DL (2007) Humanized mice in translational biomedical research. Nat Rev Immunol 7(2):118–130
Tadesse T, Gearing M, Senitzer D et al (2014) Analysis of graft survival in a trial of stem cell transplant in ALS. Ann Clin Transl Neurol 1(11):900–908
Uchida N, Buck DW, He D et al (2000) Direct isolation of human central nervous system stem cells. Proc Natl Acad Sci USA 97(26):14720–14725
Vescovi AL, Gritti A, Galli R et al (1999a) Isolation and intracerebral grafting of nontransformed multipotential embryonic human CNS stem cells. J Neurotrauma 16(8):689–693
Vescovi AL, Parati EA, Gritti A et al (1999b) Isolation and cloning of multipotential stem cells from the embryonic human CNS and establishment of transplantable human neural stem cell lines by epigenetic stimulation. Exp Neurol 156(1):71–83
Vukicevic V, Jauch A, Dinger TC et al (2010) Genetic instability and diminished differentiation capacity in long-term cultured mouse neurosphere cells. Mech Ageing Dev 131(2):124–132

Kolibri 360 GmbH
Ferdinand-Jühlke-Straße 7
99095 Erfurt, Deutschland
produktsicherheit@kolibri360.de